# International Scientific-Technical Conference on Innovative Engineering Technologies, Equipment and Materials 2016 (ISTC-IETEM 2016)

IOP Conference Series: Materials Science and Engineering
Volume 240



Kazan, Russia
7 – 9 December 2016

ISBN: 978-1-5108-4892-4
ISSN: 1757-8981

**Printed from e-media with permission by:**

Curran Associates, Inc.
57 Morehouse Lane
Red Hook, NY 12571



Printed by Curran Associates, Inc. (2017)

For permission requests, please contact the Institute of Physics
at the address below.

Institute of Physics
Dirac House, Temple Back
Bristol BS1 6BE UK

Phone: 44 1 17 929 7481
Fax: 44 1 17 920 0979

techtracking@iop.org

**Additional copies of this publication are available from:**

Curran Associates, Inc.
57 Morehouse Lane
Red Hook, NY 12571 USA
Phone: 845-758-0400
Fax: 845-758-2633
Email: curran@proceedings.com
Web: www.proceedings.com

# Table of contents

## Volume 240

**International Scientific-technical Conference on Innovative Engineering Technologies, Equipment and Materials 2016 (ISTC-IETEM-2016)**

**7–9 December 2016, Kazan, Russian Federation**
**(previous edition: ISTC-IETEM-2015)**

**Accepted papers received: 11 September 2017**
**Published online: 28 September 2017**

### Preface



### Papers

IOP Conf. Series: Materials Science and Engineering **240** (2017) 011001 doi:10.1088/1757-899X/240/1/011001

International Scientific Conference" Innovative engineering technologies, equipment and materials - 2016 " has been hosted by the exhibition center "Kazan Fair" within the framework of the 16th international specialized exhibition "Mechanical engineering. Metalworking. Kazan »and 11th specialized exhibition" TechnoWelding " on the 7th and 9th of December, 2016.

The conference program was including 5 sections of work:

Section 1 - "Innovative technologies in the design and manufacture of engineering products. Practice and prospects of creation and application "

Section 2 - "Innovative development of mechanical engineering and economics"

Section 3 - "Mathematical and physical modeling of information, technical, technological and managerial systems and processes"

Section 4 - "Innovative development of small and medium-sized enterprises"

Section 5 - "Innovative welding technology in the industry"

Participants and guests of the conference discussed the scientific and technical achievements in the field of development, design and manufacture of engineering products. See the results of studies of the country's leading universities in the field of robotics, new materials (including on the basis of production waste), mechanical engineering, energy, manufacturing economy.

Presidium of the session "Innovative development of mechanical engineering and economics" headed by Director of the Institute of Engineering Prof. Nail Faikovich Kashapov. The Presidium also included: Deputy. Director of the Institute for educational activities Ildar Hafizov Ilsurovich and Director of the Integrated Learning Center Energy Efficiency Gilmanshin Iskander Rafailevich.

Research and Innovation have been presented by the staff of the Engineering Institute of KFU, Institute of Naberezhnye Chelny, Institute of Chemistry named after A.M. Butlerov, Institute of Management, Economics and Finance.

Active participation in the conference were graduate students and students of the Engineering Institute. The experience of live communication with scientists, practitioners, representatives of the real economy will allow buduyuschim experts to determine its scientific direction for a successful solution of important problems of import substitution and enhancing competitiveness assigned to the engineering industry of our country.

IOP Conf. Series: Materials Science and Engineering **240** (2017) 011002 doi:10.1088/1757-899X/240/1/011002

# Peer review statement

All papers published in this volume of *IOP Conference Series: Materials Science and Engineering* have been peer reviewed through processes administered by the proceedings Editors. Reviews were conducted by expert referees to the professional and scientific standards expected of a proceedings journal published by IOP Publishing.

IOP Conf. Series: Materials Science and Engineering **240** (2017) 012001 doi:10.1088/1757-899X/240/1/012001

# Scientific bases and principles of obtaining carbon-metallic material by catalytic pyrolysis of ethanol

**G M Akhmadiev**

Russia, Republic of Tatarstan, Naberezhnye Chelny institute (branch) of the Kazan (Volga) Federal University, 423812, Naberezhnye Chelny, Mira Ave., 68/19

GMAhmadiev@kpfu.ru, ahmadievgm@mail.ru

**Abstract**. The method of producing of carbon-metal material in the form of mixture of carbon fibers and encapsulated in unstructured carbon particles of nickel in diameter from 10 to 150 nanometers catalytic pyrolysis of ethanol comprising the catalyst in the form of nickel oxide and magnesium applied on the surface of inert substrate and placed in theclosed sealed volume maintained at constant temperature is fed through the inlet manifoldethanol vapor andis removed through outlet manifold gaseous pyrolysis products, wherein ethanol vapor diluted with inert gas in a ratio ethanol: inert gas of 1:4 ...5. The inert gas dilution use ethanol vapor argon. The catalyst is used in the form of alloy of nickel oxide (II) and magnesium in a weight ratio of NiO: MgO 4:1. The temperature is maintained constantly during synthesis of the form in the range of 600 to 750 °C. Synthesis was carried out under atmospheric pressure. The catalyst applied on the surface of inert substrate in the form of graphite foil is used in pulverized or granular state.

**Introduction**. Nowadays the technology of producing carbon-metallic material by catalytic pyrolysis of ethanol is used in the chemical industry. Scientific production experience demonstrates that carbon-metal material in the form of mixture of carbon fibers and encapsulated in unstructured carbon of nickel particles with diameter from 10 to 150 nanometers is produced by catalytic pyrolysis of ethanol at atmospheric pressure. Chemically active catalyst in the form of nickel oxide and magnesium deposited on the surface of graphite foil as inert substrate in pulverized or granular state is placed in the closed sealed volume which regularly maintain constant temperature of 600 - 750 ° C. Through this input collector is fed vaporof ethanol and gaseous pyrolysis products are removed from the output manifold. Ethanol vapor is diluted with inert gas such as argon at weight ratio of ethanol: inert gas of 1:4 ...5. The time of active chemical synthesis is from 1 to 180 minutes. Thetechnologydeveloped bytheauthors relates to chemical technology of heterophase reactions of solids with gas or thermal decomposition. Carbon-metal material is mixture of carbon fibers and encapsulated in unstructured carbon ofnickel particles with diameter from 10 to 150 nanometers. Carbon atomic fibers are characterized substantially by constant diameter, ratio of length to diameter as 10:1 with crystal structure of the ordered carbon atoms having an outer diameter in the range between 18 and 62 nanometers. Ratio ofcarbon atomic fibers and encapsulated particles ofnickel is 3..4:1[1].

**The purpose of this paper** is to analyze the producing carbon-metallic material by catalytic pyrolysis of ethanol.

It is known the method of carbonization of carbonaceous materials comprising feeding of bundles of treated carbonaceous material in the reactor containing thermally insulated housing with muffle, moving of bundles through the reactor chamber while supplying simultaneously hot and cold inert gas and removal of pyrolysis products with additional influence of temperature on gas during pyrolysis and selection of the treated bundles in the intermediate amount when cooling (UK patent GB 2,184,819 МПКD01F 9/22, D01F 9/32, 1987). The disadvantage of this technical solution [2] is low productivity of chemical carbonization process. It is associated with inability to maintain constant temperature environment while moving in the cavity of the ovenbundles with the material to be treated. The technological process is also associated with restrictions of temperaturewiththe danger of overheating of the product which leads to the deterioration of its properties and inability to obtain encapsulated in unstructured atomic carbon of nickel particles. Another disadvantage is the need for processing of raw material before carbonization and unsuitability of this method to obtain the desired products by catalytic pyrolysis.

**Material and research methods**. The most promising technological direction is the way to handle original gaseous hydrocarbon products. In this particular case pre-treatment of raw material in the majority of the cases are minimized. The U.S. patent №5165909, МПК D01F 9/10, 1992[5] describes the process ofcontinuous production of carbon atomic fibers which comprises contacting formation of fibers on catalyst containing metallic particles with gas, containing atomic carbon. However according to the known method ([RF patent № 2296827, МПК DOIF 9/127, 9/133 DOIF], 2005 [4] the process has technological disadvantage of formation of carbonatomic fibers and encapsulated in carbon fibers ofnickel particles. In this case there is technological impossibility to obtain encapsulated inunstructured carbon ofnickel particles. It is also known the method of [6] obtaining inorganic nanotubes from renewable raw materials (application for invention № 2008113245МПК S07S 1/00. 2009), the method comprises the following stages: a) synthesis of alcohol (s) by fermentation, b) dehydration of alcohol, c) introduction into the fluidised bed in the second reactor of the catalyst powder at temperature of 450-850 ° C which contains at least one metal, wherein catalyst is supported on inert solid support with diameter of catalyst pellets less than 300 mu.m; d) reduction of alkene obtained in stage b) into contact with powdery catalyst of the stage c) to form carbonhydrogen nanotubes andhydrogen on the surface of the mentioned catalyst by catalytic decomposition of the mentioned alkene, and e) recovering carbon nanotubes obtained in the d) stage. However the above mentioned technical method, contact of ethanol vapor occurs with suspended catalyst layer, which leads to instability of conditions of nanotubes growththroughout the volume of the machine and inability to obtainencapsulated in unstructured carbon of catalyst particles.

**Results and discussion**. The scientific basis for producing carbon-metallic material is catalytic pyrolysis of ethanol. Therefore in this case it is appropriately the result isknown as technical solution adopted for the prototype the method of producing carbon nanotubes with encapsulated particles of nickel and cobalt (RF patent number 2,310,601, МПКC01V 31/02, 2007). According to this known solution - the method [3] is prepared solution of cobalt acetylacetonate or nickel in benzene or in mixture with ethanol. For this purpose theglass vessel is filled with liquid. Then it is filled with bubbler benzene. Then the represented installation is sealed and is filled with nitrogen from the cylinder. The quartz reactor made with possibility of heating in high temperature furnace heated in nitrogen stream is fed from the container. Then the vessel with reaction mixture is suppliedwith nitrogen under pressure and is sprayed solution of cobalt acetylacetonate or nickel through the capillary in the reaction zone of the reactor with producing of appropriate catalyst. Thereafter the reactor is supplied with bubbler benzene resulting in the decomposition of benzene on the catalyst.

IOP Conf. Series: Materials Science and Engineering **240** (2017) 012001 doi:10.1088/1757-899X/240/1/012001

This method utilizes as carbon source benzene which is more costly raw material than ethanol, since ethanol is produced of renewable sources. In produced material nickel particles of encapsulated in carbon nanotubes rather than unstructured carbon and have not the form nearly to the spherical shape. Thus the use of volatile catalyst solution of nickel or cobalt acetylacetonate in benzene or its mixtures with ethyl alcohol complicate considerably the process of control of chemical technological synthesis process compared with using powdered or granular catalyst.

**Conclusion.** Thus the proposed method of producing carbon-metal material in the form of mixture of carbon fibers and encapsulated in unstructuredcarbon nickel particles with diameter from 10 to 150 nanometers of catalytic pyrolysis of ethanol. New technical solution lies in the fact that catalyst in the form of nickel oxide and magnesium is applied on the surface of inert substrate and is placed in the closed and sealed volume. While maintaining constant temperatureethanol vapor is fed through the inlet manifold and through the outlet manifold gaseous pyrolysis products are removed. The distinguishing features are that ethanol vapor is diluted with inert gas in a ratio ethanol: inert gas of 1:4 ... 5. Argonis used as inert gas for diluting ethanol vapor. The catalyst is used in the form of alloy of nickel oxide (II) and Mg in a weight ratio of NiO: MgO 4:1. The temperature is maintained at constant value that synthesis in the range of 600 to 750 °C. Chemical technological synthesis technology is carried out at atmospheric pressure. The catalyst applied on the surface of inert substrate in the form of graphite foil is used in pulverized or granular state in chemical processes.

**References:**

[1] Rukhova M O, Tkachev A G, Tugolukov E N, Ruhov A V, Kotelnikov S A 2013 *The method for producingcarbon-metal material by catalytic pyrolysis of ethanol* Patent for invention No 2516548
[2] The UK Patent GB 2,184,819 МПК D01F 9/22, D01F 9/32, 1987
[3] Russian patent number 2,310,601 МПКC01V 31/02, 2007
[4] RF Patent number 2296827, МПК DOIF 9/127, DOIF 9/133, 2005
[5] US patent number 5,165,909, МПК D01F 9/10, 1992
[6] The invention application number 2008113245 МПКC07S 1/00., 2009

# Management of processes of electrochemical dimensional processing

**I D Akhmetov, A R Zakirova and Z B Sadykov**

Kazan Federal University, 18 Kremlyovskaya street, Kazan, 420008, Russian Federation

achilnur91@mail.ru

**Abstract**. In different industries a lot high-precision parts are produced from hard-processed scarce materials. Forming such details can only be acting during non-contact processing, or a minimum of effort, and doable by the use, for example, of electro-chemical processing. At the present stage of development of metal working processes are important management issues electrochemical machining and its automation. This article provides some indicators and factors of electrochemical machining process.

## Introduction

Electrochemical machining is widely used for the treatment of extra-fragile, and highly heat-resistant alloys, high purity tungsten, for the manufacture of electrical contacts of power relays, magnets made of samarium-cobalt alloys. Production of details of these materials by traditional methods is practically impossible due to the occurrence of a number of technological challenges, including the emergence of mesh cracks, chips, failure to tolerances on physical parameters. Forming such details can only be acting on the workpiece during non-contact processing, or a minimum of effort, and doable by the use of electrochemical or elektroabrasive processing under the local effect of an electric current, which is carried out by controlling the electric field in the gap.[1]

## Processes indicators and factors of electrochemical dimensional processing

Management of processes of electrochemical dimensional processing and its automation at the present stage of development of processing of metals is very important. At automation of process of electrochemical dimensional processing perhaps two approaches: ensuring stability of process or to enter amendments into process during processing of each detail. The first path is very difficult as during processing changes of structure, impurity, a causeticity, temperature of electrolyte, an interelectrode gap, resistance of chain units and other parameters are inevitable.

The second path is flexible production systems - the technologies having huge potential in the environment of small and medium business. In this regard many higher educational institutions enter into curricula in the engineering directions discipline "Flexible production systems". Flexible production systems are characterized by an opportunity to consider changes in process, in particular in the course of electrochemical processing. [2]

IOP Conf. Series: Materials Science and Engineering **240** (2017) 012002 doi:10.1088/1757-899X/240/1/012002

Currently, in scientific and technical literature is available enough materials describing relations between various parameters of process that allows to use these dependences for the fissile monitoring and management of processes of electrochemical dimensional processing.

However electrochemical processing is characterized by a number of the interdependent phenomena that complicates the analysis and the choice of criteria for process control. Methods of mathematical statistics can be applied to the analysis and the choice of criteria.

**Research electrochemical dimensional processing with the fixed electrodes**

The research of electrochemical dimensional processing with the fixed electrodes was conducted for the choice of significant criteria of management of processes. Process of dimensional electrochemical processing can be carried out at the fixed relatively each other electrodes and with movement of one of them in processing. Process of dimensional electrochemical processing with the fixed electrodes is applied when their movement in processing is difficult or inexpedient. With the fixed electrodes it is necessary to refer simplicity and reliability of a design of an inventory in which there are no composite mechanisms and systems of automatic keeping track of by giving of electrodes and the composite rapid systems of protection against short-circuits to electrochemical dimensional processing method advantages. The method with the fixed electrodes is perspective at exercise of a number of the difficult and unique technological operations where removal of metal is insignificant, including in case of manufacture of details with a contour of various configuration in a sheet, during the marking, flash removal, etc. [1]

On section exemplars with sizes of 10*10*20mm from steel 12X13. The cathode was manufactured of L62 brass by the sizes 3*3mm. $NaNO_3$ electrolyte. For a research were a number of parameters and limits of their variation is chosen, there are shown in Table 1.

**Table 1** Parameters and limits of their variation for a method of electrochemical dimensional processing with the fixed electrodes

| Name and units of measure of a factor | Coded identifications | Variation levels | | |
|---|---|---|---|---|
| | | Lower level | Basic level | Upper level |
| Initial interelectrode gap, mm | $X_1$ | 0,05 | 0,3 | 0,5 |
| Electrolyte pH value | $X_2$ | 2,5 | 5 | 7,5 |
| Electrolyte concentration,% | $X_3$ | 5 | 7 | 10 |
| Electrolyte temperature, $^0$C | $X_4$ | 19 | 30 | 40 |
| Electrolyte speed, m/sec | $X_5$ | 0,2 | 6 | 12 |
| Treatment time, sec | $X_6$ | 3 | 31 | 60 |
| Operating voltage, V | $X_7$ | 7 | 17 | 27 |

Efficiency of process (У1) was estimated by print depth. Accuracy of processing (У2) was estimated by a difference of the sizes of the cathode and the sizes of a print on the anode. The surface roughness after processing (У3) was estimated by height of roughnesses. As a result of data interpretation the regression equations were received:

$$У1=142,23-16,69X_1-6,59X_2+20,31X_3+18,21X_4+97,75X_5+110,53X_6+84,41X_7-$$
$$-56,39X_1*X_5+72,08X_5*X_6+56,73X_5X_7+61,27X_6X_7,$$

$$У2=1160,77+271,71X_1+21,21X_2+136,29X_3+39,66X_4+79,52X_5+404,59X_6+426,38X_7+$$
$$+192,49X_1*X_5+90,62X_1*X_6+151,51X_1*X_7-84,27X_2*X_3-71,23X_2*X_5+67,82X_6*X_7,$$

$$У3=54,49-9,47X_1-5,73X_2-4,99X_3-7,40X_4-7,13X_5+7,16X_6-4,69X_7+$$
$$+6,32X_1*X_2+6,68X_1*X_4+6,63X_1*X_5+7,18X_3*X_4-6,27X_5*X_6-6,21X_6*X_7.$$

## Conclusions

Follows from the analysis of the received equations of regression that the main criteria influencing electrochemical dimensional processing with the fixed electrodes are tension on electrodes, outlet velocity of electrolyte. Management of processes maintaining of a constant current density for the account by management of tension allows to increase efficiency of process and accuracy of copying, at slight increase of a roughness of the processed surface.

It is also revealed that the print sizes at defined values of tension and speed of electrolyte can be less than the sizes of the cathode. This phenomenon can be used when obtaining necessary accuracy of the sizes of a print at electrochemical dimensional processing.

## References

[1] Zakirova A.R. Tehnologija izgotovlenija naklonnyh otverstij v tonkolistovyh zagotovkah. Metalloobrabotka. 2004. № 6., p. 19-21. /Manufacturing technology inclined holes in the web workpiece/
[2] Flexible manufacturing systems in the formation of engineering competencies Zakirova A.R., Makhanko A.V. 52-64 Innovations in education ed. by L. Shlossman. - Том. Volume 4

IOP Conf. Series: Materials Science and Engineering **240** (2017) 012003 doi:10.1088/1757-899X/240/1/012003

# New electrode-tool for the combined kerf of electrically conductive materials

**I D Akhmetov, A R Zakirova, Z B Sadykov and I I Khafizov**

Kazan Federal University, 18 Kremlyovskaya street, Kazan, 420008, Russian Federation

achilnur91@mail.ru

**Abstract**. The article explains the choice of combination electrodiamond cutting as the most suitable method for separating solid, hard-processing deficient alloys. When processing by this method, the electrode-tool comes into electrical contact with the side surfaces of the cut channel, giving rise to excessive metal removal is not controllable. To solve this problem, it introduced a new electrode-tool.

## Introduction

Rapid development of mechanical engineering, instrument making, the aviation, space-rocket industry, caused broad use of hard-processing, natural scarce materials and their alloys, such as a tungsten, molybdenum, niobium, etc. As original materials in production of load-bearing frames of products where it is necessary to provide high requirements for accuracy and quality of the processed surfaces at the acceptable efficiency of processes. Use of these materials allows to increase strength and production characteristics of products that, in turn, allows to create more functional mechanisms possessing small dimensions at higher functionality.

Despite the advantages of using a super-hard materials in engineering, processing is often complicated. In modern industry, for cutting natural and scarce, hard metals are used as traditional, such as machining, cutting and new species, such as ultrasound, laser, plasma, and others. Laser and plasma cutting materials are the most modern and advanced methods and more in practical use in the manufacture of hard-processing materials.

These methods provide high performance of the production process and the possibility of cutting the curved contour of the workpiece, but have some drawbacks that prevent their use in case of stringent requirements on the quality of processing. For example, these techniques have a thermal effect on the workpiece, disrupt its structure in the affected area. There is also the likelihood of the modified layer in the material at a considerable depth, there is a likelihood of a grid of cracks due to the stress-strain state in the thermal strain. To eliminate the effects of the above laser and plasma cutting requires additional surgery to remove the altered layer, resulting in the loss of scarce material increases the complexity and labor-intensive process.

## Benefits of the combined electrodiamond method of cutting

The most acceptable from the point of view of cost savings is the use of a scarce material combined electrodiamond cutting method, since the cutting is carried out with a thin disc with a diamond coating. The main advantages of the combined electrodiamond treatment over other types of processing are: the ability to process the conductive material, regardless of their hardness, high precision machining, high quality machined surfaces.

At kerf the problem caused by an electrical link between an electrode tool and lateral areas channels of a cut exists a diamond disk. There is excessive not controlled removal of material of preparation from groove sidewalls that results in defects of the planes of a cut. [1].

Now for cutting hard-processing materials diamond disk tool where lateral areas of disks follow-up cover with an insulant, except surfaces with diamond grains are manufactured. Diamond grains are fixed by a conducting nickel covering. However, in processing the nickel covering comes into an electrical link to the processed preparation that leads to receiving a V-shaped groove, instead of the required P-shaped groove.

For elimination of a problem of excessive not controlled removal of metal from preparation authors of the patent No. 142793 "The device for the combined cutting of electrically conductive materials" suggest to apply follow-up on lateral and face areas of disks of a combined electrode tool the electro isolating material which thickness decreases from the periphery to the center. This offer yielded a particular positive take, but completely did not solve problem [2].

The reason in design feature of these disks as they are made of whole thin-sheet material owing to what the possibility of an electrical link between an electrode tool and sidewalls of the channel of cutting remains. Disks with such design do not allow to localize process of electric removal to dot and to carry out process on minimum gaps [1, 2].

**New solution of the problem of uncontrollable removal of metal at electrodiamond processing**

Authors propose in essence new solution of the problem of uncontrollable removal of metal from lateral areas of the channel of a cut. In the offered option there is no constant electrical link between an electrode tool and the processed preparation owing to what lateral areas of the channel of a cut turn out plane-parallel (patent no. 2597843 "An electrode tool for the combined kerf of electrically conductive materials").

The essence of the offer consists in manufacture an electrode tool from the composites of a carbon fiber 1 (Figure 1) located in the radial direction from the center of a circle. Carbon fibers go beyond edges of a circle and provide the current distributor. From two parties the carbon layer is covered with an organic silicon layer 2 which in the course of manufacture of a circle bakes. The face surface of a circle and a part of lateral areas are covered with a diamondiferous layer 3. Thickness of a diamondiferous layer of external lateral areas of an electrode tool makes from 0,1R to 0,2R.

The fundamental difference of our method from existing is that side etching which is present at all previous works is excluded. The electrical link between an electrode tool and the processed preparation 4 occurs in essence according to the new scheme. In the method offered by us electric current moves pointwise, on a microgap, i.e. processing on more weak modes is provided (on microcurrents). Transfer of current is carried out through nanocarbon fibers. Thus, the way of management of power transmission immediately on the processed area where there has to be a removal is found.

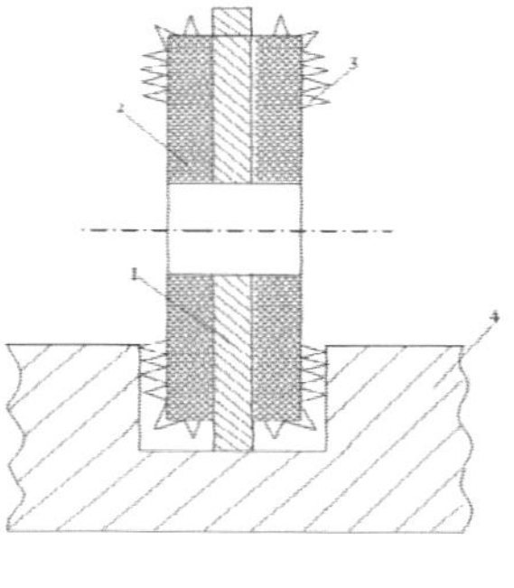

IOP Conf. Series: Materials Science and Engineering **240** (2017) 012003 doi:10.1088/1757-899X/240/1/012003

Figure 1 New electrode tool (a look in section)

**Conclusions**

For example, cutting a high-clear tungsten core (a tungsten falls into to a class of the hard-processing materials having high hardness and a brittleness) with a diameter of 6 mm and 120mm long on plates (tablets) with a diameter of 6 mm with a length of 1,2mm is considered. All process of kerf happens in the environment of flowing electrolyte, as electrolyte use a 15% aqueous solution of Sodium nitritum ($NaNO_3$). To the processed preparation and the rotating electrode tool give tension from a source of technological current. The rotation frequency of an electrode tool of 2800 rpm, tension 8V, diameter of a disk is 100mm, thickness is 1mm. At rotation of an electrode tool and its giving on incision in preparation processing is carried out owing to electrochemical disolution of material of preparation, electroerosive and cavitational destruction, mechanical cutting and removal of products of processing.

After cutting by this electrode tool receive the following results: not the ploskoparallelnost of lateral areas of the channel of a cut in limits of tolerance of 0,01mm, a roughness of the processed surfaces makes 0,03–0,05mm.

Thus, in a new electrode tool a parallelism of lateral areas of the channel of a cut is reached by an exception of an electrical link between an electrode tool and lateral areas of the channel of a cut. A positive take of a new electrode tool is the exception of padding operations of processing for elimination not of a parallelism of canal surfaces of a cut that excludes loss of scarce material and does not demand padding expenses [3].

**References**

[1] Akhmetov I D, Zakirova AR and Sadykov Z B 2016 The analysis and selection of methods and facilities for cutting of naturally-deficit materials *IOP Conf. Ser.: Mater. Sci. Eng.* 134 **012002**

[2] Khafizov I I, Zakirova A R and Sadykov Z B *The device for the combined cutting of electrically conductive materials* Pat. 142793 No 2015111479

[3] Sadykov Z B, Zakirova A R, Akhmetov I D, Hafizov I I and Kashapov R N *An electrode tool for the combined kerf of electrically conductive materials* Pat. 142793 No 2015111479

# Development of the technology of manufacturing a ceramic tip of a resectoscope

**I D Akhmetov[1], R N Kashapov[1] and L N Kashapov[1]**

[1] Kazan Federal University, 18 Kremlyovskaya street, Kazan, 420008, Russian Federation

kashramil.88@mail.ru

**Abstract**. The development of medical technology is strongly correlated with the development of new materials and the technology of their processing. At the moment, the most common type of surgery is resection of the prostate gland, for its use a resectoscope is used. In this work, the problem of increasing the mechanical characteristics of ceramic tips of resectoscopes is solved. The material of yttrium stabilized zirconia has been tested and manufacturing technology based on CAD / CAM and heat treatment has been proposed

## Introduction

Endoscopic operations have become widespread in clinical practice, while gaining increasing recognition in professional medical circles. Treatment with this method covers a wide range of diseases, whose treatment was once possible only by traditional surgical intervention. To ensure the accuracy of surgical manipulations and ensure the effectiveness of the operation, special endoscopic instruments are used. For this, resectoscopes are actively used in urology and gynecology.

In urology, resectoscopes are used for transurethral electroresection of the bladder neck with its sclerosis, resection of the prostate gland with its adenoma and cancer, resection under the control of papilloma vision and tumors of the bladder etc. In gynecology, resectoscopes are used to diagnose and perform various kinds of surgical interventions in the uterine cavity, in which the instrument is inserted through the cervix (cervical canal). At the same time, during the operation, the abdominal opening is not performed and the anatomical integrity of the organism is not disturbed. Directly, resection (removal of tissues) and coagulation of blood vessels is carried out by high-frequency electric current. Resectoscopy is a minimally invasive operation, the postoperative rehabilitation period is significantly reduced.[1],[2],[3]

The use of resectoscopes ensures high accuracy of medical manipulations and controllability of the operation as a whole. The principle of action resectoscope allows you to get a highly effective result, and in addition, significantly reduces the rehabilitation period.[4]

The modular design of the instrument (Figure 1a) consists of an external tube, an internal tube with an insulating ceramic tip for insulation and protection against heating of the cutting element (Fig. 1b), an operating element with a handle and an electrode attachment, an obturator with a smooth distal end, which is inserted into the inner tube for atraumatic injection of the resectoscope into the cavity.

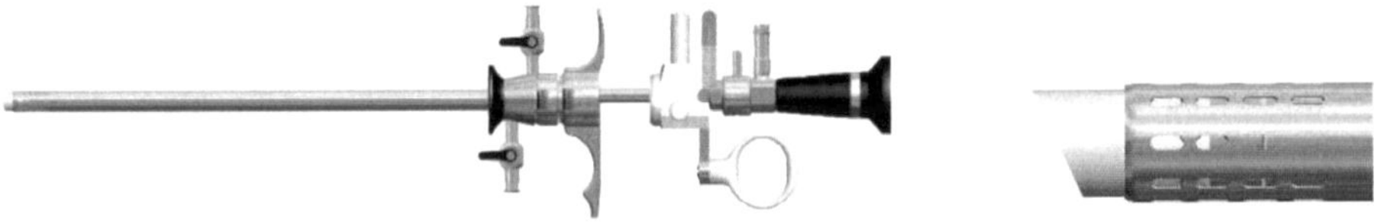

a) Resectoscope b) Ceramic tip

Figure 1. Design of resectoscope.

Developers are constantly improving the designs of resectoscopes. New resectoscopes practically exclude the risk of thermal burns of healthy tissues and postoperative complications. Modern models of the device provide a full visualization of endosurgical manipulations, safe excision of damaged tissues and accurate coagulation. [1]

After analyzing the design of the resectoscope, it was established that the ceramic tip for insulation and protection from heating the cutting element is not always reliable - it is brittle enough and in the process of operation it breaks down. During the operation, this can lead to irreversible consequences. In addition, these tips are imported from abroad and have a high cost. To date, these tips are made of aluminum oxide.[1],[2],[5],[6],[7]

The purpose of the research is to increase the life of the tip by replacing it with new material. Thus, it is necessary to solve two main tasks: 1) to choose a material with high dielectric, strength and thermophysical properties; 2) choose technology and processing modes. [1],[8]

**Main part**

In the studies, two types of tips were used, shown in Figures 2a and 2b.

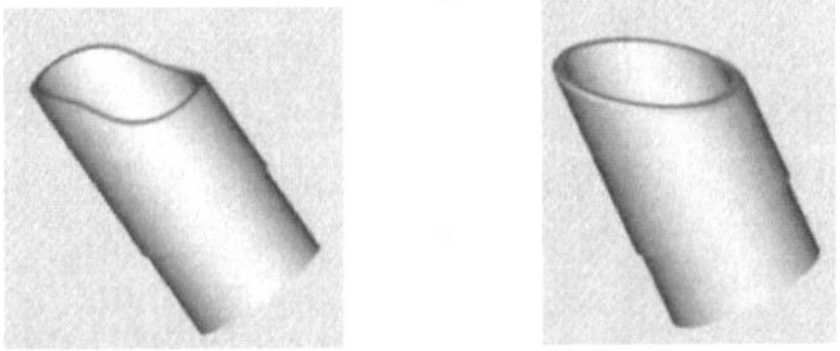

a) option 1 b) option 2

Figure 2. 3D models of tips

Both tips are a tube 15.5 mm long and an external diameter of 8 mm.

When choosing the material, certain requirements for the material were taken into account, namely high bending strength (750-1000 MPa), Young's modulus - 200 GPa, crack resistance - 8 MPa m½, coefficient of thermal expansion - 10 * 10-6 * K-1, нigh dielectric properties, breakdown voltage - 1000V, and ease of processing.

Such characteristics are possessed by zirconium oxide, namely ceramics based on zirconium dioxide ZrO2 partially stabilized with yttrium oxide (Y2O3).

Zirconium oxide in sintered form is difficult to process, so milling is carried out before heat treatment. Before sintering, zirconium oxide is similar in properties and characteristics to ordinary chalk. The processing of such material is feasible only by using the CAD / CAD system. CAD / CAM technology has a number of advantages: high precision of work, short terms of manufacturing of parts, saving of working place, automatic creation of trajectory of grinding, preliminary account of material shrinkage during sintering, high productivity.

It is also important that zirconium oxide is allowed for medical use, which will greatly simplify the process of launching the product into production.

Due to the fact that the tips have a thin wall, and the material itself is brittle enough (before sintering), a CAM-center was used for manufacturing. Namely, the model developed with the help of CAD-technology was grinded on an automated 5-axis CNC milling machine Coritec 450i.

To work with the CAM system, we need three-dimensional models of tips in the STL format that were created in the CAD system.

Then the 3D models of the tips were imported into the cad system. At this stage, first the necessary zirconium oxide blank is selected in the Sum 3D program, the shrinkage factor is specified (because during sintering, zirconium oxide shrinks and only afterwards acquires the required dimensions), the tips are placed, support is built, the necessary grinding strategy is selected with allowance for the modes of machining (cutter size, feed rate, tool rotation speed, etc.) and machining path, then the g-code is generated.. In the same program, using the "Simulation" function, a visual check of the resulting trajectories is performed.

After completing the work in the SUM 3D program, the finished g-code is transferred to the machining center - the 5-axis milling machine Coritec 450i.

When drilling, drills of three sizes were used: 2.5, 1.02 and 0.6 mm. Drills 2.5 and 1.02 mm are required for rough and medium processing, and 0.6 mm for final finishing. As a blank was used a standard disc of zirconia partially stabilized with yttria (Y2O3) G 526 thickness of 12mm and a heat shrinkage factor of 1,256. The processing of the two tips was carried out for 20 minutes. After milling, the tips are extracted from the zirconium oxide blank and sintered.

Sintering was performed on a P 330 Nabertherm controller. For the sintering of parts, a sintering regime for small structures was chosen. In this mode, the temperature rises to 900 ° C for 90 minutes. When this temperature is reached, the exposure is held for 30 minutes, then within a hundred minutes the temperature increases to 1550 ° C. When the 1550 ° C mark is reached, the temperature rise stops and a 2-hour exposure at 1550 ° C is carried out. Further, the products cool down to 200 ° C for 2 hours.

Obviously, the most optimal and effective technology for processing zirconium oxide is CAM-technology. Its use excludes manual work and settings on the milling machine, allows you to adjust the parameters and machining configuration without loss of time.

Figure 3a shows a photo of a zirconium oxide billet. Figure 3b shows the finished tips after extraction from the preform and subsequent sintering.

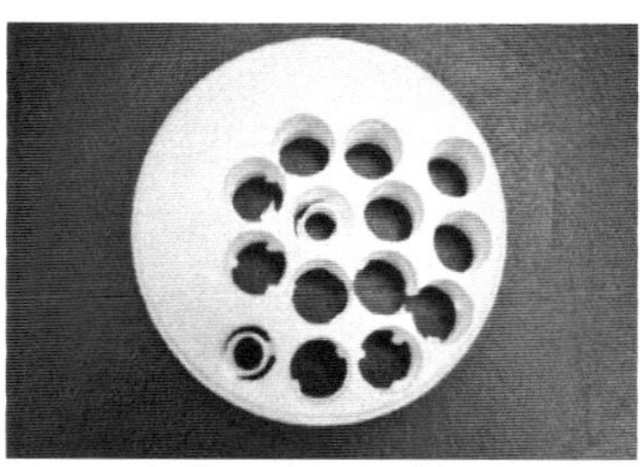

a) zirconium oxide

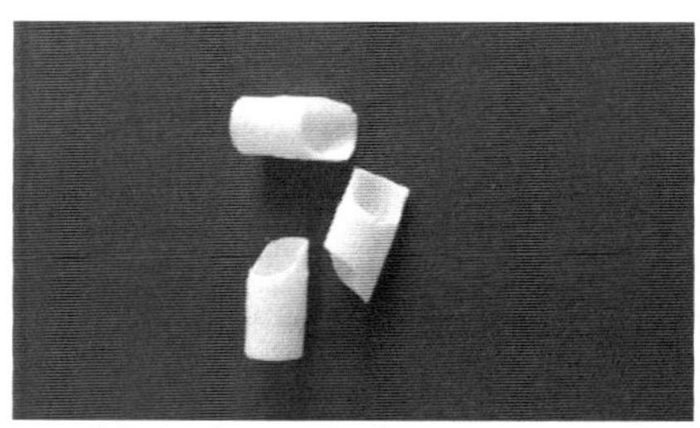

b) ready-made tips

Figure 3. Machined parts

The resulting finished products were tested for dielectric strength and performance. Tests on the samples were carried out using a megawatt F 4102 / 2-1M at a voltage of 1000V. The tests proved the dielectricity of the new ceramic tip. [8]

Thus, the suitability for using a zirconia tip was established.

**Conclusions**

As a result of the carried out researches the technology for manufacturing ceramic tips for resectoscopes was chosen. This led to a simplification of manufacturing technology and increased reliability of the tips.

The tool has become much easier to manufacture and more reliable to use.

The developed technology for the manufacture of ceramic parts makes it possible to manufacture ceramic tips for resectoscopes of «Aleps» firm, which are used in gynecological and urological operations.. The use of the new technology made it possible to fabricate tips with the necessary spatial configuration and sufficient dimensional accuracy. The material used was more durable than the material in the imported instruments. The use of tips for resectoscopes obtained by the new technology makes it possible to increase the reliability, service life of the product and reduce the cost of the final instrument.

**References**

[1] Belov S V, Danileyko Y K, Shulutko A M, Semikov V I, Gryaznov S E, Osmanov E G, Gorbacheva A V, Patalova A P, Osiko V V and Salyuk V A 2017 Use of high-tech electrosurgical tools to improve the efficiency and safety of operations on the thyroid gland *Medical equipment* No 1 (301) pp 8-11

[2] Kashapov L N, Rudyk A N and Kashapov R N 2014 Applying 3D-printing technology in planning operations of cancer patients *IOP Conf. Ser.: Mater. Sci. Eng* Vol 69 **012016**

[3] Fattakhova A N, Khramova A Y, Belinite M A, Kashapov R N and Kashapov N F 2014 Human and animal cells adhesion on the modified glass *IOP Conf. Ser.: Mater. Sci. Eng* Vol 56, Issue 6 pp 456-58

[4] Kashapov R N, Korobkina A I, Platonov E V and Saleeva G T 2014 The method of manufacture of nylon dental partially removable prosthesis using additive technologies *IOP Conf. Ser.: Mater. Sci. Eng Vol* 69 **012026**

[5] Gavrilova V A, Kashapov N F and Kashapov R N 2011 Plasma application of protective polymer-powder coatings for ultrasonic sensors *Meditsinskaia tekhnika* Issue 5 Pp 43-46

[6] Gavrilova V A, Kashapov N F and Kashapov R N 2011 Plasma application of protective polymer-powder coatings to ultrasonic sensors *Biomedical Engineering* Vol 45 Issue 5 Pp198-00

[7] Kashapov L N, Kashapov N F, Kashapov R N and Pashaev B Y 2016 The application of additive technologies in creation a medical simulator-trainer of the human head operating field *IOP Conf. Ser.: Mater. Sci. Eng* Vol 134, Issue 1 **012011**

[8] Akhmetov I D, Kashapov N F and Kashapov R N 2014 Modernization of clamping unit of draft of endosurgical monopolar forceps *IOP Conf. Ser.: Mater. Sci. Eng* vol 69 **012002**

# Development and investigation of the stressed-deformed state of the demountable foundation for support

**L S Sabitov[1], N F Kashapov[1], I R Gilmanshin[1,4], Yu M Strelkov[2] and D M Khusainov[3]**

[1] Kazan Federal University, 18 Kremlyovskaya street, Kazan, 420008, Russian Federation
[2] Kazan CHPP-1, Engineering Center "Energoprogress"
[3] Kazan State University of Architecture and Civil Engineering
[4] Kazan National Research Technical University named after A N Tupolev, Kazan, Russian Federation

sabitov-kgasu@mail.ru

**Abstract.** The development is demountable foundation for support, including separate reinforced concrete blocks in the form of prisms mounted on the surface of the base and pulled together by horizontal strands, and anchor devices for fixing the supports. The reinforced concrete blocks are made in the form of hollow prisms consisting of walls and square bottoms, and the strands are made in the form of bolts that tighten the walls along the top and bottom, while the anchoring devices for fixing the supports are made in the form of anchors on the bottom of the central prism and horizontal spacers between the support and the walls of the prism in its upper part. Numerical studies of the foundation have been carried out and its optimal sizes have been found in the PK Lira SAPR.

The invention relates to the field of construction, namely, to demountable foundations of poles for lightning, cell tower equipment, antennas, billboards and etc.

It is known the foundations for support, for example, billboards (see RF Patent No. 2243596. MPK G09F15 / 00, bulletin No. 36 dated December 27, 2004). The well-known foundation is installed in an open trench and is made in the form of a monolithic or prefabricated reinforced concrete plate with anchor bolts to fix the base of pole [1].

The disadvantage of the foundation is the large transport dimensions, which makes it difficult to transport and install.

It is known a different foundation for lightning pole, including a demountable reinforced concrete plate with anchor bolts for fixing the base of the pole. The foundation is installed directly on the surface of the base (www.kaal.nl, kaal master more than masts).

The disadvantage of the foundation for the pole is its large dimensions, with a high support height and weak bases. Large dimensions make it difficult to transport and install.

The closest to the invention is the demountable foundation (Fig. 1), which includes separate reinforced concrete blocks in the form of rectangular prisms mounted on the surface of the base and pulled through in two directions through horizontal strands, while a part of the blocks is provided with anchor devices for fixing the support in the form of bolts, protruding beyond the surface of the upper facet of reinforced concrete blocks (www.cellblocksinc.com)

Fig.1. Installation of Cellblock type foundation.

The disadvantage of the known solution of the demountable foundation is a large mass during transportation and installation, the difficulty in installing end-to-end long strands and the large costs of anchoring devices for supporting the support.

The purpose of the invention is to reduce the transport and installation weight of the units, simplify the installation of cords and reduce the cost of anchoring devices for supporting the supports.

The goal of the invention is achieved by the fact that in a demountable foundation for a support including separate reinforced concrete blocks in the form of prisms mounted on the surface of the base and pulled together by horizontal strands, and anchoring devices for fixing the supports, reinforced concrete blocks are made in the form of hollow prisms consisting of the walls and square bottoms, and strands are formed as bolts, tightening the wall over the top and bottom, thus, anchoring devices for faixing the supports are in the form of anchors at the bottom of the central prism and horizontal spacers between the support and the prism walls at an upper its part.

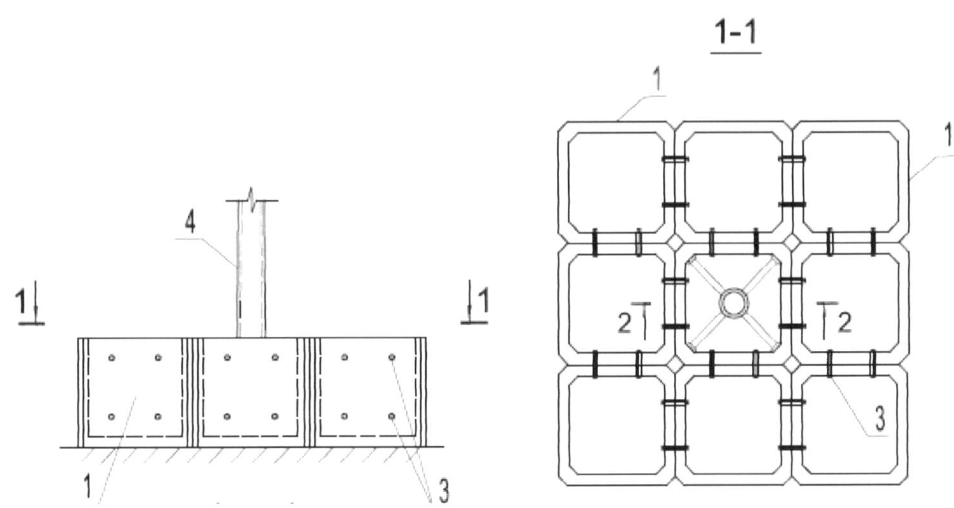

IOP Conf. Series: Materials Science and Engineering 240 (2017) 012005 doi:10.1088/1757-899X/240/1/012005

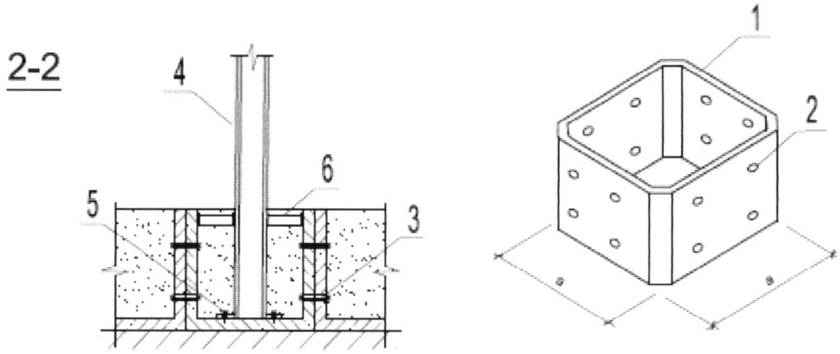


Fig.2. The proposed demountable foundation of ballast type.

Fig. 1 shows a general view of demuntable foundation under support; Fig. 2 shows a top view of the demountable foundation section 1-1 on Fig. 1; Fig. 1.3 shows a section 2-2 on Fig. 2; Fig. 4 shows the axonometry of a reinforced concrete block in the form of a hollow square prism with the dimensions of the bottom "a".

The demountable foundation for support includes reinforced concrete blocks in the form of hollow square prisms 1, with holes 2 for bolts 3. The support 4 is fixed by the base with anchor bolts 5 on the bottom of the hollow prism, and in the upper part by horizontal spacers 6 between the support and the walls .

The demountable foundation under the support is realized as follows: the reinforced concrete blocks in the form of hollow square prisms 1 are installed on the base and are tightened together by bolts 3 installed in the holes 2 in the walls of the prisms. The support 4 is mounted on the bottom of the central prism and fixed with anchor bolts 5, and at the top of the prism by spacers 6. To increase the load-bearing capacity of the foundation for overturning, the cavity of the prisms is filled with loose inert material, for example, crushed stone. The foundation is dismantled in the reverse order. First, crushed stone 7 is removed from the prism cavities, then spacers 6 and anchor bolts 5 of the support are released. After dismantling the support 4, the bolts 3, which tighten the prism walls, are released.

**Stress - deformed state of the foundation**

The calculation was carried out in the software package "Lira-SAPR". The design scheme of the building is a spatial finite-element model "foundation - ground base", consisting of voluminous 6 and 8 nodal physically nonlinear soil elements, volumetric 6 and 8 nodal linear elements simulating a reinforced concrete foundation, and core elements of one-sided elastic bonds modeling the boundary layer of supporting the foundation on the ground. To ensure the possibility of applying external loads to the foundation block located in the center, embedded parts are conventionally adopted in the form of cores with rigidity much greater than the stiffness of the elements of the main scheme. The connection of the embedded parts and the surfaces of the foundation block is made using absolutely rigid bodies [4].

Sand is used as soil. A class of concrete foundation is adopted B25. The characteristics of the finite elements are shown in Table. 1.

Table 1

Physical and mechanical characteristics of finite elements

| № п/п | Name | Characteristics | | |
|---|---|---|---|---|
| | | KE soil | KE foundation | KE One way communication by Z |
| 1 | 2 | 3 | 4 | 5 |
| 1 | Type KE | 271 - 274 | 34, 36 | 262 |
| 2 | $E_0$, MPa | 10 | 30000 | - |
| 3 | v | 0.3 | 0.2 | - |
| 4 | $\rho$ , t/m$^2$ | 1.6 | 2.5 | - |
| 5 | $\sigma^+_{max}$ , MPa | 0.001 | - | - |
| 6 | $\sigma^-_{min}$ , MPa | -0.05 | - | - |

| | | | | |
|---|---|---|---|---|
| 7 | C, MPa | 0.001 | - | - |
| 8 | $\alpha$ | 30 | - | - |
| 9 | $R_z$, t/m | - | - | 1000 |

Two variants of models consisting of 9 and 25 cells of foundation blocks are considered.

**Method of computer modeling**

The models are calculated in a physically nonlinear formulation. The following parameters have been adopted as the main variable parameters: bending moment value M (5 tm, 7,5 tm and 10 tm), backfilling density $\rho$ (1000 kg / m$^3$, 1300 kg / m$^3$ and 1600 kg / m$^3$). Two consecutive downloads are accepted: loading 1 - vertical loads including own weight of foundation, backfill, loading 2 - bending moment. Each download was included 3 equal steps of the load application. The calculation scheme of the foundation and the scheme for applying loads are shown on Fig. 3.

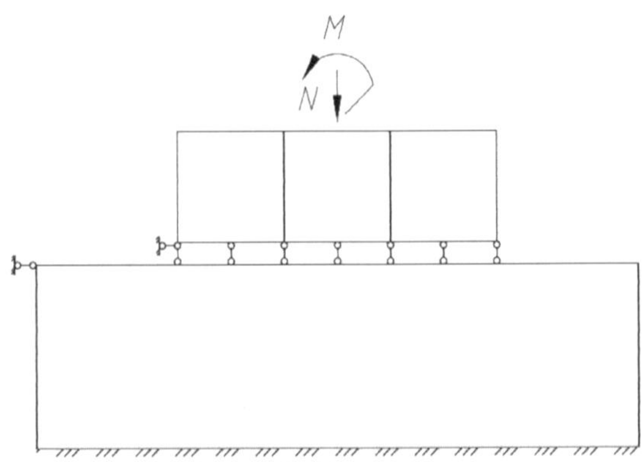


Fig. 3 Calculation scheme

**Results of computer modeling**

The main results of computer modeling are taken in accordance with Fig. 4 and are presented in Table 2.

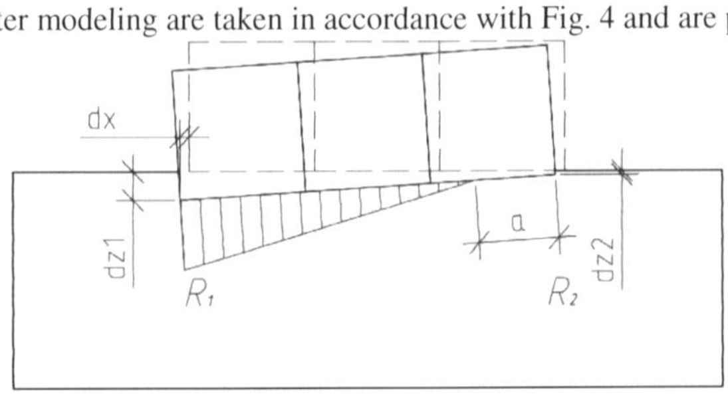


Fig.4 Scheme of results

Table 2

The main results of calculation

| № п / п | The magnitude of the moment original/actual M, tm | Back fill density, t/m$^2$ | Results | | | | | | | | Note |
|---|---|---|---|---|---|---|---|---|---|---|---|
| | | | Stress in the ground, MPa | | | Displacement, mm | | | Medium stress in concrete, MPa | | |
| | | | $R_1$ | $R_2$ | a, мм | $dz2$ | $dz1$ | $dx$ | $N_1$ | $N_3$ | |
| 1 | 2 | 3 | 4 | 5 | 6 | 7 | 8 | 9 | 10 | 11 | 12 |
| Foundation 3x3 | | | | | | | | | | | |
| 1 | 1 | 1.6 | -0.009 | -0.003 | 0 | -1.82 | -3.1 | -0.05 | 0.44 | -0.53 | Reinforcement is not |

IOP Conf. Series: Materials Science and Engineering **240** (2017) 012005 doi:10.1088/1757-899X/240/1/012005

| | | | | | | | | | | | |
|---|---|---|---|---|---|---|---|---|---|---|---|
| | | | | | | | | | | | required |
| 2 | 3.5 | 1.6 | -0.04 | 0.001 | 510 | -0.94 | -6 | -1 | 1.5 | -1.3 | Reinforcement required |
| 3 | 3.5 | 1.3 | -0.039 | 0.001 | 600 | -0.71 | -5.57 | -0.93 | 1.5 | -1.3 | Reinforcement requiredc |
| 4 | 3.5 | 1 | -0.04 | 0.0009 | 750 | -0.51 | -5.53 | -0.94 | 1.49 | -1.32 | Reinforcement required |
| 5 | 5 / 3.3 | 1.6 | -0.045 | 0.001 | 600 | -0.83 | -6.8 | -1.5 | 1.69 | -1.45 | Reinforcement required Destruction! |
| 6 | 7.5 /2.5 | 1.6 | -0.034 | 0.001 | 300 | -1.12 | -4.8 | -0.55 | 1.6 | -1.67 | Reinforcement required Destruction! |
| 7 | 10 / 3 | 1.6 | -0.1 | 0.001 | 1770 | +3000 | -166 | 41 | 5,2 | -2 | Destruction! |
| Foundation 5x5 | | | | | | | | | | | |
| 7 | 5 | 1.6 | -0.027 | -0.001 | 0 | -1.45 | -3.6 | -0.1 | 2.65 | -1.05 | Reinforcement required |
| 8 | 7.5 | 1.6 | -0.03 | -0.003 | 160 | -1 | -4.31 | -0.3 | 4.11 | -1.53 | Reinforcement required |
| 9 | 10 | 1.6 | -0.037 | 0.001 | 540 | -0.67 | -5.4 | -0.8 | 5.56 | -2.01 | Reinforcement required |

Note.
1. Destruction occurs on the ground.
2. The value dz with the sign "-" corresponds to the downward movement, dx with the sign "-" corresponds to the movement to the left.
3. The stress values with the sign "-" correspond to compression.
4. For concrete, the mean values of the principal stresses are indicated.

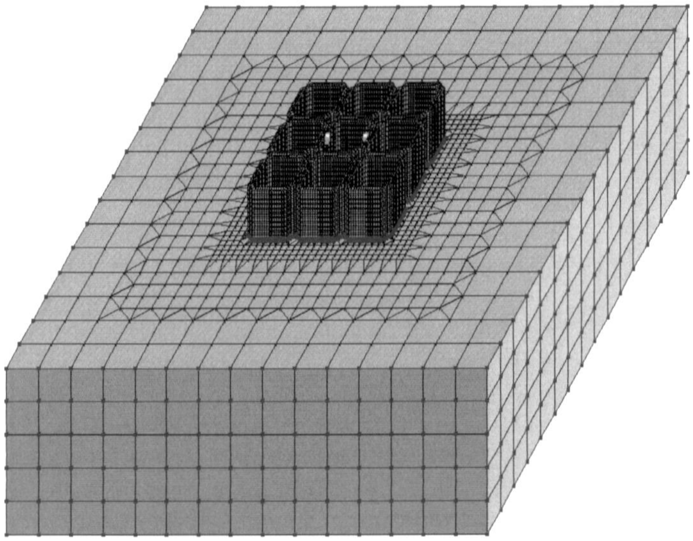

Fig. 5 The general view of foundation 3x3m model

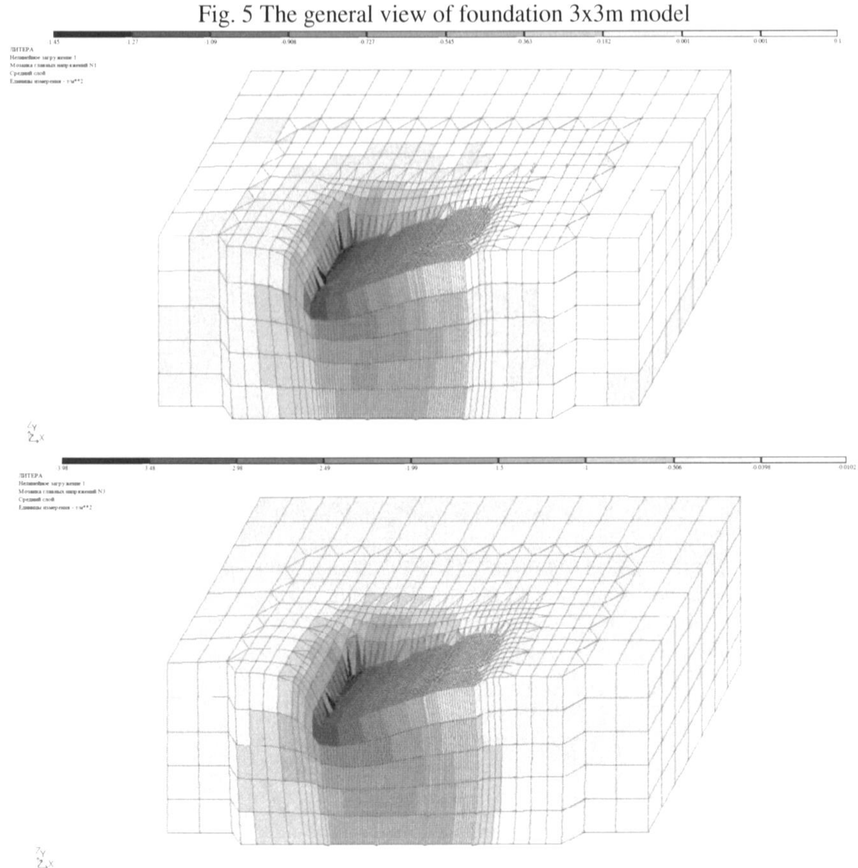

Fig. 6 Mosaic of main stresses N1, N3 in the ground

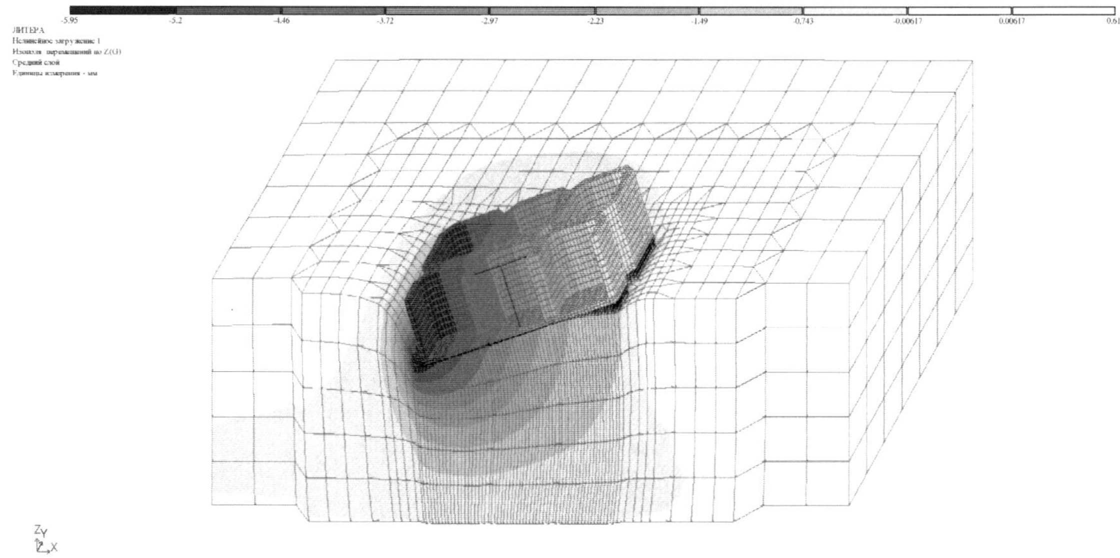

*Fig. 7 The isopy of displacements along Z (G)*

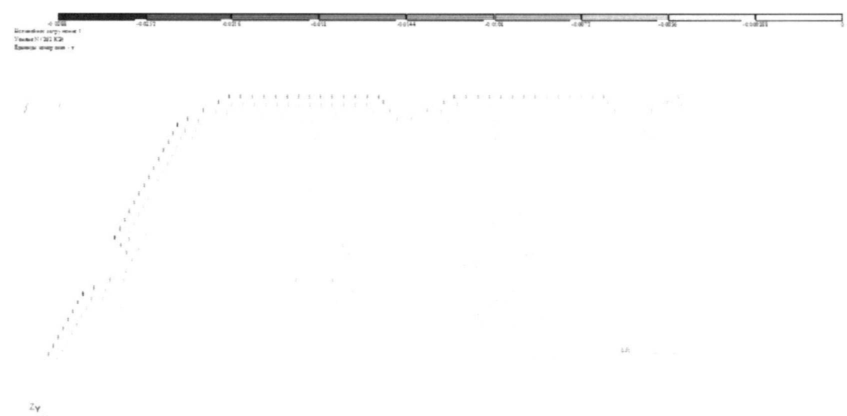

*Fig. 8 Force N in the boundary layer*

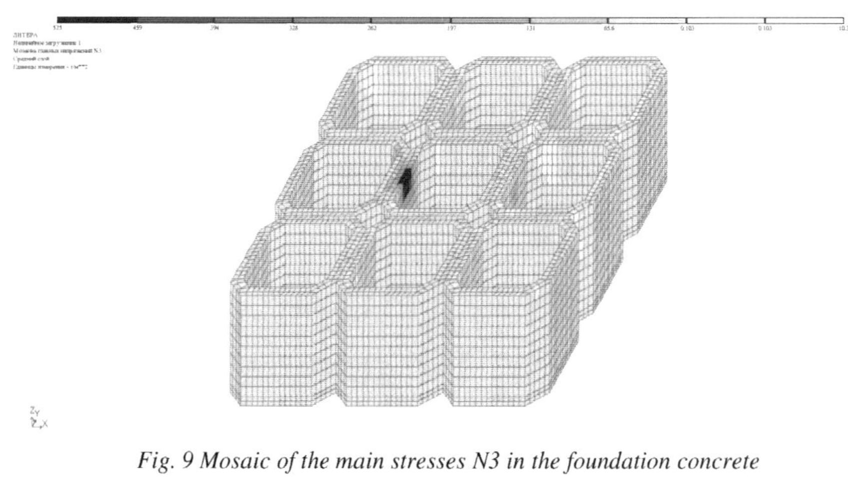

*Fig. 9 Mosaic of the main stresses N3 in the foundation concrete*

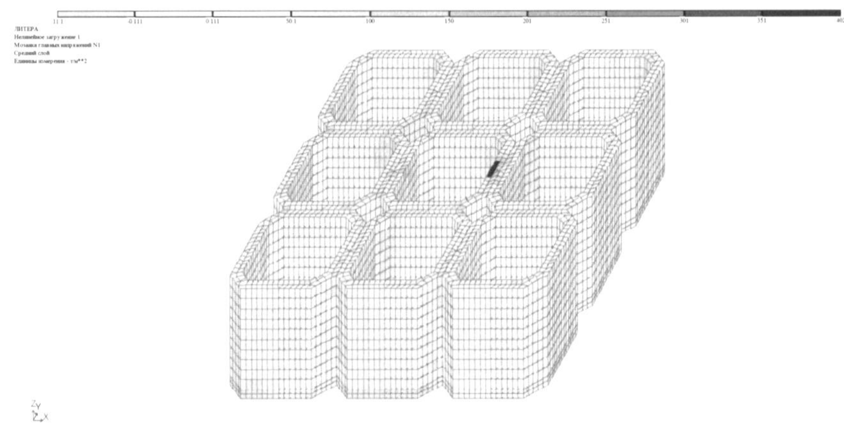

*Fig. 10 Mosaic of main stresses in concrete foundation*

**Conclusion:**

1. The proposed new demontable foundation allows to reduce its weight during transportation and installation, as well as to reduce the consumption of materials and labor for fixing reinforced concrete prism blocks and supports.
2. Numerical studies have made it possible to obtain an actual stressed-deformed state of the foundation and soil. On the basis of this, the base dimensions of 3x3m and 5x5m were obtained for the support of the power transmission line.

## References

[1] Akhunova Z R, Khusainov D M, Kuznetsov I L, Strelkov Yu M, Sabitov L S 2016 Demountable foundation under support *Materials of the international scientific and technical conference "Innovative machine-building technologies, equipment and materials -2016"* Part 2 pp 227-30
[2] *Cell Blocks Inc* [Electronic resource]: URL: www.cellblocksinc.com
[3] *Kaal Masten BV* [Electronic resource]: URL: www.kaal.nl
[4] Sokolov B S, Radaykin O V, Abuberb S S 2016 Application of computer simulation for analysis of the stressed-deformed state of bent concrete reinforced elements in the zone of joint action of bending moments and shear forces *In the collection: Modern problems of calculating reinforced concrete structures, Buildings and structures for emergency impacts* pp 388-93

# Optimal design of a combustion chamber of gas turbine engine by a Combustion chamber 1D-2D computer program

**Y B Aleksandrov**[1] **and B G Mingazov**[1]

[1]Kazan National Research Technical University named after A N Tupolev, Kazan, Russian Federation

alexwischen@rambler.ru

**Abstract**. The paper shows a method of modeling and optimization of processes in combustion chambers of gas turbine engines using a computer program developed by a team at the Department of Jet Engines and Power Plants (DJEPP) of Technical University named after A N Tupolev KNRTU-KAI.

## Introduction

Modern gas turbine engines and power plants are among the most complex technical objects. They are characterized by a large number of structural and circuit design. Their operating conditions are characterized by a variety of operating modes under different external conditions, whereby there are conflicting requirements imposed to them concerning power, specific fuel consumption, weight, dimensions, pollutant emissions, noise, etc. All this is a very complicated process of creating a modern gas turbine engine.

This process has now been inconceivable without the use of computer technology and mathematical models at all stages of the design. The use of mathematical models provides an opportunity to explore a wide range of issues related to the creation of GTE and thus can significantly reduce their full-scale tests.

One of the most difficult for the design of the engine components is the combustion chamber (CC), which takes into account the development of gas-dynamic processes, the chemical interactions of combustible components of fuel with the oxidant (air), thermal mass transfer processes. Working conditions, safety and environmental performance of the whole engine directly depend on the efficiency of this engine unit.

## Main part

The process of designing and creating the combustion chamber can be divided into a series of sequential steps shown by block diagrams in Fig. 1. Let us consider a phased combustion chamber design process.

Stage 1. The stage in which the original data of the combustion chamber (flow rate and pressure of air and gas and other) and design of the compressor and turbines serve as material for creation of a 1D model. There are two possible courses of action: 1) use of a prototype, 2) creation of a constructive image of a combustion chamber on the basis of statistical data. After calculation of this model it is possible to obtain: distribution of secondary air, mixing of fuel with air, and graphs of dependence temperature and nitrogen oxides, fuel combustion completeness and other parameters. The important point is fast results, allowing to sort out a significant number of options for modifications of the combustion chambers and determine the 1D image of the combustion chamber, which will give information on the optimal design of the combustion chamber. This stage is fully realized in the created computer program «Combustion chamber 1D-2D» [1]. The scientific basis for the program creation were the approaches outlined in works of Mingazov BG [2]. The calculation of combustion processes for one-dimensional model can be carried out quickly enough, which will help to organize the design optimization for various operating and design parameters at the entrance. This is a very important advantage of one-dimensional program in practical calculations of a combustion chamber.

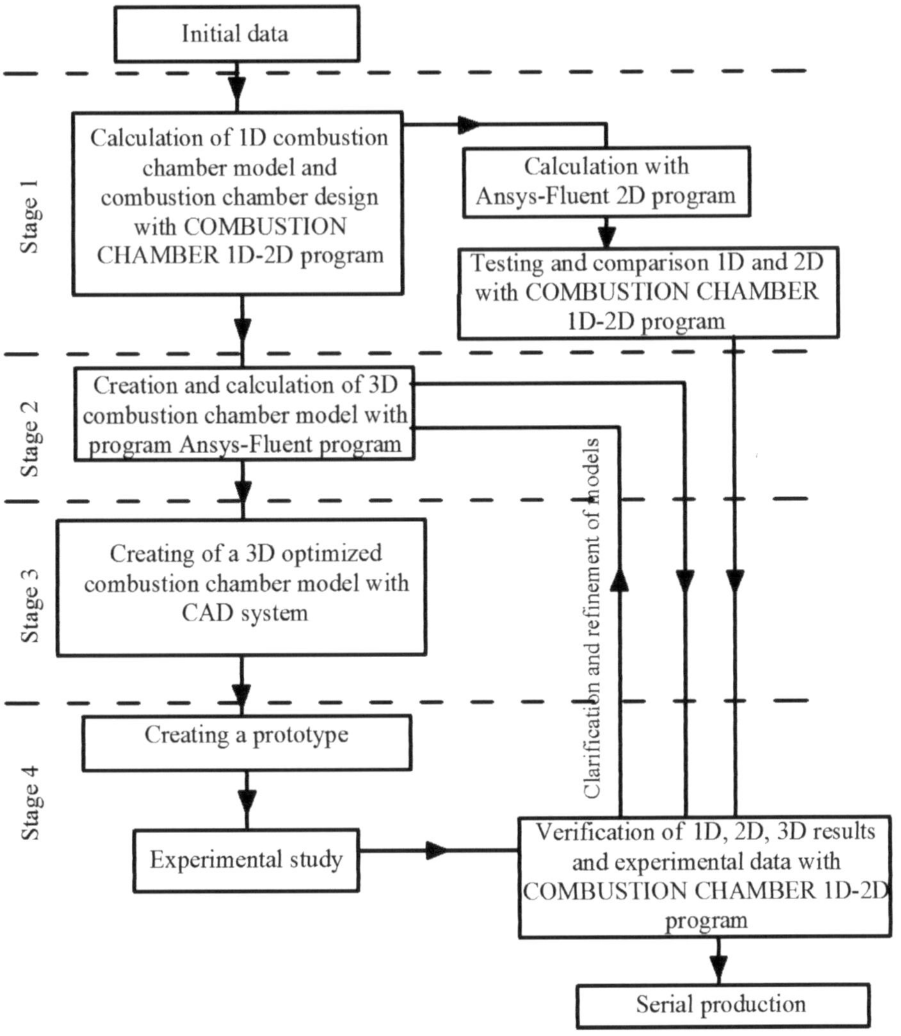


Fig. 1. Stages of the combustion chamber design using the program «Combustion chamber 1D-2D»

Stage 2. Based on the results obtained in the first stage of the work a 3D model is created. This model takes into account the additional design features assigned by a designer, such as design of a burner, twist flow, district unevenness of cooling zones holes and secondary air supply. Results of the calculation in CFD (Computational fluid dynamics) software are the most complete and serve to optimize the final design of a combustion chamber. It should be noted that this is the most time-consuming step. At the end of the stage make additional results calculation is needed to compare 3D with 1D and 2D.

IOP Conf. Series: Materials Science and Engineering **240** (2017) 012006 doi:10.1088/1757-899X/240/1/012006

Stage 3 design layout of the combustion chamber image. Calculation 3D optimized design is transferred to CAD (Computer-aided design / drafting) program, where all necessary design and technological documentation for the prototype is prepared.

Stage 4: Creating a prototype of the combustion chamber and conducting the experiment; it is the most expensive and a fairly long work stage. The above mentioned 1D, 2D and 3D optimization approaches allow to minimize costs and time of the final quarter of the stage. The experimental data is verified (compared) with the calculations of 3D data and in case of discrepancies calculations and improvements are carried out to refine the results, such as elaboration of models of turbulence and combustion (hybrid approach). For final image approval you must complete a combustion chamber prototype and carry out a research. The experiment takes precedence over numerical calculations. If the prototype parameters satisfy requirements specification, the design is transferred to serial production.

In practice of this calculation a layered approach is applied and 1-, 2- and 3-dimensional models are used. Recently, with the growth of computer performance 4-dimensional combustion models have become a reality (taking into account non-stationary processes). To clarify, one can make a rough comparison of speed calculation made on average computers - calculation of combustion with 1-dimensional model takes less than one second, with two-dimensional models – ten of minutes, with three-dimensional models from several hours to days, with 4-dimensional model calculations are relevant only on supercomputers.

From the point of view of the consistent implementation of multi-level calculations of intra-chamber processes, the one-dimensional model of the combustion chamber should be used initially. The calculation of combustion processes with one-dimensional model can be carried out quickly enough, which will organize the design optimization in various operating and design parameters at the entrance. This is a very important advantage of one-dimensional program in practical calculations of combustion chambers.

**One-dimensional and two-dimensional approach for calculation and optimization of a combustion chamber**

When calculated with 1D and subsequently 2D axisymmetric models, some simplifications will be entered into the model:

- flow model assumes a quasi-dimensional approach and the impact of flow swirls from blade twists on the formation of mixture and combustion is being taken into account using the mixing ratio which contains a twist parameter stream n φ;
- calculation of a fuel plume is measured on influence of fuel supply pressure and fuel injector nozzle size the average size of fuel droplets;
- air distribution along the length of the combustion chamber is performed by supplying air through the orifice zones taking into account impact of consumption and mixing coefficients;
- formation of local compositions of mixture in each section is calculated on vaporized fuel and residual and secondary air mixed with the flow;
- mixture burns on the surface of a combustion mechanism in a turbulent flow with turbulence intensity along $\rho \approx 0.4$ chamber length.

Despite the simplifications in 1D or 2D models, they allow to calculate a variety of options and identify areas to optimize the air distribution in a short period of time.

Result of a 2D calculation, unlike 1D is obtained for a longer time, but it is possible to analyze the flow parameters change, not only along the flow, as implemented in 1D, but also in transverse directions, which provides an additional advantage when analyzing a combustion chamber.

Initial data for 2D calculation are: air consuming, air pressure and temperature through the CC; fuel gas consuming and temperature. There can be variants, for example, instead of the air consuming sometimes inlet air speed is specified. These data can be obtained from a preliminary 1D calculation.

A one-dimensional and two-dimensional calculation of the combustion chambers is implemented fully in the "Camera" software. It has automated laborious process of creating a model of the combustion chamber geometry for 2D calculations. As an example, in Fig. 2 shows these files: a demo version of the 3D geometry of the combustion chamber of a gas turbine engine NK-16-18ST for visual representation (Fig. 2a.) and computed 2D axisymmetric option of applying of a contour fill depending on temperature (Fig. 2b.).The color scale on the left, displays the value. Color changes from red to blue varying rom largest to smallest respectively.

In the designated area of the combustion chamber calculations are performed sequentially using an iterative method: gas dynamics with the "cold" mode, combustion processes with the "hot", calculation of nitrogen oxides.

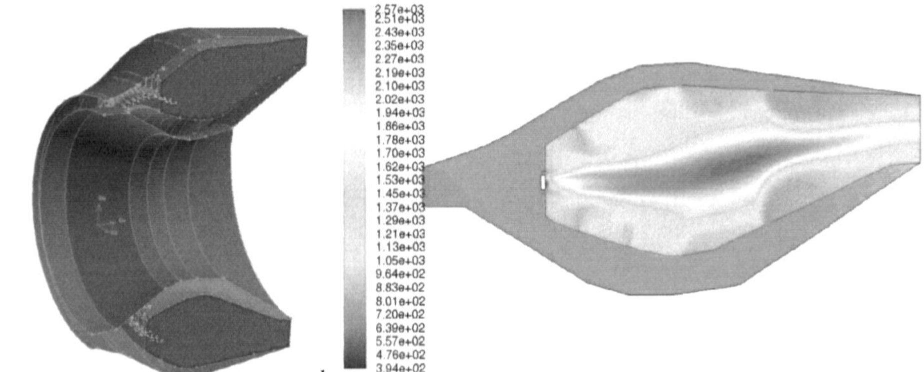


a b

Fig. 2. a - spatial model of the combustion chamber, 180° sector; b - result of 2D calculation, contour temperature filling with "hot" mode

For a complete picture of the correspondence of data the results of 1D and 2D calculations are being compared. Comparison can be performed on any desirable parameters calculated along combustion chamber length. As an example, Fig. 3 shows a comparison of the average temperature along the cross section of the flame tube.

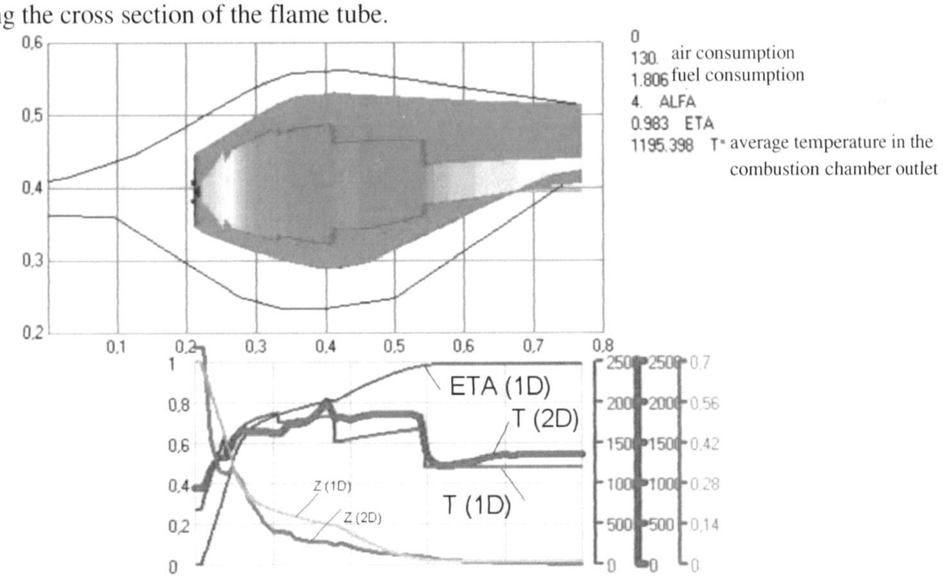


Fig. 3. Graphic dependences of 1D- and 2D- calculations comparison

Thicker lines in Fig. 3 indicate the design parameters, chosen by a user, calculated in the program “Fluent” (2D calculation), thinner lines were defined using the "Camera" program (1D calculation). By comparing the temperature dependences (T) there is a slight discrepancy along the combustion

IOP Conf. Series: Materials Science and Engineering **240** (2017) 012006 doi:10.1088/1757-899X/240/1/012006

chamber flame tube length. The obtained results indicate the possibility of using 1D model on the first stage.

**Three-dimensional approach to a combustion chamber calculation**

A combustion chamber is a relatively complex structure in terms of the implementation of its spatial computer model and construction of computational grid in its area. High-quality grid in such a model will contain at least one million cells and require considerable time to complete its calculation, for example, using a standard dual core computer it will take at least three days. In this regard, its design is simplified as much as possible, for example, remove the rivets, bolts, and all that makes a small contribution to the formation of gas-dynamic flows and heat transfer processes in a CC. For a preliminary calculation of gas dynamics and combustion chamber, a segment is often chosen so that it is symmetrical with respect to other ones which surround it on two sides [3]. In his segment a nozzle model is detailed.

As an example there is shown a combustion chamber of a gas turbine installation made in a 3D setting with a comparison of various parameters, distributed along the length of the combustion chamber, in one-dimensional and two-dimensional formulation.

The boundary conditions were inlet air flow in the combustion chamber diffuser G = 125.82 kg/s with a homogenous distribution of the flow parameters, fuel gas temperature T = 724.9 K, fuel gas consumption Gg = 1.388 kg/s, gas temperature Tg = 353 K. At the output there was set a condition of the static pressure constancy $p^* = 19{,}156\ kg/cm^2$.

To calculate the 3D model a 1/42 combustion chamber sector was selected containing 1.5 million grid elements. Due to the presence of a swirling flow pattern from a swirler, a viscosity model RNG k-e was used in the 3D model with standard wall functions, which provides the most appropriate picture of the current and provides stability when performing calculations. Burning of gas fuel was calculated using laminar micro-flame models for unmixed mixture. This approach is equivalent to the definition of a turbulent flame front as an ensemble of laminar micro-flames (flamelets). Laminar flame parameters are preliminary calculated and stored in a table, and then the laminar flame is introduced into turbulent flow using statistical methods. As a set of chemical reactions of methane oxidation the Kee mechanism (18 mixture components and 58 chemical reactions) is considered [4].

Parameters distribution along the combustion chamber was controlled by a specially designed control surfaces (3D) or lines (2D), using the program of the automated processing and analysis of the calculated parameters of gas dynamics and combustion processes in a combustion camera [5].

As a result of the calculation there has been obtained comparable data distributed over the length of the combustion chamber and shown in Fig. 4.

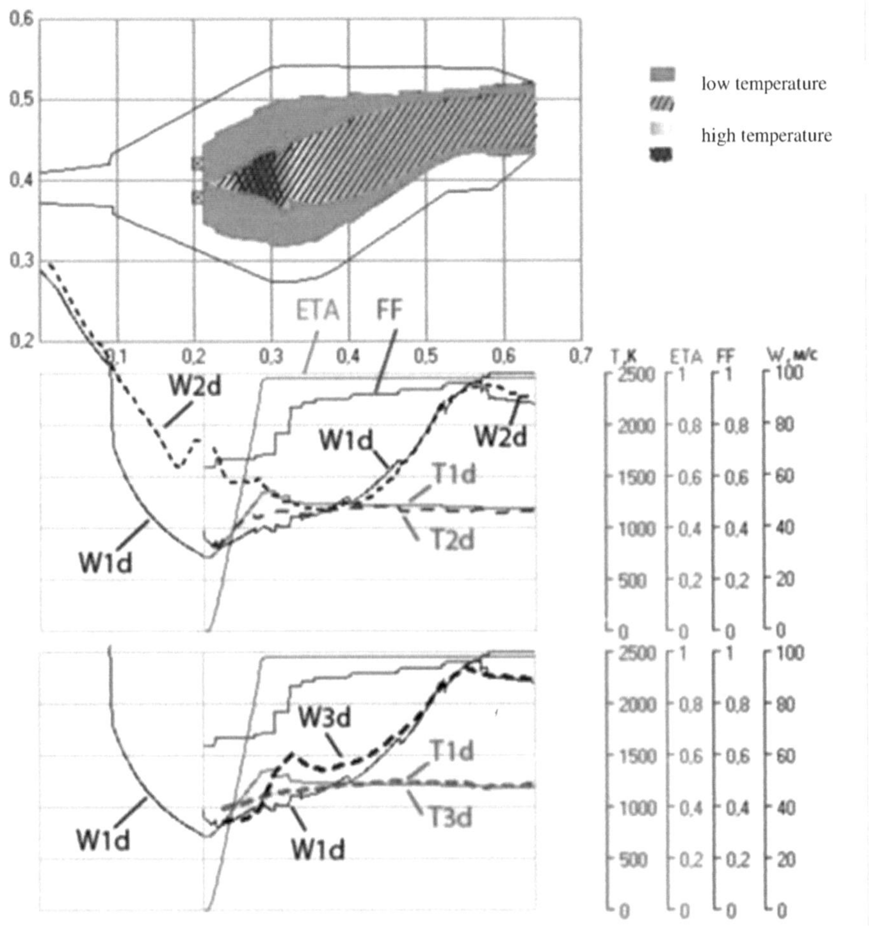


Fig. 4. Distribution of temperature (T) and velocity (W) along the length of a combustion chamber

Based on the results of one-dimensional calculation by a color field Fig. 4 shows the temperature distribution from blue to red, respectively, from the minimum to the maximum gas flow temperature values. Blue background in the outer zone of the flame tube - corresponds to the air temperature not involved in burning and mixing processes. Colors and shading show the area of the flame tube where air, fuel and fuel combustion are uniformly mixed. FF parameter – a relative area, it shows the distribution of areas of zones openings in the flame tube. ETA shows completeness of fuel combustion. Graphical dependence of the temperatures, averaged at cross sections, are shown in Fig. 4. They demonstrate their sufficient convergence at 2D and 3D setting, the only differences are observed in the primary air inlet area in the combustion zone and 1D. This can be explained by a simplified scheme model of a mixture formation. At flow speed rates differences are manifested in a greater degree and can be explained by the simplified scheme model of a mixture formation of 1D and 2D models, in comparison with the model made in 3D setting.

## Conclusion

The foregoing methodology allows to carry out comprehensive studies in design and finishing in GTE combustion chambers creation; a developed program of one-dimensional calculation «Combustion chamber 1D-2D» can be successfully applied in a preliminary design stage to predict the characteristics of a combustion chamber.

IOP Conf. Series: Materials Science and Engineering **240** (2017) 012006 doi:10.1088/1757-899X/240/1/012006

## References


[1] Aleksandrov Y B, Mingazov B G and Tokmovtsev Y V *The program of one- and two-dimensional calculation and optimization of combustion chambers of gas turbine engines* (Combustion chamber 1D-2D) (Certificate of state registration of the computer program №2016611770, application № 2015662645, received 22.12.2015, registered 10.02.2016, Moscow)

[2] Mingazov B G, Aleksandrov Y B, Kosterin A V and Tokmovtsev Y V 2015 *Combustion processes and automated design of combustion chambers* GTE and GTU teaching guide (Kazan: KNITU-KAI Press) 160 p

[3] Sabirzyanov A N, Aleksandrov Y B and Yavkin V B 2016 The rationale for reducing the number of sectors of the annular GTE combustion chamber in the modeling of work processes. *Problems and prospects for the development of engine building: materials of the reports of the international. Scientific-technical conference* of 22-24 June 2016, Vol 1 (Samara: Samara University Press) p 223-4

[4] Peeters T 1995 *PhD thesis on Numerical Modeling of Turbulence Natural-Gas Diffusion Flames* (Delft: Delft Technical University) p 263

[5] Aleksandrov Y B, Mingazov B G and Tokmovtsev Y V *Program for automated processing and analysis of gas dynamics and combustion processes in combustion chambers of gas turbine engines* (Servis 3D) (Certificate of state registration of the computer program №2016611856, application №2015662627, received 22.12.2015, registered 11.02.2016, Moscow)

IOP Conf. Series: Materials Science and Engineering **240** (2017) 012007 doi:10.1088/1757-899X/240/1/012007

# Key concepts for production of high-quality parts

**V I Astashchenko[1], N N Zapadnova [1], G F Mukhametzianova [1] and A N Shafigullina[1]**
[1]Naberezhnye Chelny Institute (branch), Kazan Federal University
Sjujumbike street 10a, Naberezhnye Chelny, 423800 Tatarstan, Russia

misharin_82@mail.ru

**Abstract.** The paper reveals the negative effect various stress raisers have on the part cyclic resistance. It proposes the innovative solutions to eliminate the segregation occurrences in steel, mitigate the negative effects of crack-like defects, and towards the design features required in part fillets.

## 1. Introduction

The performance capability and operating life of vehicle parts depend on many factors. The high-priority task for experts from the different fields (mechanical engineer, metallurgist, field engineer, etc.) is to find root causes of low product durability, develop and implement innovative engineering and process solutions aimed at producing high-quality and competitive vehicle parts.

The objective of this paper is to develop and substantiate the design, metallurgical, process and engineering solutions towards higher field reliability and durability of vehicle parts.

The paper presents the most representative and effective solutions to increase the fatigue strength of steel parts used for different purposes – ball joints for vehicle drive axles, crankshafts, crosses and gear wheels.

## 2. Body text

The design solutions adopted at the design and road bench test stage play an important role in ensuring the reliability and durability of products. A notable example is provided by the experimental studies of the impact of fillet radius (Rf) at the transition from hub Ø105 mm to flange (Fig. 1) in the vehicle ball joint on its durability [1]. The bench tests showed that with a fillet radius of 0.5 mm the average durability (28 parts tested) was $0.076 \times 10^6$ cycles, with a radius of 2.5 mm it was $0.142 \times 10^6$ cycles, with a radius of 4.0 mm it was $0.268 \times 10^6$ cycles, with a radius of 8.0 mm it was $0.524 \times 10^6$ cycles. These results revealed that the part durability (N) depends on the fillet radius (Rf) in its weakest section:

$$\lg N = 5 + 0.1\,Rf$$

The fillet radius should be minimum 4.3 mm to ensure the target vehicle mileage (350 000 km) and rule out a destruction failure of a ball joint. The fatigue strength of parts can be additionally increased by 21-34% if parts are toughened by thermal residual stresses arising from accelerated cooling after tempering from a temperature of 630-650°C [2]. After water cooling the compressive residual stresses occur on the part surface due to a significant temperature difference (~500°C) between the product surface and core. At the product design stage a similar approach should also be used for other part regions where stress is concentrated (keyslots, grooves, collars, thread, holes, etc.).

As stress raisers in parts are unavoidable their negative impact can be minimized by shot peening. The proper shot size and properties together with its effect on the surface help to create high compressive stresses in the problem areas and thus increase the durability of parts significantly for cyclic load operation [3].

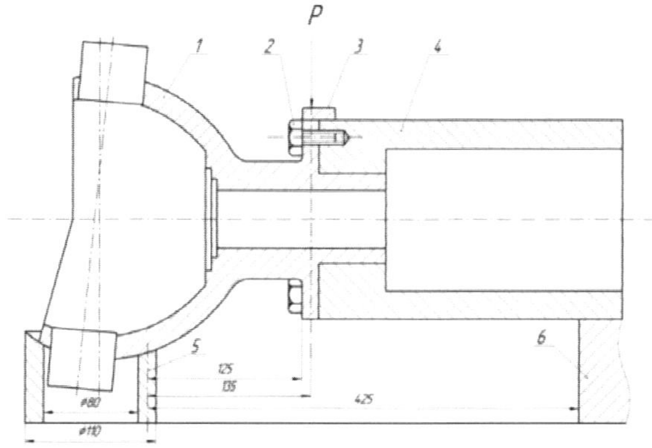


Figure 1. Load distribution in a ball joint: 1 – ball joint; 2 – fastening bolts; 3 – plate; 4 – pipe; 5 – pivot ring; 6 – rectangular-shaped bearing.

Among various negative metallurgic factors affecting reliability and durability of steel products the most critical ones are undoubtedly segregation occurrences and a degree of alloy contamination with endogenous and exogenous inclusions. The greatest danger comes from chemical and structural steel heterogeneity like a segregation square and especially in cases of segregate emersion to the highest stressed regions in a product. Figure 2 shows the failed gear. The cause of its failure was the segregation square which had been shifted towards the part tooth root at the hot working stage of the metal work-piece [4].

The method of investigating metal flow at hot forging was developed to eliminate unwanted micro- and macrostructures in steel products arising during process metal conversion in the machine production [5]. This method helps to find different shifting metal regions in the die cavity and, if necessary, control flows even in complex-shaped products by designing an accurate working die contour.

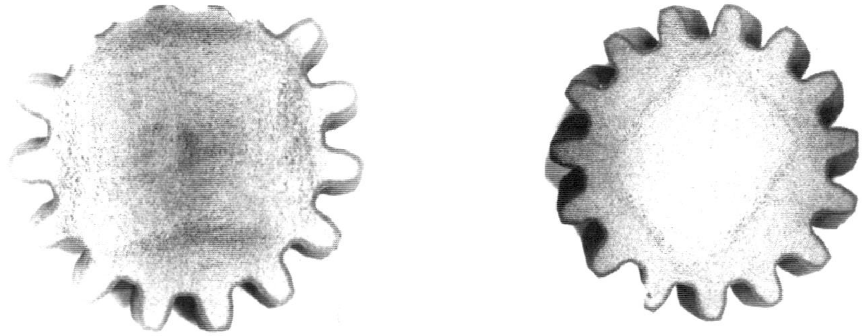

Figure 2. Macrostructure of part sections with the segregation square found at the gear tooth root (a) and center (b).

The crack-like stress raisers resulting from process metal conversion decrease the steel resistance to dynamic and static loads significantly. Such defects include folds, laps, undercuts, marks, etc.

Mechanical engineers should focus their efforts on complete removal of such stress raisers from a part surface or their elimination by shot peening. The greatest effect is achieved by using combined cold working process: first, impacting with large-sized shots, then relieving the stresses, and finally impacting with small-sized shots [3, 6]. The shot size for each stage depends on the type and size of a defect.

## 3. Conclusions

1. The design solutions were proposed to help make a rational choice of fillet radius in the mating areas of a part for high field reliability of products.

2. The method of investigating metal flow at hot forging was proposed to eliminate unavoidable metallurgy-related stress raisers in the parts. Subsequently, it prevents segregation occurrences and other steel defects from coming up to the surface.

3. It is recommended to use shot peening in order to increase the filed durability of products with inherited stress raisers of different origin.

## References

[1] Faskhiev Kh A 2009 Povyshenie dolgovechnosti sharovoi opory perednego mosta gruzovogo avtomobilia [Improvement of durability in ball joint of truck front axle] *Gruzovik* vol 5 pp 11-3

[2] Astashchenko V I, Shveeva T V and Rodkin I M 2012 Tekhnologicheskie resheniia pri peredele metalloprokata dlia povysheniia svoistv stal'nykh izdelii [Process solutions for conversion of rolled steel to improve properties of steel products] *Tekhnologiia metallov* vol 8 pp 9-13

[3] Saverin M M 1955 Drobestruinyi naklep [Shot peening] (Moscow: Mashgiz Publ.) p 312

[4] Astashchenko V I, Shveev A I and Shveeva T V 2011 Kontrol kachestva i nasledstvennost stroeniia stali pri tekhnologicheskom metalloperedele [Quality control and structural heredity in steel for process metal conversion] (Moscow: Academia Publ.) p 239

[5] Astashenko V I, Ishchenko V I, Shveev A I, Shveeva T V, Martiugin V S and Makhonin V V 2013 Sposob issledovaniia techeniia metalla pri goriachei obyemnoi shtampovke izdelii [Method of investigating metal flow at hot forging of work-pieces] Patent RF No 2506138

[6] Gliner R E and Astashchenko V I 2015 Vvedenie v tekhnologiiu poverkhnostnogo uprochneniia metalla [Introduction to surface hardening process] (Kazan, Kazan University Publ.) p 342

# Management in the system of waste utilization of production and consumption

**U I Azimov[1], I R Gilmanshin[1,2] and D R Krainova[1], I A Galeev [1]**

[1]Kazan Federal University, 18 Kremlyovskaya street, Kazan, 420008, Russian Federation
[2]Kazan National Research Technical University named after A.N.Tupolev, 420111, Kazan, Russia

Is-er@yandex.ru

**Abstract.** The main problems of waste management in accordance with the legislation are considered in the article. The economic benefits of separate waste collection are listed. The necessity of transition to a new level of waste management in the Republic of Tatarstan is determined.

The intensive increase in the volumes of formation and accumulation of production and consumption wastes (including communal and industrial waste), with a parallel increase in the share of artificial polymers in the morphological composition of the waste, caused a crisis and predetermined the problem of ensuring environmental safety in the sphere of waste management. The systemic nature of the problem is linked to the issues of the economic efficiency of the allocation and utilization of very significant amounts of valuable secondary resources that are part of the waste. Also, the issues of formation of understanding of sustainable development, moral and ecological education of the population are actualized [1, p.112].

In a sustainable development strategy, waste should be considered as a residue of production and consumption, suitable for further use or processing.

In fact, the majority of waste categories, both industrial and domestic, contain 40-60% (and sometimes 80%) of the most valuable types of secondary resources, including waste paper, glass, polymeric waste, metals, etc. Dynamics Changes in the quantitative and qualitative composition of waste categorically predetermine the need to adopt new approaches and solutions in the management of waste management.

Federal Law №458 of December 29, 2014 "On production and consumption wastes" establishes the basic principle of the theory and practice of waste management. Waste is treated as an object of ownership, which accordingly provides for the right to license activities in the field of handling them.

Priorities in the field of waste management:

- waste treatment;
- recycling;
- waste neutralization.

Proceeding from the presented principle of organization of the waste management procedure, the main tasks of the management at the regional level are the application of scientifically grounded technologies in securing the recycling of secondary resources extracted from waste through development, both investment attractiveness and activation of small business.

The economic and environmental benefits of using secondary raw materials in comparison with production based on primary resources are presented in Table 1.

Table 1

Economic and environmental benefits of using recycled materials

| Indicators | Benefits by type of cost, % | | | |
|---|---|---|---|---|
| | Aluminum | Steel | Glass | Paper |
| Saving energy | 90-97 | 47-74 | 23-71 | 4-32 |
| Expenditure of additional types of resource components | 90 | - | 80 | 35-40 |
| Saving of water resources | - | 40 | 58 | 50 |
| Reduction of air emissions | 95 | 85 | 74 | 20 |
| Reducing the discharge of pollutants into water bodies | 97 | 76 | 35 | - |
| Reduction of waste generation | - | 97 | - | 80 |

From the indicators presented in the table it becomes obvious that recycling of secondary resources with the selection of the latter from wastes determines not only the economic efficiency of large-capacity production, but also of small and medium-sized businesses, determining the opening of new jobs and ensuring employment of the population.

For the branch economy, the selection of secondary resources from waste products, in particular from domestic waste, is the opportunity to provide raw materials to enterprises. For example, providing the paper mill in Naberezhnye Chelny with high-quality waste paper is a condition for the stability of the production process, a way to diversify the industry, and increase profitability. And as a result of implementing complex measures for reclamation of existing landfills, it is possible not only to eliminate hazardous and carcinogenic emissions of landfill gas into the atmosphere, but the generation of heat and electricity in cogeneration mode [2, p.42].

The transition to a new level of system organization of the procedure for handling waste products of production and consumption should be justified by motivational principles that ensure the activation of economic activity in expanding the attractiveness of the recycling procedure for secondary resources allocated from production and consumption wastes.

As an effective mechanism of motivation in the management of secondary resources, one should focus on improving the system of institutional economy. The most effective principle of motivation is economically justified indicators for the purchase of secondary resources, or the establishment of economic and legal indicators of the tariff scale for garbage disposal, depending on the depth of selection of secondary raw materials from waste generated by waste owners.

A classic example of this mechanism is the establishment of a reasonable price for the purchase of secondary metal (today black metal is 8-10 rubles / kg) determined 100% collection of secondary raw materials. Also, since 2016, in connection with the increase in the cost of purchasing secondary cardboard and paper products to 5-8 rubles per kilo, the collection of recyclables has increased dramatically both by large operators and by individual entrepreneurs.

The model of consumer behavior in the choice of the technological method of managing waste management can be described as a function of the usefulness of the procedure for separate waste collection in the form of an additive relationship:

$$U(\bar{x}) = \sum_{j=1}^{n} a_j \times x_j^{b_j}$$

IOP Conf. Series: Materials Science and Engineering **240** (2017) 012008 doi:10.1088/1757-899X/240/1/012008

where the price estimate of the efficiency indicator of the accepted waste management process - $U(\bar{x})$ is produced as a generalized indicator of the additive sum of partial indices $\bar{x} = (x_1, x_j, x_n)$. As private indicators, it is necessary to establish indicators of economic, ecological efficiency, changing social security with the representation of parametric coefficients $a_j, b_j$in relative numerical ratios.

Thus, a rational variant of management in the field of waste management of production and consumption is the transition to separate collection directly within their owner, as well as the construction of an integrated waste management system in all life stages, including disposal and disposal.

**References**

[1] Azimov Y I, Gabirov R F 2009 The concept of recycling in the tasks of minimizing the consumption of natural resources *II International Ecological Congress (IV International Scientific and Technical Conference) "Ecology and life safety of industrial transport complexes"* pp 113-14

[2] Gilmanshin I R, Kashapov N F, Azimov Y I, Gilmashina S I, Ganeeva D A, Valishov R D 2015 Utilization of biogas from landfills of solid household waste by building an energy complex based on alternative energy facilities *Kazan economic bulletin* 2 (16) pp 41-45

[3] FL №458-FL of 29.12.2014 "About production and consumption wastes"

IOP Conf. Series: Materials Science and Engineering **240** (2017) 012009 doi:10.1088/1757-899X/240/1/012009

# Modeling of technological processes of heat supply, as a tool for assessing the state of objects of the property fund and engineering networks of enterprises

**U I Azimov**[1], **I R Gilmanshin**[1,2]

[1]Kazan Federal University, 18 Kremlyovskaya street, Kazan, 420008, Russian Federation
[2]Kazan National Research Technical University named after A.N.Tupolev, 420111, Kazan, Russia

Is-er@yandex.ru

**Abstract.** The approach to the mathematical modeling of technological processes of production, manufacture and consumption energyresources on the property fund facilities and engineering networks is presented. This approach is defining the information support system analysis of the kinetic changes of thermodynamic parameters sequentially occurring thermal processes in the flows of heat transfer agent in a closed structures of heating energy working in the recycle mode of the heat flow. It is determined the possibility of setting and solving problems of energyefficiency on the objects with close cycle operating and working in the fluctuation mode of the environmental parameters.

Technological network processes of heat and power supply of objects of property fund representing multi-stage dynamic systems of heat flows in network structures of heat consumption, accompanied by kinetic change of thermodynamic parameters of the heat carrier, change of thermal potential in closed circuits of circulating heat supply, the mathematical model is written in the form of a recurrent system of differential equations determining the change in the thermodynamic potential of thermal energy In networks those Provision of food.

Local identification of model equations determines the condition of high reliability of information support for the development of control systems for heat and power supply processes at the consumption objects.

Mathematical modeling of technological processes of power supply of objects of property fund and engineering networks of enterprises determines the possibility of solving the problem of energy and resource saving, optimization of energy consumption regimes, provision of sanitary and hygienic standards in premises in conditions of fluctuations in temperature parameters of the environment.

A systematic information approach is being implemented to solve the problem of efficient use of fuel resources in the conditions of stabilization of power supply modes at power consumption facilities.

The information approach of the mathematical modeling procedure can be represented by recording the balance equations of the dynamics of the change in the thermodynamic parameters of thermal networks.

The task of managing the intensity of energy production in conditions of optimal energy supply to consumers is recorded by a system of differential equations of kinetic variation that determine the parameters of the coolant in the structural chain of technical equipment in heat production, distribution and consumption of thermal energy with provision of sanitary hygienic thermal regimes in the premises.

The intensity of accumulation of the thermal potential is determined by the relation:

$$Q = \int_{\tau_0}^{\tau_K} m_{i;j} \cdot I(\tau) d\tau \qquad (1)$$

where $m_{i;j}$- Flow rate of coolant ($m^3$/hour);$I(\tau)$ – enthalpy of the heat agent (kJ/$m^3$); here $I(\tau) = c_v \cdot t(\tau)$; $c_v$ – bulk heat capacity of coolant;$t(\tau)$ – temperature of heat agent ($°K$).

In this task, the state of the control object is considered as a multidimensional, vector variable of the change in the energy potential of the coolant flow Q (kJ), the components of which determine parametric parameters, such as the volume flow of the heat carrier m (m3 / h), and the thermodynamic temperature T (K).

The vector state of the control object in the form of a network power supply system at the objects of the property fund and engineering networks of enterprises and, in particular, housing and communal services objects is determined by the change in the thermal potential of the heat carrier Q in a continuous flow in a certain range of values or a finite set of values.

The variable parameter of the technological process of heat supply should be determined such a current indicator as the temperature of the coolant $t_i$ on the i-th section of the supply network, while such an indicator as the volume flow of the coolant in the relevant sections of the power network should be defined as a given and constant value $m_i$

The generalized form of the mathematical model of thermal networks of heat production and heat supply is written in the form of a set of recurrent differential equations of the dynamic change in the energy potential of the coolant flow estimated by the heat energy index as the coolant temperature. This approach determines the possibility of developing a system for optimal management of energy consumption objects based on the principles of system analysis of the thermodynamic state of the flow at certain stages and sections of the network structure of its operation and regulation by parametric feedback indicators.

The structural diagram of thermal heat and power networks of heat supply at the objects of the property fund and engineering networks of enterprises, presented in Fig. 1, determines the solution of the problem of minimizing the consumption of thermal energy in conditions of ensuring the normative operating modes of the facility based on temperature characteristics.

IOP Conf. Series: Materials Science and Engineering **240** (2017) 012009 doi:10.1088/1757-899X/240/1/012009

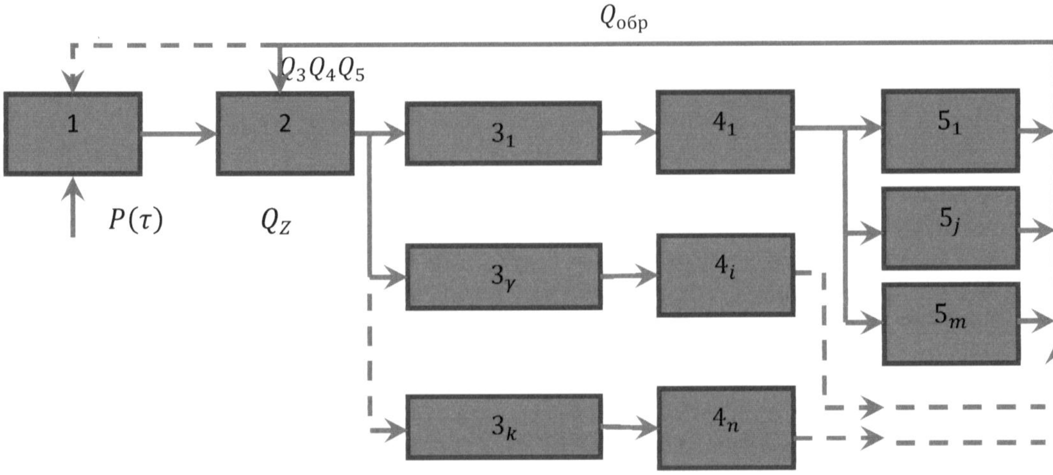


Fig.1 Structural diagram of heat and power supply in consumption networks: G - heat agent (natural gas, superheated steam); $Q_{ij}$ - heat flow; $Q_{обр}$excess-return flow of the coolant; $P(\tau)$ - the heat agent consumption; 1. heating boiler; 2. Boiler heat treatment of thermal conditions of the coolant; 3. the coolant supply line; 4. Nodal objects of heat consumption; 5. local heat consumers.

The balance ratio of the heat-energy flow pulse in accordance with the one shown in Fig. 1 technological structure scheme of heat and power supply in consumption networks is recorded in a generalized form in the form of a set of recurrence relations of instantaneous change in the heat content of the heat flow in a closed cyclic circulating heat and power supply system with respect to:

$$\Delta Q_1 = \left(\left(\left(\Delta Q_2 + \Delta Q_\Sigma\ \right) - \sum_{\gamma=1}^{k} \Delta Q_\gamma\right) - \sum_{\gamma=1}^{k}\sum_{i=1}^{n} \Delta Q_{ij}\right) - \sum_{\gamma=1}^{k}\sum_{i=1}^{n}\sum_{j=1}^{m} \Delta Q_{\gamma ij} \quad (3)$$

where:$\Delta Q_1$,$\Delta Q_2$, $\Delta Q_\Sigma$ ,$\Delta Q_\gamma$, $\Delta Q_{ij}$, $\Delta Q_{\gamma ij}$, Respectively, the energy flow of the formation energy in the boiler room, the output in the boiler, the flow of the heat carrier in the return line, the heat loss in the main network pipeline, the decrease in the thermal load at the γi nodal heating site, at the γij local heat consumption section.

The generalized recurrence equation for modeling the technological process of heat exchange in the structural system of heat supply to consumers of complex nodal objects of the property fund and engineering networks is a recurrence relation written in the form of a set of initial equations:

$$Q_1(\tau) = \mu \int_{\tau_0}^{\tau} P(\tau)\Delta H d\tau \ \rightarrow min$$

$$=> \int_{\tau_0}^{\tau} m_\tau(\tau)\cdot c_v \cdot t_2(\tau)d\tau - \sum_{\gamma=1}^{k}\int_{\tau_0}^{\tau}\int_{L=0}^{L_\gamma} K_\gamma \cdot f_\gamma \cdot \left(t_\gamma(\tau) - Q(\tau)\right) dL \cdot d\tau$$

$$-\sum_{\gamma=1}^{k}\sum_{i=1}^{n}\int \Delta Q_{\gamma J}(\tau) d\tau$$

$$-\sum_{\gamma=1}^{k}\sum_{i=1}^{n}\sum_{j=1}^{m}\left[\int_0^{\tau} V_{\gamma ji}\cdot c_v^{н} d\tau - \int_0^{\tau} Z_{\ \gamma ji}\cdot S_{\gamma ji}\cdot \left(T_{\gamma ji}(\tau) - U(\tau)\right) d\tau\right]$$

The identification of the generalized relationship is carried out by parametric indicators of the change in the value of the heat flux at the local points of the consumption node.

The recorded mathematical model of the technological process of production and consumption of heat energy resources on the objects of the property fund in the form of a recurrence relation determines the possibilities for setting and solving a number of management tasks:

- Deep elaboration of information support is provided for all parametric indicators both for all local analysis points and for the cumulative system as a whole.
- It is possible to evaluate all local sites and the structure as a whole in the parametric indicators of consumption and losses. The problem of eliminating bottlenecks is being solved.
- the theoretical and practical basis of development and implementation of ACS-TP operating in the mode of providing consumers with standard sanitary and hygienic life-support conditions is laid down; minimization of energy and resource saving, energy losses, minimization of payments by consumers for energy services.
- The presented approach to modeling the processes of use and consumption of material and energy resources in production systems can be considered as an opportunity for a deep level of information support in the development of an optimal control system.

**References**

[1] Azimov Yu I, Savdur S N, Kirpichnikov A P, Kostromin A V 2015 Network modeling of fertilizer production of agricultural enterprises *Bulletin of Kazan Technological University* T 18 No 12 pp 122-24

[2] Azimov Y I, Savdur S N, Mukhametgaleev D M 2015 Modeling of the technological process of biochemical sewage treatment of the polymer industry *Theoretical Foundations of Industrial Policy Formation* pp 378-97

[3] Basyrov I R 2001 Expansion of Petri nets, problem-oriented to modeling multi-assortment production systems *Uchenye zapiski KFEI: Sat. Sci. Works* Issue 16 pp 177-82

[4] Gilmanshin I R, Ferenets A V 2009 Energy saving in housing and communal services: construction of a complex of a centralized automated system for collecting, monitoring and analyzing household energy consumption *News of higher educational institutions. Problems of energy* No 9-10 pp 82-88

IOP Conf. Series: Materials Science and Engineering **240** (2017) 012010 doi:10.1088/1757-899X/240/1/012010

# Adaptive system of supplying lubricant to the internal combustion engine

**E P Barylnikova[1], A T Kulakov[2] and O A Kulakov[3]**

[1] Kazan (Volga) Federal University, Russia, 420008, Russian Federation, Kazan, Kremlin St., Bld. 18,
[2] Kazan (Volga) Federal University, Russia. 420008, Russian Federation, Kazan, Kremlin St., Bld. 18,
[3] NC "Kamazavtosport", 423800, Russian Federation,NaberezhnyeChelny, Avtozavodsky St., Bld. 2

692401@mail.ru, alttrak09@mail.ru, oleg.kulakov@kamaz.ru

**Abstract.** This paper assesses the impact of reducing the pressure in the lubrication system on the failures of the crankshaft bearings. The method of adapting lubricating system of the diesel engine as the wear in operation and depending on the operation modes.

In the operation of the internal combustion engine (ICE), the connecting rod bearings are experiencing oil starvation because of the gap in the supply flow channels. This increases the likelihood of such costly failure as rotation of the inserts [1] due to the lubrication system pressure reduction [2,3] (Table. 1, Fig. 1).

Table 1

The pressure in the engine lubrication system, KAMAZ-740 on different runs

| Fequency, min$^{-1}$ | Operating vehicle mileage, th. km | | | | | | |
|---|---|---|---|---|---|---|---|
| | 20 | 40 | 60 | 80 | 100 | 120 | 140 |
| 2600 | 0.39 | 0.34 | 0.35 | 0.29 | 0.24 | 0.23 | 0.23 |
| 600 | 0.21 | 0.7 | 0.16 | 0.12 | 0.07 | 0.09 | 0.07 |

The reason for reducing the pressure in the engine lubrication system to nominal mode to P = 0.2 ÷ 0.3 MPa after 100 ÷ 140 thousand km. may be insufficient performance of the oil pump [3, 4].

Starting with the engine KAMAZ-740.11 is used the pump of increased productivity [2], which made it possible to slow down the pressure reducing operation (Fig. 1).

The efficiency of bearings is determined mainly by lubrication conditions and in its turn wear of the bearing lead to reduce the pressure in different parts of the lubrication system [2]. Therefore the degree of wear of bearings of the internal combustion engine (ICE) determines the pressure at different points of the lubrication system.

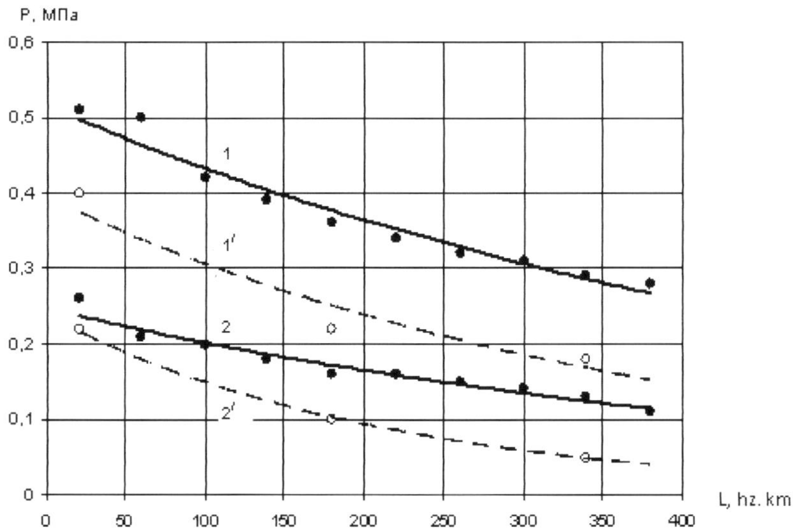


Fig. 1. Dependence of pressure in the engine lubrication system by KamAZ use: unprimed - KAMAZ-Euro; primed - KAMAZ-740
(*1* - for n = 2200 $min^{-1}$ and *2* - when n = 600 $min^{-1}$)

To ensure constancy of the flow of oil to the connecting rod bearings its continuity and availability in its centrifugal trapped it is necessary to preserve constant oil pressure for the excess crankshaft axis is not less than 0.2 MPa as in the initial period of engine operation [3].

With increasing of the engine crankshaft rotation speed increasepressure loss of inertial forces. The gap in the crank bearing increases because of wear as well as under influence of the load due to the increase of the relative eccentricity. This increases oil flow through the clearance in the main bearing. This oil flow through the channel of the crankshaft to the axis is constantly decreasing. All this leads to the fact that oil pressure on the axis of the crankshaft will decrease. At the same time increase flow gaps in the connecting rod bearings because of the growth gap dimensions. The magnitude of the pressure on the crankshaft is reduced continuously and can be zero that is the limit value. It is found that new engines value of engine pressure is 0.2-0.3 MPa. Allowable pressure can be taken as 0.05 MPa, 0 MPa is limit.

In the method developed by the ICE on vehicleservice limitthe degree of wear of bearings [5] is proposed to measure the pressure in the oil line on the oil supply channel portion to the crank bearing on the crank shaft after the main bearings during the engine operation. The value obtained at any time must be greater than zero.

Usingthe proposed method of evaluation of the technical condition of the internal combustion engine can reliably determine the condition of the bearings of the crank mechanism as well as to predict the time of trouble-free operation of the engine and the remaining service life.

In order the lubrication system is self-adjusted according the ICE wearis developed adaptive lubrication system which differs from the traditional following.

There are valves in the oil pump body: pressure reduction and lubrication systems (differential). Pressure reducing valve is designed to limit the maximum oil pressure at the outlet of the pump and adjusted to the opening pressure 0.85-0.89 MPa and valve lubrication system has to maintain operating pressure in the main oil line 0.40-0.45 MPa. In this case the control pressure is supplied from the master cylinder oil passage. Its main purpose is to control draining oil from the high pressure area, preventing from decreasing and increasing of the set pressure [1].

The disadvantage of this device is that the valve setting does not take into account changing of oil flow through the users particularly through the connecting rod bearings to the extent of wear during operation.

The adaptive tuning lubrication system allows for differential pressure valve taking into account changing flow of oil through the connecting rod bearings as they wear in service due to the fact that the channel for supplying the pilot pressure to the differential valve from the crankshaft axis through the sleeve (Fig. 2).

At the beginning of operation of the engine the adaptive system will support oil pressure at the outlet of the oil pump 3, equal to the sum of the overpressure 0.2 MPa and losses from centrifugal forces and losses to customers. This amount is 0.5 MPa. With increasing pressure losses because of increased flow of oil through the main and connecting rod bearings due to wear of parts, to maintain the desired excess pressure of 0.2 MPathe differential valve lubrication system 2 will be covered. The pressure at the outlet of the oil pump will increase thus the pressure will rise in the lubrication system. Similarly the device will operate with increasing rotation frequency of the crankshaft 16 leading to an increase in pressure loss by centrifugal forces.

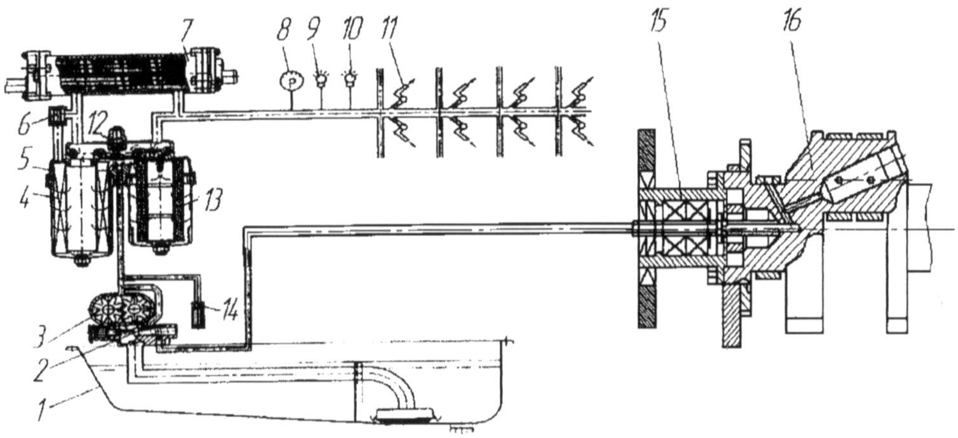


Fig. 2. Schematic diagram of the adaptive system of supply of lubricant to the internal combustion engine:
1 - oil pan; 2 - differential valve lubrication system; 3 - oil pump; 4 - full-flow filter element; 5 -oil filter; 6 - relief valve; 7 - water-oil heat exchanger; 8 - oil pressure indicator; 9 – indicator of oil pressure; 10 - switch of oil overheating; 11 - piston cooling nozzles; 12 - thermostatic valve; 13 - partially in-line filter element; 14 - reducing valve; 15 - coupling; 16 - crankshaft.

Thus the valve setting will take into account the change in the oil flow through the bearings as they wear during operation, that will help to change the pressure in the internal combustion engine lubrication system and maintain optimal lubrication conditions.

## Literature

[1] Denisov A S, Kulakov A T 2007 *Ensuring the reliability of automotive engines* Saratov: Saratov State Technical University 422 p
[2] The engine of KAMAZ 740.11-240 1977 *Operating Instructions* 740.11-3902001RE, Naberezhnye Chelny 120 p
[3] Kulakov O A, Kulakov A T, Denisov A S, Gafiyatullin A A 2010 Change the connecting rod bearing lubrication conditions during operation of the engine *Friction and lubrication in machinery* No 10 pp 13-22
[4] Kulakov A T, Makushin A A, Kulakov O A 2008 Experimental study of changes in the connecting rod bearing lubrication conditions during operation *Friction and lubrication in machinery* No 4 pp 17-20

[5] Makushin A A, Kulakov O A, Kulakov A T, Barylnikova E P, Gafiyatullin A A 2013 The method for diagnosing the degree of wearout of the internal combustion engine bearings. The patent for the invention No 2560972

IOP Conf. Series: Materials Science and Engineering **240** (2017) 012011 doi:10.1088/1757-899X/240/1/012011

# Mathematical modeling of the crystallization process to obtain a fine fraction aluminates

**D A Bashmakov, D I Israphilov and A D Samigullin**

Kazan Federal University, 68/19 Mira av., Naberezhnye Chelny, 423810, Russian Federation

BashmakovDA@yandex.ru

**Abstract**. Considered process of melt crystallization of the material in a twin-roll mold. Resolved an important problem of heat conduction through the wall of variable thickness, using the Fourier method and theorem Duhamel. Obtain theoretical dependence of the thickness of a frozen layer on the physical parameters of the melt and technological parameters of the crystallization process.

## 1. The first section in your paper

Development and implementation of technologies for the processing of low-grade ores of nonferrous metals and extraction of precious metals from scrap and waste around the world is becoming an important area of activity of metallurgical companies. In connection with the above, an urgent task becomes optimizing production to reduce the cost of deeper extraction of precious metals from ores and waste.

In order to exclude the long-term operation of cooling and energy-consuming step of crushing pigs, proposed to use modified twin roll crystallizer. For easy clarifying character of crystallization process, two-roll scheme during molding considered on example of the right roll (Figure 1). Second roll will be same in operating.

Full rotational cycle may be divided into four zones marked as I, II, III and IV.[1]

First zone - crystallization time $t_1$ layer thickness $\delta_1$, that is the contact time of roller surface element with a liquid melt, which is located in a container-crystallizer.

Second area - further crystallization time $t_2$, during it imposition of liquid film thickness $\delta_2$ is completely solid.

Third zone - cooling time of "rolls wall - hard layer" system till full crystallized layer at which the destruction of the material under a permanent load by shear knifes 8. This zone is limited to the area of passage crystallized surface layer element by the moment of complete solidification imposition films to chopping knifes.

In the fourth zone only wall of the roll is cooled till come to entrance of container-crystallizer zone.

IOP Conf. Series: Materials Science and Engineering **240** (2017) 012011 doi:10.1088/1757-899X/240/1/012011

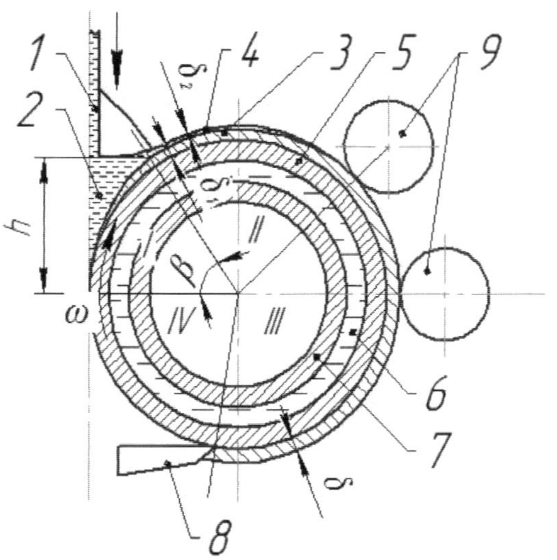


Figure 1. Scheme of continuous operation roll crystallizer.
1 - melt filing; 2 - container-crystallizer; 3 - crystallized layer; 4 - film liquid melt; 5 - drum, 6 - coolant, 7 - internal drum 8 - blade, I - first cooling zone, II - second cooling zone, III - third cooling zone IV - cooling zone of the roll, $\delta_1$ - thickness of crystallized layer at the outlet of the first zone, $\delta_2$ - layer thickness of liquid melt at the outlet from the first zone, $\delta$ - thickness cooled melt, ω - angular rotation velocity of the rolls, $\beta$ - angle of the first zone, $h$ - the depth of container - crystallizer

## 2. Heat transfer in thin-layer crystallization on the rolls.

Considered adjoin problem of heat conduction through variable thickness of wall "rolls wall - hard layer" with a convective heat exchange from two sides [2], solid layer with release heat of phase transition.

Heat transfer problem for the drum wall comes to the solution of the heat equation [3]:

$$\frac{\partial T_C}{\partial t} = \alpha_{CT} \frac{\partial^2 T_C}{\partial x^2}, \; 0 \le x \le l;$$

where $T_C$ - layer temperature, $\alpha_{CT}$ - heat transfer coefficient from freezing layer to the rolls wall, $x$ - coordinate of the rolls wall layer beginning from inside of rolls wall, washed by coolant, $l$ - thickness of the rolls wall, $t$ - contact time of cooled roll surface with the melt.

In boundary conditions [4, 5]:

$$\frac{\partial T_C}{\partial x} = \frac{\alpha_2}{\lambda_{CT}} \left( T_{CT} - T_2 \right), x = 0;$$

$$T_{CT} = T_0\,(t),\; x = l;$$

$$T_C = f(x),\; t = 0;$$

where $\lambda_{CT}$ - rolls wall heat transfer coefficient, $\alpha_2$ - rolls wall heat transfer coefficient to the coolant, $T_2$ - coolant temperature, $T_{CT}$ – outside rolls wall temperature.

To solve this problem of the inhomogeneous heat on roller wall used Fourier method [6] and Duhamel theorem.

The task of freezing melt layer in our formulation was solved by multivariable L.G. Loitsiansky, common to the heat conduction problem with moving boundary E.M. Smirnov [7].

After the transformation equations to new dimensionless variables and then integrate a series expansion in powers of the parameters, we obtain a system of equations for the determination of $T = T_0\,(t)$ and $\delta_1 = \delta(t)$, describing the process of freezing the liquid product on the surface of the roll wall.

IOP Conf. Series: Materials Science and Engineering **240** (2017) 012011 doi:10.1088/1757-899X/240/1/012011

As a result of solving the problems described above to determine the dependence of the film thickness, freezed on the surface of the roll, during his stay in container-crystallizer [8, 9].

$$\delta_1 = 6{,}5 \cdot 10^{-5} \frac{\left(\lambda_P \dfrac{T_{\kappa p} - T_2}{c_{\kappa p}\rho a_P}\tau\right)^{\frac{2}{3}}}{\left(\lambda_P\left(\dfrac{1}{\alpha_2} + \dfrac{\delta_{cm}}{\lambda_{CT}}\right)\right)^{\frac{1}{2}}\left(\alpha_1 \dfrac{T_1 - T_{\kappa p}}{T_{\kappa p} - T_2}\right)^{\frac{1}{6}}};$$

$$\delta_2 = 0{,}94\left(\frac{\mu u}{\sigma}\right)^{\frac{1}{6}}\left(\frac{\mu u}{\rho g \sin\dfrac{\beta}{2}}\right)^{\frac{1}{2}}\frac{\rho}{\rho_{m\theta}}.$$

Where: $\mu$ - dynamic viscosity of the of liquid melt, $u$ - linear outer surface of the rotational speed of the roll, $\sigma$ - the surface tension of liquid melt, $\rho$ - melt density at a temperature close to the temperature of crystallization, $g$ - acceleration due to gravity, $\beta$ - angle of first zone, $\rho_{m\theta}$ - frozen melt density.

Experimental data of conducted experiments on the two-roll mold shows a good correlation for the given formula.

Calculation of twin roll crystallizer conducted to determine the required flow of coolant (water) and rotational speed of the rolls when aluminates crystallize.

As the input task of data accepted a requirement the cooling on the existing two-roll mold aluminate 0,034 m3 during 120 seconds with the ability to specify the installation location of the shearing knife. Results of calculations presented in graphical form in Figures 2.

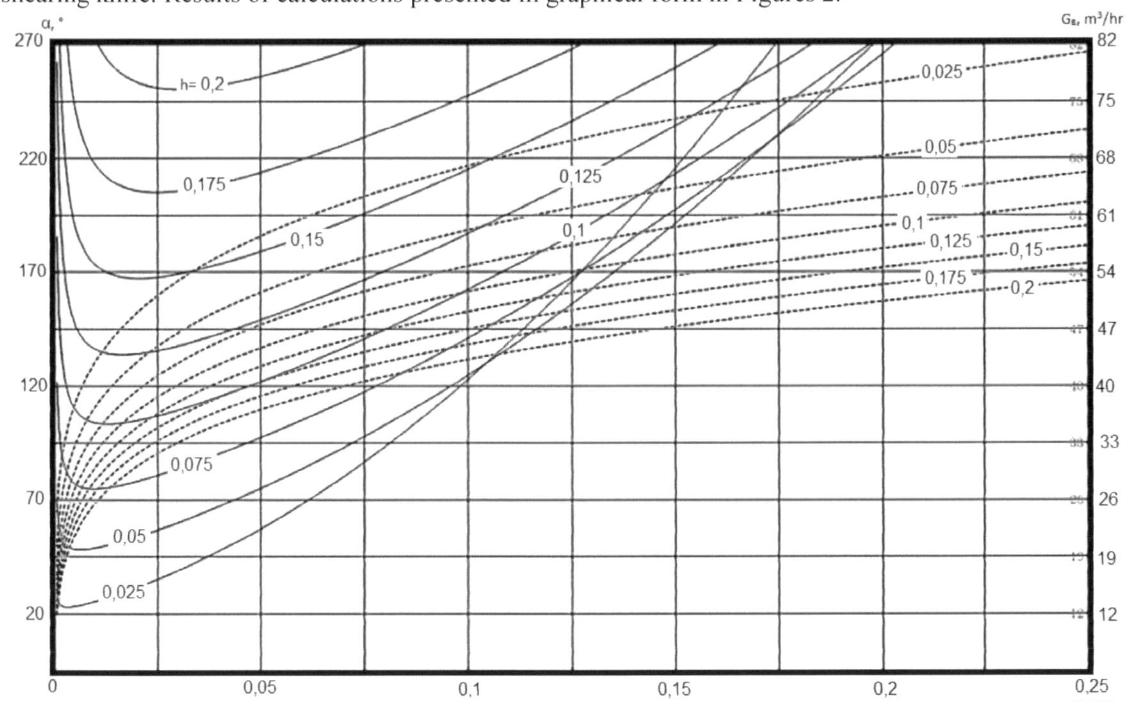


Figure 2. The nomogram for determining the parameters crystallizer work.
- The installation angle of the blades, α°; ---- Coolant flow in m³/h; h - depth container-crystallizer, m.

The result of this work is a method of calculating the two-roll crystallizer, allowing to select appropriate modes of operation depending on the crystallizing melt. All this allows choosing the optimal operating conditions crystallizer work mode for the specific requirements of the final product.

This work was supported by the research grant of Kazan Federal University.

**References**

[1] Kravchenko V M 1950 Triple system benzene-toluene-naphthalene *Journal. nat. chemistry* **24** pp 1033-1056

[2] Anosov V J, Pogodin S A 1947 *The basic principles of physical and chemical analysis*. (Publishing AN USSR, Moscow) p 228

[3] Wang H 1979 *Basic formulas and data on heat transfer for Engineers. Directory.* (Atomizdat, Moscow) p 216

[4] Israphilov I H, Israphilov D I, Bashmakov D A, Galiakbarov A T, Samigullin A D 2015 Calculation of thermal processes in bottom electrode *Contemporary Engineering Sciences* Vol 8 no 1 pp 13 – 20

[5] Israphilov I H, Israphilov D I, Bashmakov D A, Galiakbarov A T, Samigullin A D 2014 Numerical analysis of temperature distribution in bottom electrode of DC arc furnace in process *Contemporary Engineering Sciences* Vol 7 no 26 pp 1483 – 1491

[6] Timmermans J 1959 *The physicochemical constants of binary systems in concentrated solutions*. (Interscience Publishers Inc., N.Y.) V 1

[7] Gelperin N I & Nosov G A 1975 *Basic techniques of melt crystallization* (Chemistry, Moscow) p 182

[8] Kuzmenko Yu P, Kasymbekov B A & Myasnikov S K 1984 *Phase equilibrium liquid - solid body in the system n-hexadecane - n-octadecane*, Dep. In TSSNF TSNIITENEFTEHIM, IGIC them. NS Kurnakova USSR, Moscow, USSR

[9] Israphilov I H, Galiakbarov A T & Israphilov D I 2014 *The bottom electrode for metal smelting* RU Patent 142995 FSAEI HPE KFU, Russian Federation 25 Feb 2014 2014107020/07

IOP Conf. Series: Materials Science and Engineering **240** (2017) 012012 doi:10.1088/1757-899X/240/1/012012

# Study of sound-absorbing properties of glass-fiber reinforced materials used in engineering

**V E Egorova, R R Habibova and L N Shafigullin**
Kazan Federal University, Russia, Republic of Tatarstan, Tukaevsky district, Naberezhnye Chelny, Pr. Mira 68/19

Nika29.94@yandex.ru, regina_16-94@mail.ru

**Abstract.** Modern engineering makes high demands to the noise level in the passenger compartment or cabin of KAMAZ. An effective means of dealing with noise is to use sound absorbing materials produced by the automotive industry. To increase sound-absorbing capacity of materials and structures using glass fibre reinforced polyurethane foams (PUF) obtained by the technology FiberCompositeSpraying.

One of the promising areas of development of mechanical engineering is the development and production of sound-absorbing materials and products.Modern production makes high demands to reduce background noise created by the nodes and aggregates: engine, transmission, suspension, steering, etc.

From the standpoint of human physiology and ecology noise is a sound irritant, which is generally characterized by a combination of different intensity and frequency components. Noise is a factor of the sensor, informative overload and is a typical stressor.Acoustic discomfort negatively affects the well-being and health of people [1]. Sound levels in the truck cab must not exceed 70 dBA, and in the cabin of cars and buses should not exceed 60 dBA[2].Increased noise adversely affects human health, leading to an increase in occupational diseases (deafness, pneumatic hammer disease, etc.), increases the likelihood of cardiovascular disease..In this context, the struggle with the noise generated by the equipment is an urgent practical problem. Silencers, heat-proofing screens are most frequently usedfor struggle against noise in the vehicle cabin.

As a sound-absorbing engineering materials polyurethane composite materials (CM) which have high sound absorption, low specific gravity, strength and thermal conductivity are widespread but they were not tested of sound absorbtion.

In this regard, the study of sound-absorbing properties of the glass-filled polyurethane foam material is an actual material science problems.

Experimental study of sound absorption properties were performed on glass-filled polyurethane materials, consisting of polyurethane systems based on the polyol component A (SPECFLEX 753) and the isocyanate component B (SPECFLEX 138) in the ratio of components A: B - 1.8: 1, [3].As the fillerglass fiber roving of brand EDR 24-2400-386 (2400 tex) [4] is used in an amount of 25 parts by weight 100 pbw matrix component. Products from a glass-filled polyurethane were produced by innovative «FiberCompositeSpraying» technology (FCS) [5]. The main process equipment for manufacture of products used: ABB robot 721 68; Zamboni KraussMaffei; Press SIEMAG.

We carry out the test of four pieces of glass foam:

- № 1, 2, 3 - 2 mm;

- № 4 - 5 mm;

Sound absorption coefficient was determined on acoustic tube Kundt type 4206 the firm "Brüel&Kjær" according to GOST 16297-80 "Sound insulation and sound absorbing materials. Test methods " [6]. We carried out three measurements of samples sound absorption coefficient (Ks) parts 1-4. ThetestresultsareshowninFig. 1-4.

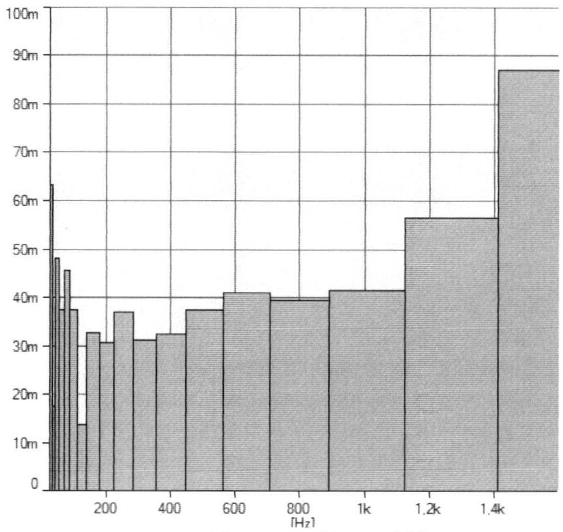


Figure 1 –KspartNo1

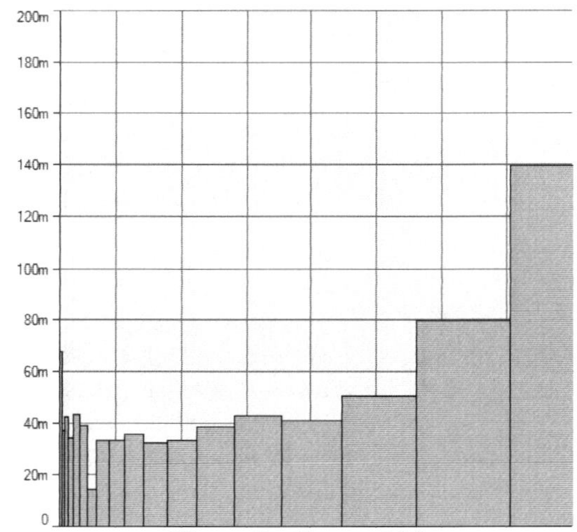


Figure 2 - Kspart №2

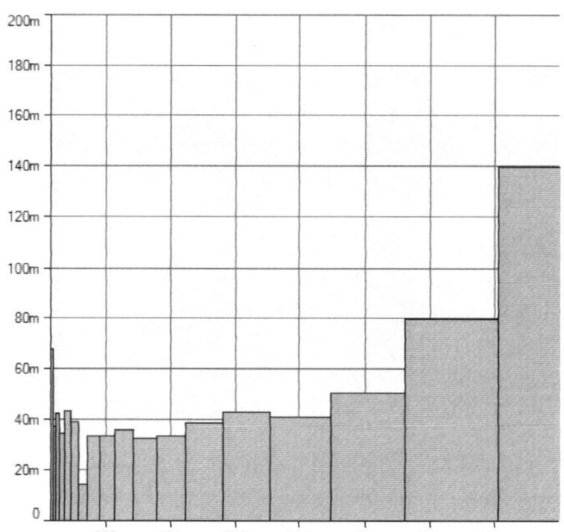


Figure3 - Kspart №3

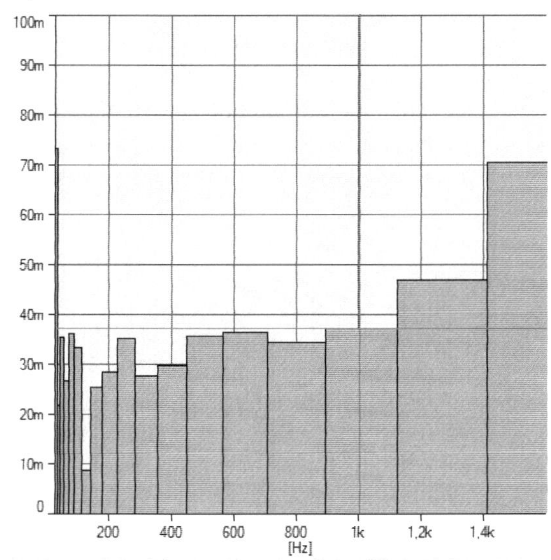


Figure 4 - Ks parts №4

The samples have low KZ in the test frequency range,KZ maximum are reached in the mid-range - 1600 Hz.

Analysis of the data showed a satisfactory sound-absorbing properties of the glass-filled polyurethane.

IOP Conf. Series: Materials Science and Engineering **240** (2017) 012012 doi:10.1088/1757-899X/240/1/012012

**References**

[1] Kochergina KA, Garmonov SY, Mavleev A I and Kochergin A V 2010 Save useful information system and the protection of human hearing in production Bulletin of Kazan Technological University No 7 pp 162-70
[2] 4616-88 Sanitary norms "Sanitary rules on occupational health car drivers"
[3] URL: http://www.dow-izolan.com/ru/products/ (date: 06.07.2016)
[4] URL: http://poly-tex.ru/fiberglass/glass-roving/ (date: 06.07.2016)
[5] URL: http: //www.kraussmaffei.com/ru/home.html (date: 06.07.2016)
[6] GOST 16297-80. Sound insulation and sound absorbing materials. Test methods (instead of GOST 16297-70); Introduced to 01.01.1981. - M .: Publishing House of Standards, 1980.

# Mathematical modeling of capillary flow of viscous fluid

**V L Fedyaev[1,2], E R Galimov[2] and F M Il'yasova[2]**

[1] Institute of Mechanics and Mechanical Engineering of Kazan Scientific Centre of Russian Academy of Science, Kazan, Russian Federation
[2] Kazan National Research Technical University named after A N Tupolev, Kazan, Russian Federation

morenko@imm.knc.ru

**Abstract.** We consider flow of viscous fluid in the capillaries under assumption that the contact angle of the curve varies over the length of the capillary. For the two laws of change we obtain estimates of the distance of meniscus from its initial position and reduction of its movement speed. It was found that the results of the evaluations are consistent with the existing physical concepts.

## 1. Introduction

Fluid flow in capillaries (capillary flow) is often found in nature and technology; it plays an important role in the plant and animal life, when filtering the fluid media in the grounds, in the drying processes of dispersed bodies, media purification from mechanical impurities, impregnation of porous materials and in many other cases. Fundamental principles of capillary fluid flow theory are presented in [1-6 and etc.]. Let's name some of them.

In general case, the fluid transfer in the capillaries occurs by both molecular (diffusive) and convective mechanisms. First is caused by the difference in concentration of a substance along the length of the capillary. It is important to note that because of braking action of the capillary solid walls the diffusion coefficient is much smaller than that in a free fluid and this fluid transfer mechanism will be the most significant in very thin capillaries.

When capillary lateral dimensions are relatively large the convective transfer is dominating its intensity is determined, firstly, by the pressure drop along the length of the capillary, secondly, by the temperature gradient along it, thirdly, by the action of the surface tension forces at the interface between fluid and gaseous medium (meniscus), mass forces, in particular, gravitational forces.

The surface tension force is the force acting on a contour unit length of the media interface and is characterized by the surface tension coefficient σ also called surface tension. The magnitude of the surface tension coefficient depends on the material of media in contact, temperature, presence of adsorbed (surface active) agent on the fluid surface and other factors.

## 2. Part of mathematical modeling

Taking into account the surface tension, in the absence of mass forces at the contact boundary of immiscible viscous fluid and gaseous medium the following relations are written [3]:

$$p_{nnl} + p_\sigma = p_{nng}, p_{nsl} + p_s = p_{nsg}, \tag{1}$$

They characterize the balance of loads acting on the interface surface;

$$v_{nl} = v_{ng} = w_n, v_{sl} = v_{sg} = w_s.$$

IOP Conf. Series: Materials Science and Engineering **240** (2017) 012013 doi:10.1088/1757-899X/240/1/012013

Here, the subscripts $l, g$ mean the parameters of the fluid and gaseous medium, respectively; $p_{nn}$, $p_{ns}$ – normal and tangential components of stress $\vec{p}_n$; $v_n, v_s$; $w_n, w_s$ – components of motion velocity of medium, interface surface; $p_\sigma, p_s = (grad\ \sigma)_s$ – capillary pressure, tangential force per unit area.

In the simplest case, when $p_{nnl} \approx -p_l, p_{nng} \approx -p_g$, the surface tension along the contact boundary is constant $((grad\sigma)_s = 0)$, the capillary – tube of radius $r$, meniscus has a shape of a spherical segment, capillary pressure [5]

$$p_\sigma = \pm 2\sigma/\tilde{R}.$$

Here, $\tilde{R}$ – mean radius of the interface curvature, $\tilde{R} = r/\cos\theta$, $\theta$ – limiting wetting angle; plus sign is selected for a convex fluid surface, minus sign – for a concave $(p_\sigma < 0)$.

In accordance with the Young's equation [4-6]

$$\cos\theta = \left(\sigma_{sg} - \sigma_{sl}\right)/\sigma, \tag{2}$$

where $\sigma_{sg}, \sigma_{sl}$ – surface tension at the interfaces solid body – gas, solid body – fluid.

In work [6] it is shown that the contact angle determined according to (2) gives the angle value at the three phase boundary of the media in contact on a microscopic scale. The macroscopic contact angle depends on the system geometry which is, in turn, under a great influence of the mass forces, structure and material of the solid body surface, medium temperature.

When studying the capillary flow of fluid in the above-mentioned cases it is usually considered that capillary is a tube, probably, with varying along the length cross-section [5], the media in a capillary are under action of gravitational force, along the movement of meniscus the limiting wetting angle remains constant.

Analyzing the available result you can conclude that during the spraying of polymeric powder materials [7-10] at the stage of coalescence of fluid particles situated on the treated body surface, their spreading [8], the medium movement is similar to the capillary one. For better understanding of peculiarities of the given processes taking into account the specificity of particles coalescence we consider the capillary flow of fluid between two plates situated at a small distance from each other $d = 2r$. Let's assume that the meniscus cross-section is close to a circular arc of radius $R$ its surface is under the influence of pressure $p_g$ from the gas side, pressure $p_P$ – fluid. Let's also assume that during the meniscus movement the limiting wetting angle varies according to a particular law.

Assuming that the surface of the channel walls is wettable the contact angle $0 < \theta < \pi/2$, we represent the dependence $\zeta = \cos\theta$ in the form:

$$\zeta = \zeta(\bar{h}) = 1 - (1 - \bar{\zeta}_K) \cdot \bar{h}. \tag{3}$$

Here $\bar{h} = h/h_K, h = h(\tau)$ – meniscus distance from the original position occupied by it in the capillary at the time $\tau = 0$;

$$h_K = \sigma\zeta_K/\Delta p_c^*, \zeta_K = \cos\theta_K,$$

$\theta_K$ – characteristic limiting wetting angle for the considered system solid body-fluid-gas, parameter $\bar{\zeta}_K \geq \zeta_K$, $\Delta p_c^* = p_g - p_p^*$ – pressure drop of media in contact on the meniscus surface.

The average velocity of the meniscus movement will be:

$$\tilde{v}_M = dh/d\tau = h_K \cdot d\bar{h}/d\tau.$$

Considering as it is customary in the capillary hydrodynamics that the given movement of fluid is laminar, close to the stationary (quasi-stationary) to describe the average velocity of its movement in the absence of the mass forces we use the Poiseuille solution:

$$\tilde{v}_\Pi = r^2(p_\sigma - \Delta p_c)/(3\eta h_K \bar{h}),$$

where $\eta$ – fluid viscosity, $\Delta p_c = p_g - p_p$, when it flows plane-parallel between the plates

$$p_\sigma = \sigma\zeta(\bar{h})/r.$$

Equating the velocities $\tilde{v}_M$ and $\tilde{v}_\Pi$, after a series of transformations we obtain the equation:

$$d\bar{æ} - \bar{\delta}_P \cdot d\bar{æ}/\bar{æ} = \bar{\delta}_\eta d\tau. \tag{4}$$

Here function $\bar{æ} = \bar{æ}(\bar{h}) = \bar{\delta}_P - \bar{\delta}_\zeta \bar{h}$, parameters

$\bar{\delta}_P = \sigma/r - \Delta p_c$, $\bar{\delta}_\zeta = \sigma(1 - \bar{\zeta}_K)/r$, $\bar{\delta}_\eta = (\sigma(1 - \bar{\zeta}_K)/h_K)^2/(3\eta)$.

Since for $\tau = 0$ the value $\bar{h} = 0$, $\bar{æ} = \bar{\delta}_P$, integrating (4) we find the dependence $F(\bar{h}, \tau) = 0$:

$$\bar{h} + \bar{\sigma}_P \ln(1 - \bar{h}/\bar{\sigma}_P) = -\bar{\sigma}_\eta \tau, \quad (5)$$

where $\bar{\sigma}_P = \bar{\delta}_P/\bar{\delta}_\zeta = (\sigma - r\Delta p_c)/(\sigma(1 - \bar{\zeta}_K))$, $\bar{\sigma}_\eta = \bar{\delta}_\eta/\bar{\delta}_\zeta = r\sigma(1 - \bar{\zeta}_K)/(3\eta h_K^2)$.

Hence, the average velocity of the meniscus movement in the capillary

$$\tilde{v}_M = \tilde{v}_M(\tau) = (\bar{\delta}_P r^2/3\eta h)(1 - \bar{h}/\bar{\sigma}_P). \quad (6)$$

If the ratio $\bar{h}/\bar{\sigma}_p \ll 1$, then from (5) and (6) we obtain the estimate of the functions $h(\tau)$, $\tilde{v}_M(\tau)$:

$$h \approx (2r(\sigma - r\Delta p_c)\tau/(3\eta))^{1/2}, \quad (7)$$

$$\tilde{v}_M \approx (r(\sigma - r\Delta p_c)/6\eta\tau)^{1/2} \approx h/(\sqrt{2}\tau).$$

It is seen that immediately after the beginning of fluid movement in the capillary the remoteness of the meniscus increases proportionally to $\sqrt{\tau}$ at that velocity of its displacement decreases with time, the movement is slowed down. According to (6) the meniscus stops when $\bar{h} = \bar{\delta}_P$ i.e. at the distance from the movement start

$$h = h_{OK} = \frac{r\zeta_K(1 - \Delta\bar{p}_c)}{(\Delta\bar{p}_c(1 - \bar{\zeta}_K))},$$

where $\Delta\bar{p}_c = r\Delta p_c/\sigma$– dimensionless pressure drop and $0 < \Delta\bar{p}_c < 1$ $0 < \Delta\bar{p}_c < 1$.

In accordance with (7) time to the meniscus stop will be approximately,

$$\tau = \tau_{OK} \approx 1.5h_{OK}^2\eta/(r\sigma(1 - \Delta\bar{p}_c)). \quad (8)$$

From formula (8) it follows that the time to the meniscus stop is smaller the smaller are the fluid viscosity, the capillary width, the greater is the surface tension. The given regularities are quite consistent with the existing physical notions.

In the considered case the contact angle $\theta$ increases (see (3)) along the meniscus movement. However, the situation where on the contrary the angle $\theta$ decreases starting from $\theta_K$ is of great interest. In this case, we assume that dependence $\zeta = \zeta(\bar{h})$ is described by the relationship:

$$\zeta = \zeta(\bar{h}) = \zeta_K + (\bar{\zeta}_K - \zeta_K)\bar{h}.$$

By analogy with the previous case we introduce the auxiliary function $æ^* = æ^*(\bar{h}) = \delta_P^* + \delta_\zeta^* \bar{h}$, где $\delta_P^* = \zeta_K\sigma/r - \Delta p_c$, $\delta_\zeta^* = \sigma(\bar{\zeta}_K - \zeta_K)/r$, parameter $\delta_\eta^* = (\sigma(\bar{\zeta}_K - \zeta_K)/h_K)^2/(3\eta)$, using which we obtain the equation for finding $æ^*$ :

$$dæ^* - \delta_P^* dæ^*/æ^* = \delta_\eta^* d\tau. \quad (9)$$

Taking into account that at the initial time $\tau = 0$ $(\bar{h} = 0)$ function $æ^* = \delta_P^*$, integrating (9), we determine:

$$\bar{h} - \sigma_P^* \ln(1 + \bar{h}/\sigma_P^*) = \sigma_\eta^* \tau,$$

where $\sigma_P^* = (\zeta_K - \Delta\bar{p}_c)/(\bar{\zeta}_K - \zeta_K)$, $\sigma_\eta^* = r\sigma(\bar{\zeta}_K - \zeta_K)/3\eta h_K^2$.

At that the average velocity of the meniscus movement

$$\tilde{v}_M = \tilde{v}_M(\tau) = \left(\frac{r^2\delta_P^*}{3\eta h}\right)\left(\frac{1 + \bar{h}}{\sigma_P^*}\right),$$

approximately $(\bar{h}/\sigma_P^* \ll 1)$

$$h \approx (2r\sigma(\zeta_K - \Delta\bar{p}_c)\tau/3\eta)^{1/2}, \quad (10)$$

$$\tilde{v}_M \approx (r\sigma(\zeta_K - \Delta\bar{p}_c)/6\eta\tau)^{1/2} \approx h/\sqrt{2}\tau.$$

In this case, in contrast to the previous one when changing the variable $\zeta$ from $\zeta_K$ to $\bar{\zeta}_K(0 < \bar{h} < 1)$ the velocity of the meniscus movement is not equal to zero, the meniscus does not stop. After the lapse of control time $\tau_K$ from the beginning of the meniscus movement it will shift approximately at a distance $h_K$ determined by the formula (10) and velocity of its movement is $\tilde{v}_K \approx h_K/\sqrt{2}\tau_K$. For fixed $r$, $\zeta_K$, $\Delta\bar{p}_c$, time $\tau_K$ distance of the meniscus from the initial position increases with increasing of the surface tension $\sigma$ and decreasing of the fluid viscosity $\eta$ proportionally to $\sqrt{\sigma/\eta}$.

## 3. Conclusions

IOP Conf. Series: Materials Science and Engineering **240** (2017) 012013 doi:10.1088/1757-899X/240/1/012013

In conclusion note that within the applied approach it is possible to take into account the impact on the dynamics of the meniscus movement in the capillary of the mass forces, change over time of the pressure drop of the media in contact, temperature.

**References**

[1] Landau L D 1944 *Continuum mechanics* (Leninrad: Association of state book and magazine publishers Gostekhizdat)
[2] Shchelkachev V N 1949 *Reservoir hydraulics* (Leningrad: Gostoptekhizdat)
[3] Levich V G 1952 *Physical-chemical hydrodynamics* (Moscow: Publishing house of the USSR AS)
[4] Betchelor Dzh 1973 *Introduction to fluid dynamics* (Moscow: Mir)
[5] Kogan V B 1977 *Theoretical foundations of unit operations of chemical engineering* (Leningrad: Khimiya)
[6] Adamson A 1979 *Physical chemistry of surfaces* (Moscow: Mir)
[7] Galimov E R 2012 *Polymeric powder coatings for special purposes* (Kazan: Publishing house "Ofset servis")
[8] Yakovlev A D 1987 *Powder paints* (Leningrad: Khimiya)
[9] Gavrilova V A, Kashapov N F and Kashapov R N 2011 Biomedical Engineering 45 pp 198-200.
[10] Kashapov L, Kashapov N and Kashapov R 2013 J. Phys.: Conf. Ser. 479 012005.

This work was performed with financial support from Ministry of Education and Science of the Russian Federation in the framework of the Federal Target Program "Research and development in priority development fields of science and technology sector of Russia for 2014-2020" the agreement No. 14.583.21.0057 on 28th of July 2016 (Project ID RFMEFI58316X0057).

IOP Conf. Series: Materials Science and Engineering **240** (2017) 012014 doi:10.1088/1757-899X/240/1/012014

# Formation of polymeric powder coatings

**V L Fedyaev[1,2], E R Galimov[2], N Ya Galimova[2], I R Gimranov[2] and M M Marat[2]**

[1] Institute of Mechanics and Mechanical Engineering of Kazan Scientific Centre of Russian Academy of Science, Kazan, Russian Federation
[2] Kazan National Research Technical University named after A N Tupolev, Kazan, Russian Federation

morenko@imm.knc.ru

**Abstract.** The initial stage of the formation of polymeric powder coatings is considered. The equations for estimating the convergence velocity of the particles and the obstacle at impact, as well as other characteristics were obtained when depositing the particles on the body surface. Using fictitious bubble and the ideal model of the porous medium we found dependencies that describe the coalescence dynamics of the deposited softened powder particles, determined at the time of their full agglutination.

## 1. Introduction

In general, during the formation of the polymeric coating the initial stage of particles deposition on the body surface, then their agglutination, coalescence, spreading over the surface (sintering, melting) and also curing and polymerization are distinguished [1, 2]. Each of these stages can be the subject for a separate research. In this paper we consider without going into details the first of these stages largely determining the quality of the polymeric coatings.

## 2. Deposition of polymeric particles on the treated body surface

Among the large number of technologies, methods and techniques of application of polymeric powder coatings we focus on the jet spraying [2-5]. For this method it is characteristic that powder particles moving in a stream of air, flue gases or their mixture under the influence of the inertial forces, fields of the body forces fall on the treated body [6-11].

In order to simplify the modeling of the sprayed polymeric particles we assume that in the unit volume $V$, adjacent to the unit area $S_m$ of the treated body the particles are of the same average diameter $d$, fly with the same velocity $\vec{v}_d$. At that the total number of impacts of all particles with the surface $S_m$ over the time $\tau$ will be:

$$N_m = \tau \sum_d (N_d v_{dn})/\ell,$$

where $v_{\text{dn}}$ – projection of the vector $\vec{v}_d$ on the normal to the body surface, $N_d$ – number of particles with the diameter $d$, which are situated in the considered unit volume $V$, $V_d$ – the total amount of particles with the diameter $d$, quantitatively $V_d$ is equal to the concentration of particles with the size $d$; $\ell$ – unit length.

At low impact velocity of a single particle in the quasi-stationary mode in accordance with Newton's second law [9]

$$\int_0^s P(s)ds = (4\pi\rho_p\ r_p^3/3)\int_v^{v_p} v dv, \tag{1}$$

where $r_p, \rho_p, v_p$ – radius, material density, particle impact velocity, respectively; $P = P(s)$– impact force; $s, v$– current convergence, velocity of the colliding particles and obstacle.

Assuming that $P(s) \sim 2P_* (s/s_0)(1 - s/2s_0)$, from (1) we find:

$$v = v(s) = (v_p^2 - (3s^2P_*/4\pi s_0\rho_p r_p^3)(1 - S/3s_0))^{1/2}, \tag{2}$$

Here, according to the condition $v(s_0) = 0$, the maximum convergence of the particles and obstacle

$$s_0 = 2\pi\rho_p r_p^3 v_p^2/P_*. \tag{3}$$

The maximum force on impact $P_*$ is determined by the formula [10]:

$$P_* = \delta_{pm}^{2/5}(5m_p v_p^2/4)^{3/5},$$

where $m_p$ – particle mass, $\delta_{pm} = 4(r_p)^{1/2}/(3\pi(\kappa_p + \kappa_m)), \kappa_p = (1 - \nu_p^2)/\pi E_p$, $\kappa_m = (1 - \nu_m^2)/\pi E_m$; $E_p, E_m; \nu_p, \nu_m$ – Young's modulus, Poisson's ratio of the particle and obstacle material, respectively.

If it is necessary having the dependency (2) from the equation

$$\int_0^s ds/v(s) = \tau$$

you can find the convergence law of the particle and obstacle at impact $s = s(\tau)$ and knowing $s_0$ (3), to assess the impact time $\tau_0$ and other parameters characterizing the impact.

## 3. Particle agglutination

When depositing powder particles they subsequently stick together before their liquefaction. Correspondingly, the porosity of the sprayed layer decreases. In order to describe this process mathematically the sprayed powder layer is represented as a bubble porous medium [12]. When laying particles close to the sphere in shape rhombohedrally the equivalent diameter of these pores will be

$$d_\Pi = d_p\,(6\sin^2\vartheta\sqrt{1+\cos\vartheta}/(\pi(1+\cos\vartheta)-1)^{1/3}, \tag{4}$$

where $d_p$ – particle diameter, $\vartheta$ – angle characterizing the laying of particles ($\pi/3 \le \vartheta \le \pi/2$). At that, approximately, the initial porosity of the medium

$$m_0 \sim 0.5 - 0.33\cos\vartheta. \tag{5}$$

According to [1] when particles stick together the equivalent radius of the spherical pores $a = a(\tau)$ satisfies the equation:

$$da/d\tau = -3\kappa\sigma/(4\eta). \tag{6}$$

Here, $\kappa$ – correction factor, basically, caused by the difference of the pores shape of the real material from spherical shape in the model of the fictitious medium, influence of gas pressure on the agglutination process, $\sigma$ – surface tension coefficient at the bubble boundary, $\eta$ – polymeric material viscosity.

Integrating the given equation we find

$$a = a(\tau) = a_0 - K_0\tau,$$

где $a_0$ – initial equivalent radius of the pores ($a_0 = 0.5d_\Pi$), $K_0 = 3\kappa\sigma_0/(4\eta_0)$;
$\sigma_0, \eta_0$ – average values of the surface tension coefficient and viscosity.

Hence, at initially rhombohedral laying of spherical powder particles of the sprayed layer the time of their full agglutination will be:

$$\tau_{C\Pi} \approx 1.33\,(1 - 0.22\cos\vartheta)\eta_0 r_p/(\kappa\sigma_0).$$

If you use the ideal model of a porous medium [13] then at the initial time the radius of the channels cross-sections of this medium is determined by the relationship:

$$r_{П0} = 2m_0 r_p/(3(1-m_0)).$$

Assuming that at agglutination the radius of the pores channels $r_П$ changes over time in accordance with (6) we find:

$$r_П = r_П(\tau) = r_{П0} - K_0 r.$$

Approximately, taking into account (5) time of full agglutination of particles

$$\tau_{CП} \approx 0.89(1-0.66\cos\vartheta)^2\, \eta_0\, r_p/(\kappa\sigma_0).$$

Comparing the ratio for $\tau_{CП}$ we make sure that in the case of a model of the ideal porous medium the time of the pores collapse is less than that when using the model of the bubble medium.

## 4. Coalescence and spreading of particles material over the body surface

When analyzing the coalescence of the fluid particles of the polymeric powder we consider only the behavior of particles in a layer adjacent to the solid surface of the treated body.

If the body surface is wettable then the initial ($\tau = 0$) shape of the half section of the separated particle and transformation of the cross-section over time are shown in figure 1 *a, b*, respectively. Here, *Ox* axis is directed along the body surface, *AB* – part of the upper boundary of the particle, *A'B'* – its average position, *PQ* – interphase boundary, $h_k$ – average particle height equal to $2r_p$ in the initial position, $\theta$ – limiting wetting angle ($\cos\theta = \zeta$), *BQ* – interface surface of the adjacent particles, *QQ'* – conventional boundary parallel to the *Ox* axis.

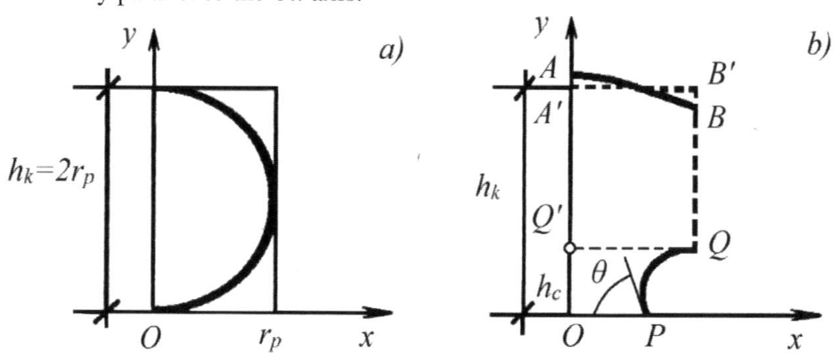


**Figure 1.** Half of the cross-section of the separated fluid particle: at the initial (a) and intermediate (b) time points.

Further we use the provisions of the capillary hydrodynamics [13-19] to solve the problem of the particles coalescence. Let's present the transformation of the cross-section (fig. 1) over time as a settling of the rectangle $A'B'QQ'$ with constant area as a result of spreading of the fluid located in *OPQQ'*, accompanied by a displacement of the points *P*, *Q* and reduction of gas cavity area up to its collapses. In this case, we assume that near *PQ* boundary the fluid moves along *Ox* axis in the horizontal direction under the action of the pressure difference and surface tension forces.

It is known [20] that at plane-parallel flow of a viscous fluid in the channel one of the walls of which is moving with a velocity $v_0$ at constant longitudinal pressure gradient $\partial p/\partial x = p_x < 0$ the fluid velocity

$$v = v(y) = v_0 y/h_c + p_x/2\eta\,(y^2 - yh_c),$$

where $h_c$– remoteness of the conventional interface of the areas *QQ'* from the axis *Ox* (fig. 1)

If $p_x \approx -\Delta p/\tilde{x}_p$, where $\Delta p$ – pressure difference at length $\tilde{x}_p (\Delta p > 0)$, then the average velocity of the fluid flow

$$\bar{v} = 0.5v_0 + \Delta p h_c^3/(12\eta S).$$

Here *S* – *OPQQ'* area, equal to $\tilde{x}_p h_c$. Provided that the fluid is incompressible its flow through the conventional boundary *QQ'* is small, $S \approx const$.

$h_c$ and $v_0$ remain so far undetermined values in relationships, characterizing the fluid flow along the solid surface. When determining them we take into consideration that

$$\bar{v} = \frac{d\tilde{x}_p}{d\tau} = -S(dh_c/d\tau)/h_c^2 = 0.5v_0 + \Delta p h_c^3/(12\eta S), \qquad (7)$$

the shear stress (specific friction force) at the conventional boundary *QQ'*:

$$\tau_{xy} = \eta \frac{dv(y)}{dy}\bigg|_{y=h_c} = \eta v_0/h_c + 0.5h_c p_x.$$

Assuming that $\tau_{xy} \sim 0$, replacing $p_x$ on the $-\Delta p/\widetilde{x_p}$, we find:

$$v_0 \approx \Delta p h_c^3/(2\eta S).$$

Substituting this value $v_0$ in (7), relative to $h_c$ we obtain the differential equation:

$$dh_c/h_c^5 = -7(\Delta p/(12\eta S^2))d\tau.$$

Integrating it on condition that at the initial time $\tau = 0$ the height $h_c = h_{co}$ is known the pressure drop $\Delta p = \Delta p_0 = const$, we determine

$$h_c = h_c(\tau) = 1/((1/h_{co}^4) + (\tau\Delta p_0/3\eta S^2))^{1/4}. \qquad (8)$$

Hence the time of complete layer deposition when gas cavity disappears, will be

$$\tau_k = \left((r_p h_{co})^4 - S^4\right)(3\eta/(h_{co}^4 S^2 \Delta p_0)).$$

The presented solutions at the given parameters $\sigma, \eta$;, the initial values $\widetilde{x_p} = \widetilde{x_{po}}$, $h_c = h_{co}$, the constant pressure difference $\Delta p = \Delta p_0$ allow you to find the area $S$, to assess the movement of the boundaries *PQ*, *QQ'*; the time of complete deposition of the particles layer, spreading of the polymeric material over the treated body surface.

At the capillary flow of the fluid in the channel in the absence of mass forces the average difference in pressure is evaluated by the dependency [13-15]:

$$\Delta p = p_\sigma - p_g + p_p.$$

Here $p_g, p_p$ - pressure in the cavity and the particle material, $p_\sigma = \sigma/R$ – capillary pressure, $\sigma$ – surface tension coefficient at the fluid-gas interface, $R = r/\cos\theta$ – curvature radius of this interface, $r$– half of the channel width. Pressure $p_\sigma$ is the sum of two specific forces $\sigma\cos\theta/d$, acting in the direction of the channel axis and applied to the ends of the meniscus.

Let's assume that when particles coalesce (see fig. 1) the force $\sigma\cos\theta/h_c$, applied at the point *P* and the force $\sigma/h_c$ – at point *Q* are acting on the *PQ*. The total specific tension force directed towards *Ox* axis,

$$p_\sigma = (1+\zeta)\sigma/h_c\,.$$

Consequently, the average difference in pressure causing the displacement of the PQ interface

$$\Delta p = \Delta p(\zeta) = (1+\zeta)\sigma/h_c - \Delta p_g(\Delta p > 0),$$

where $\Delta p_g = p_g - p_p$.

Hence the motion of *PQ* boundary will be directed towards the *Ox* axis, if

$$(1+\zeta)\sigma/h_c > \Delta p_g.$$

In the case when at the *PQ* interface the average pressure drop $\overline{\Delta p_{go}} = const$, the average curvature radius of the *PQ* interface in accordance with the Laplace equation will be:

$$\bar{R} = \sigma/\overline{\Delta p_{go}} = R_0 = const.$$

When this condition is satisfied the *PQ* interface surface displacement at the three characteristic points in time is shown in fig. 2. It should be noted that if at the initial time $\theta \sim \pi/2$ ($\zeta \sim 0$), then as cavity sizes decrease the limiting wetting angle increases (fig. 2 *b, c*) tending to $\pi(\zeta = -1)$. It is known that the values $\theta > \pi/2$ are typical for nonwettable surfaces of the solid body.

Since $\zeta = \cos\theta < 0$, the tension force $\sigma\cos\theta/h_c$ at the point *P* is directed oppositely to the movement of the interface, however, at the point *Q* it still acts in the *Ox* axis direction.

In the considered case when the cavity radius $R_0$ is constant from its geometry it follows that

$$h_c = (1+\zeta)R_0.$$

Thus, in contrast to (8), at a given value of $h_{co}$ the radius $R_0 \approx h_{co}$ ($\zeta \approx 0$), the cavity height $h_c$ depends on $\zeta = \zeta(\tau)$, the pressure difference

$$\Delta p = \sigma/R_0 - \Delta p_g.$$

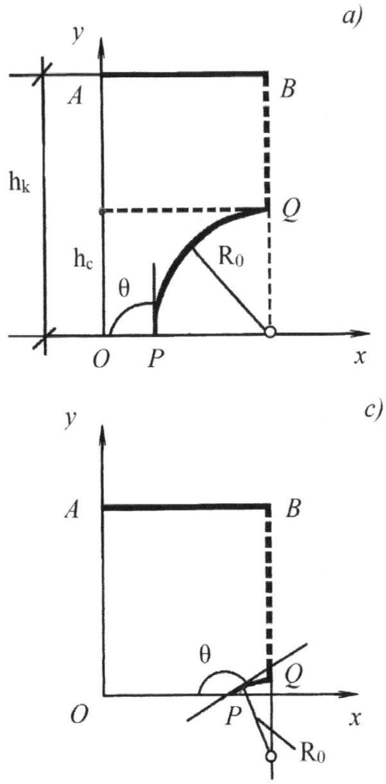


**Figure 2.** Transformation of fluid particle cross-section over time.

Hence, *PQ* interface will move in the direction of *Ox* axis, if the following condition is satisfied

$$\Delta p_g < \sigma / R_0.$$

Here the pressure drop $\Delta p_g$ can be interpreted as a control parameter when it changes both the velocity of interface movement and direction of movement also change.

## 5. Conclusions

In conclusion, note that the obtained solutions can be used to calculate the appropriate characteristics taking into account the changes of the governing parameters depending on the temperature.

## References

[1] Yakovlev A D 1987 *Powder paints* (Leningrad: Khimiya)
[2] Galimov E R, Zverev E V, Tukbaev E E, Galimova N Ya, Kuryntsev S V, Mukhin A M 2012 *Polymeric powder coatings for special purposes* (Kazan: Publishing house "Ofset servis")
[3] Fedyaev V L, Galimov E R, Siraev A R and Martines Markes L. 2016 *Current problems of Continuum Mechanics. For 25-th anniversary of IMME KSC RAS collection of scientific papers* (Kazan: Publishing house «FEN» AS RT) p 452
[4] Gots V L, Ratnikov V N and Gisin P G 1975 *Painting methods of industrial products* (Moscow: Khimiya)
[5] Volkov S S and Girsh V I 1988 *Gluing and spraying of plastics* (Moscow: Khimiya)
[6] Targ S M 1966 *Short course of theoretical mechanics* (Moscow: Nauka)
[7] Landau L D and Lifshitz E M 1965 *Mechanics* (Moscow: Nauka)
[8] Timoshenko S P and Gud'ir Dzh N 1975 *Theory of elasticity* (Moscow: Nauka)
[9] Pris K 1982 *Erosion* (Moscow: Mir)

[10] Zukas Dzh A, Nikolas T, Svift Kh F, Greshchuk L B and Kurran D R 1985 *Impact dynamics* (Moscow: Mir)
[11] Batuev G S, Golubkov Yu V, Efremov A K and Fedosov A A 1977 *Engineering methods, stuadies of impact processes* (Moscow: Mashinostroenie)
[12] Nigmatulin R I 1987 *Dynamics of multiphase media* (Moscow: Nauka) part 1
[13] Levich V G 1952 *Physical-chemical hydrodynamics* (Moscow: Publishing house of the USSR AS)
[14] Betchelor Dzh 1973 *Introduction to fluid dynamics* (Moscow: Mir)
[15] Kogan V B 1977 *Theoretical foundations of unit operations of chemical engineering* (Leningrad: Khimiya)
[16] Adamson A 1979 *Physical chemistry of surfaces* (Moscow: Mir)
[17] Bertenev G M and Zelenov Yu V 1976 *Course of polymer physics edited by prof. S Ya Frenkel* (Leningrad: Khimiya)
[18] Stromberg A G and Semchenko D P 1973 *Physical chemistry edited by A G Stromberg* (Moscow: Vysshaya shkola)
[19] Israfilov Z K and Kashapov N F 1991 *Journal of Engineering Physics 60* pp 364-368
[20] Kochin N E, Kibel' N A and Roze N V 1963 *Theoretical hydromechanics* (Moscow: Fizmatgiz) part II.

This work was performed with financial support from Ministry of Education and Science of the Russian Federation in the framework of the Federal Target Program "Research and development in priority development fields of science and technology sector of Russia for 2014-2020" the agreement No. 14.583.21.0057 on 28th of July 2016 (Project ID RFMEFI58316X0057).

# Study of the combined laser-plasma effect on metals

**AT Gabdrakhmanov, AT Galiakbarov, I H Israphilov.**

Kazan Federal University, 423800, 68/19 Mira avenue, Naberezhnye Chelny, Russia

ATGabdrahmanov@kpfu.ru

**Abstract:** Interaction processes of laser radiation with the metal surface layer during laser effect, and the main factors influencing the absorption capacity of the metal with a surface processing by a laser beam are considered. The method is proposed for increasing the absorption capacity of laser radiation by plasmatron pre-heating of the material. The results of the combined laser-plasma effect are presented.

## 1. Introduction:

The development of modern production necessitates the ever-increasing introduction of high technologies, in particular, of laser processing of materials. This processing is one of the technologies that determine the present level of production in the industrialized countries. The use of laser processing of materials ensures high quality of produced parts, specified productivity of the processes, ecological purity, as well as saving human and material resources.

At present the use of laser technologies in the instrument-making production is extremely diverse. Such technologies include welding, heat hardening, alloying, welding, cutting, dimensional processing, marking, engraving, precision micro-welding and many others [1-2]. In some cases, radiation technologies has been ahead of the competition, as using lasers can obtain technical and economic results that cannot be achieved by other technical means.

Problem of laser processing of materials is transferring the maximum amount of laser radiation energy into the material. In any case, the laser thermal effect on the material is important, not just the power of laser radiation, but the power absorbed by the material and used to getting for a beneficial result. The absorption capacity in one form or another appears in all laser processes.

Several methods are known for increasing the absorption capacity of metals [3, 4]: change in the surface roughness, the formation of oxide films during processing in an oxidizing environment, specific absorbing coating, pre-heating etc. The application of increasing roughness and applying of various coatings is not always appropriate due to an increase processing time and change the surface properties of the material.

It is known [1] that the lowest of the absorption coefficient is characterized by radiation effect of technologically advanced $C0_2$ lasers. The low level of the absorption creates at first sight a hopeless situation with the possibility to use of $C0_2$-radiation for processing of metals. However, the decisive factor is the substantial increase of absorption capacity by an increase in the temperature of the processed surface. The absorption capacity of the non-oxidized metal surface at the wavelength of laser radiation $\lambda$ = 10,6 μm can be determined by the equation of Hagen-Rubens:

$$A = 112{,}2(\sigma_0^{-1})^{1/2}$$

where A - absorptive capacity of the metal; $\sigma_0$ - conductivity of the metal at a constant current value, 1 / ($\Omega \cdot$ m). At high temperature, the electrical conductivity of metals decreases and the absorption coefficient increases.

## 2. Experimental studies

This study offers a method to increase the absorption capacity of laser radiation by pre-heating of the material with a linear plasmatron of indirect action a small capacity. The use of plasmatrons of direct action can lead to melting of the surface and the deterioration of its quality parameters [5-11].

Experimental studies have been conducted by using the developed laser-plasma hybrid installation. The samples of steels: 45, 09Г2С, 08КП, were processed. For comparison, this samples were treated only by first laser radiation, and then the combined laser-plasma effect (Fig. 1). Modes of laser radiation in the course of the study have been changing in the ranges: t=1,5...2.5 ms, E=5,3...1,4 J, dp=2 mm (where t is the time of the pulse, E-pulse energy, dp - the diameter of the focused radiation).

A microstructure of the samples studies were conducted by means of the metallographic inverted microscope called Labomed –I (Fig. 1).

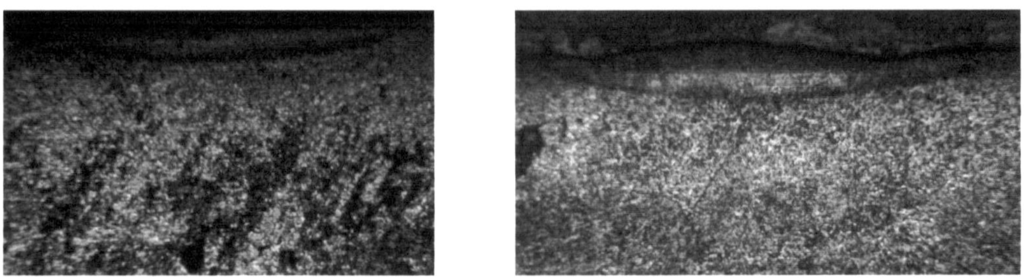

a) b)

**Figure1.** The results of metallographic: samples after the laser (a) and a laser-plasma (b) effect, treatment conditions: t = 2ms, E = 5,2 J, dp = 2 mm, an increase: × 80.

After processing the micro-hardness was measured using the method of Vickers at 100gs load on the digital hardness testing model HVS-1000. The metallographic samples cut along a track and maximally close to points of the track, were prepared for the measurements. Some micro-hardness measurements are presented in Figure 2.

Under the combined effect of plasma and laser radiation, there is an increase of microhardness by 30-40% compared to laser.

Metallographic studies of 45 steel in the field of laser and laser-plasma effect (Fig. 1 and Fig. 2) showed that the thermal effect zone of the plasma jet has the shape of a segment. When heating, by the plasma jet a metal surface heats the surface layer to different temperatures, so it has a layered structure. Depending on the microstructure and micro-hardness of steel 45 in the depth formed three layers: the melting zone, the quenching from the solid phase and the transition zone.

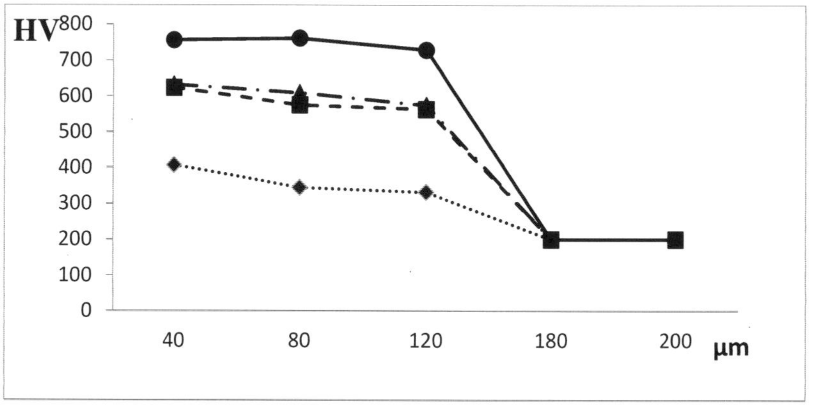


Figure 2 - Distribution of micro-hardness in depth, sample of steel 45, the processing modes: • • • - Laser t = 2ms, E = 3 J, dp = 2 mm; - - - - Laser t = 2ms, E = 5,2 J, dp = 2 mm; - • - - Laser + plasma t = 2ms, E = 3 J, dp = 2mm - - Laser + plasma t = 2ms, E = 5,2 J, dp = 2mm.

## 3. The conclusion

Tests of the designed installation were conducted to check the performance of the laser-plasma hybrid installation as compared to the laser installation on different samples. As a result of these tests, it was determined that the pre-heating of the plasma samples increases the effectiveness of laser radiation effect.

## References

[1] Grigoryants A G, Shiganov I N, Misyurov A I 2006 *Processes of laser processing* (M.: Izd-vo MGTU im. N. Uh. Bauman) 664 p

[2] Khisamutdinov R M, Zvezdin V V, Saubanov Ruz R, Israfilov I H, Rakhimov R R and Spirin A A 2016 Study of processes of steels surfaces modification with highly concentrated energy flows *IOP Conf. Ser* (Materials science and engineering) Volume 669 Issue 1 **012024**

[3] Kostecki V V, lissovsky A L 2013 Increasing the absorption capacity of the surface layer of metal during laser exposure by using absorbing coatings *Bulletin of Polotsk state University. Series: Industry. Applied science* No 3 pp 97-101

[4] Bashmakov A D, Gabdrakhmanov A T ,Israfilov Y H, Galiakbarov A T 2013 Electrostatic field as factor of increase of stability of quality of laser processing of metals *MNTK "IMTOM-2013"* Part 1 pp 325-28

[5] Kashapov N F, Sharifullin S N 2015 Hardening of the surface plasma jet high-frequency induction discharge of low pressure *IOP Conf. Ser (Materials science and engineering)* Volume 86 Issue 1 **012021**

[6] Gabdrakhmanov Az T , Israphilov I H , Galiakbarov A T , Samigullin A D , Gabdrakhmanov Al T 2016 Improving the efficiency of plasma heat treatment of metals *Journal of Physics: Conference Series* Volume 669 **012014**

[7] Gabdrakhmanov Az T , Israphilov I H, Galiakbarov A T, Bashmakov D A , Samigullin A D 2014 Pulse plasma surface thermostrengthening of machine parts *IOP Conf. Ser (Materials science and engineering)* Volume 69, Issue 1 **012037**

[8] Gabdrakhmanov Az T, Israfilov I H, Galiakbarov A T, Samigullin A D 2013 The thermal characteristics of the process of pulsed plasma processing of steel *News of the Tula State University* Vol 6 Part 1 pp 253-260

[9] Denisov D G, Kashapov N F, Kashapov R N 2015 The appearance of shock waves in the plasma electrolytic processing *IOP Conf. Ser (Materials science and engineering)* **012005**

[10] Kashapov N F, Luchkin A G 2014 The use of low-temperature plasma deposition of hardening coatings on plastics *Proceedings of the higher educational institutions. Physics.* V 57 No 3-3 pp 160-63
[11] Kashapov N F , Kashapov R N 2014 A study of plasma-electrolytic process for different ratios of the anode space to the cathode *Proceedings of the higher educational institutions. Physics* V 57 No 3-3 pp 168-70

# A shader for simultaneous ambient occlusion and edge detection for reverse engineering

**R R Gaisin[a], K V Sharifullina[b] and R N Kashapov[a,b]**

[a] Engineering institute, Kazan (Volga region) Federal University, Kremlevskaya 18, 420008, Kazan, Russia
[b] Kazan Physical-Technical Institute of RAS, Sibirsky Trakt 10/7, Kazan, Russian Federation

rusgaisin@gmail.com

**Abstract**. We have presented real-time post-processing shader, which produces the effect of ambient occlusion and edge detection at the same time. The shader is written in two versions: in Pixel Bender 2.5 (for test purposes) and in AGAL (for real-time rendering). The algorithm is simple to implement and can be easily extended to other higher-level shader languages. It uses only depth texture without screenspace normal texture and without additional noise texture, which is key component of a rendering pipeline of most of known screenspace ambient occlusion algorithms.

## Introduction

In most cases in 3D rendering, we need to select some objects to get user's attention to it. The most obvious way to do it is to draw a set of lines around the "selected" object. These lines can show many different things, including various combinations of lighting, surface discontinuities, and shape. The question is how to draw lines that simultaneously convey several of these cues.

There are different types of bounding contours. Silhouettes represent boundaries between object and background. Occluding contour is a generalization of silhouettes: these mark any depth discontinuities, not just those against the background. This type of contour adds a lot of important detail to the drawing, but still does not convey shallow features, particularly those viewed head-on.

Sharp creases of the surface - works well for polyhedral objects, and is a frequent ingredient in technical drawings. The algorithm is quite simple: just look for a dihedral angle (that is, the angle between two faces connected along an edge) smaller than a threshold.

Many shape-conveying lines based on surface differentials. Occluding contours critically depend on the normal: they are zeros of the dot product between the normal and the view direction. Ridges and valleys (crest lines) are defined as local maxima of curvature [7].

Occluding contours (also known "interior and exterior silhouettes") for any shape are locations of depth discontinuities. There are a few different ways of defining these, of which a very straightforward definition is simply those locations at which, from the current viewpoint, there is a depth discontinuity. Note that these are view-dependent lines. On the plus side, the view dependence makes it much more likely that these lines are interpreted as conveying shape, rather than as surface markings. On the other hand, this means that the lines will have to be recomputed for each frame.

IOP Conf. Series: Materials Science and Engineering **240** (2017) 012016 doi:10.1088/1757-899X/240/1/012016

On smooth surfaces, there is another definition of contours that is useful: contours are those surface locations, where the surface normal n is perpendicular to the viewing direction v. That is, places where n dot v is equal to zero.

Another important post-processing effect in computer graphics is screen space ambient occlusion (SSAO) [2]. SSAO is also useful for producing both interior and exterior silhouettes. However, primarily it was developed as algorithm of shading. This technique emulates ambient lighting by computing shadows that nearby objects would casts on each other's under the effect of an ambient shadowing. It is also can be performed in pixel shader, using the 3D scene depth buffer. For every pixel on the screen, the pixel shader samples the depth values around the current pixel and tries to compute the amount of occlusion from each of the sampled points. The occlusion factor depends on the depth difference between two sampled points.

Both this post-processing effects [3] have a deal with comparison of color of near-lying pixels. Basing on this, in present work we made universal screen post-processing shader, which can work as an edge-detection filter or as ambient-occlusion filter.]

There are three basic sets of algorithms for extracting most kinds of feature lines from 3D objects of the scene. First, there are bitmap-space algorithms that render depth map (or Lambert-shaded surface), then extract contours by doing per-pixel processing (such as color thresholding) on the framebuffer. The advantage of this kind of algorithm is that it can be fast and easy to implement. A major drawback is that it makes it difficult to control the appearance and stylization of the resulting lines.

A second class represents multi-pass hybrid algorithms, which partially perform processing in object space, but the lines show up only in the framebuffer.

Final class of algorithm operates in object space – on the model directly. These algorithms are more complex, but they can provide good control over stylization technique.

A very simple way to produce occluding contours is to render the model with white direct lighting (but without color and textures), and then perform color thresholding [8]. Any region darker than a threshold is set to black (or the line color), and anything above the threshold is set to the background color. One drawback of this algorithm is that the thickness of the lines varies and becomes thicker in low-curvature zones. A way to get fixed a line thickness is based on curvature-dependent threshold. For any constant-curvature region one can determine how thick the lines will be as a function of radial curvature of 3D surface.

Object-space approach provides two types of contour extraction. There are many implementations of object-space algorithm. Contours along the mesh edges produced via a loop over all edges to check whether each has one adjacent back face and one adjacent front face [9]. To extract contours within faces of a 3D mesh, one can use interpolated values of n dot v (n is a face normal vector, v is view direction). Unfortunately, this technique depends on geometrical complexity of the 3D scene and can be relatively slow for high-poly meshes. There is a large body of work on acceleration of these algorithms, that try to reduce time. For the contours-within-faces case, one popular technique is to construct a hierarchical data structure, where each node stores a quiver of the normals below it. The most popular is hybrid approach. It consists of three main steps: Draw front faces in white, offset towards viewer. Draw enlarged back faces. As a result, the second rendering pass will "look out" from behind the geometry rendered previously on the first pass. Algorithm is fast and easy to implement, but it is applicable for silhouettes only.

Another useful image-space algorithm uses data from the depthbuffer (a color that depends on depth) instead of rendering n dot v. This approach makes the rendering simpler, but the image processing more complex. "Main work" of this technique lays on shoulders of pixel shader rather than vertex shader. That's why it can be treated as a screen space postprocessing effect. Our shader is a bit similar.

## Main part

IOP Conf. Series: Materials Science and Engineering **240** (2017) 012016 doi:10.1088/1757-899X/240/1/012016

The key idea of the algorithm is to average dot product of screen's normal vector and a tangent vector of all visible surfaces. To compute screen space tangents we use encoded depth texture. The screen z coordinate of each generated pixel of a scene is stored in a depth texture representing depth buffer (z-buffer). The depth buffer will allow the method to reproduce the real screen space coordinates of each rendered point of the scene.

It can be done by commonly used edge detection technique with the Laplacian operator [1]. Unfortunately, it cannot produce contours of individual object: it works for the whole scene. To select the object of interest (i.e. to draw its contours) we need to render it separately. The basic algorithm is the follows:

- Initialize new screen texture. Let it be b/w.
- Clear the whole screen by black solid color.
- Render the object of interest filled by white solid color.
- Apply Laplacian operator to each pixel of the screen texture. Use finite difference approximation of Laplacian operator: L(x,y) = c(x,y+1) + c(x,y-1) + c(x+1,y) + c(x-1, y) - 4*c(x,y). Where c(x,y) is a color of a pixel of b/w screen texture. Now we can see the outline of the object of interest.
- Invert screen texture. Now object's edges are black, and void space is white.
- Initialize new screen texture and render the entire scene with textures, lighting, etc.
- Multiply b/w and scene texture. The object of interest has black outline.

On the other hand, this algorithm is also applicable for the whole scene after exclusion of the first three steps.

To encode z-values as rgb colors of the depth texture, we use well-known approach described in [4].For integer RGBA textures, we need to pack the depth value and write it into the color channel. There are plenty of ways to pack depth into texture. We use the follows approach:

In the fragment shader: just to project vertices of a scene models to the normalized screen space by means of perspective model-view-projection matrix. We use Adobe Graphics Assembly Language (AGAL) [5] to implement our shader. In AGAL syntax, it looks the following way:

```
m44 op, vt0, vc0        // project vertex coords (vt0) by MVP matrix (vc0)
m44 vt1, vt0, vc4       // project vertex (vt0) according to scene tranform matrix (vc4)
sub v0, vt1, vc5        // compute a vector (v0) from vertex to camera position (vc5)
```

The "scene transform" matrix (vc4) is the transformation matrix that transforms from model to world space.

In the pixel shader we need to use the far clipping plane of a camera. To represent the 32-bit depth buffer, we need to multiply interpolated squared distance between camera and vertex to powers of two: 20, 28, 216, 224. Let the distance from camera to the far plane of the view frustrum is “f”. As the value of expression √(f^2+f^2 ) is largest possible distance for any view frustum, we need to divide by it to normalize obtained distances. Finally, the AGAL pixel code will look like:

```
dp3 ft0.z, v0.xyz, v0.xyz // get a squared distance (ft0) by computing squared length of v0
mul oc, fc0, ft0.z                    // pack the depth (ft0) by multiplying to powers of two (fc0)
```

In this code the register fc0 has the following values:

$$fc0.xyzw = \left(\frac{1}{2f^2}, \frac{2^8}{2f^2}, \frac{2^{16}}{2f^2}, \frac{2^{24}}{2f^2}\right) \tag{1}$$

Here we took into account that the maximum depth value can be approximated this way for the simplicity of computations (no need to calculate square root):

$$\max depth = \frac{1}{\sqrt{f^2+f^2}} \approx \frac{1}{2f^2} \tag{2}$$

Now we have the depth texture that is used for further computations. Once the screenspace depth map is done, we can compute tangent vectors. To make this we simply sample depth values in adjacent pixels of the pixel in processing. Texture coordinate (UV) offsets for adjacent pixels are also known, that is why the resulting tangent vector is described by expression:

$$\vec{t} = (dx - 0.x,\ dy - 0.y,\ z - 0.z) \tag{3}$$

Here "0" is a position vector of the processed pixel; "dx" and "dy" are horizontal and vertical distances between processed pixel and its neighbor pixel. Both the "dx" and "dy" are specified in texture coordinates and corresponds to standard "u" and "v" components which are in range [0..1] fo the whole screen. In our case, we use normalized "dx" and "dy" according to width and height of the screen texture:

$$dx = \frac{du*r}{width}; dy = \frac{dv*r}{height}; \tag{4}$$

Here "r" is a sampling step in pixels, because we can sample not only nearest neighbors, but skip them and make a leap sampling. In figure 1 the processed pixel is point "0" and its left and right neighbor pixels are "L" and "R" respectively. Surface tangent vectors are respectively: "0L" and "0R".

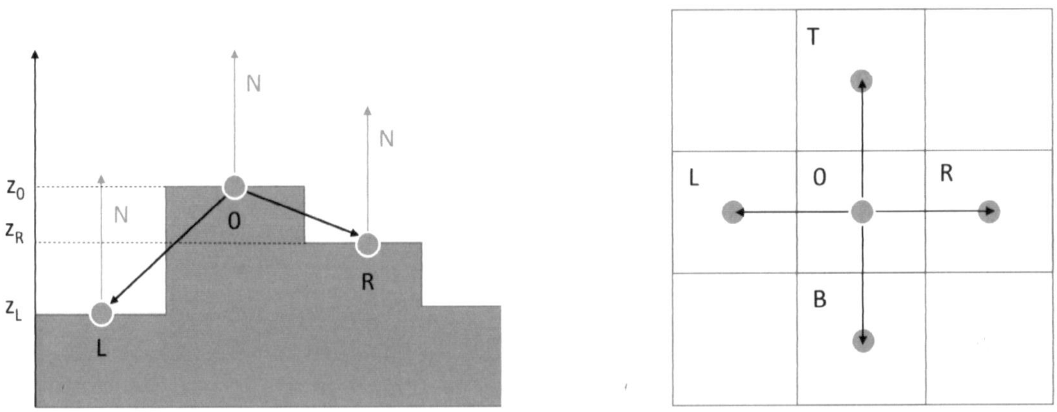


Fig. 1. Description of the idea of the shader. Left: side view of a depth map. Right: we use the same notation for top view of a depth map. "0" is a currently processing pixel. Then we compute tangent vectors for its neighbor pixels of a depth texture.

When the four tangents for the currently processed pixel of the screen had calculated, we compute a dot product of them and of the normal vector of the screen: "N" = (0, 0, 1). According to the figure 1: the lower the neighbor pixel is – the higher is the dot product. Differences between high and low depths finally will be look like different intensity of a resulting color. As a result, the most covered from the incident light places of the 3d scene will be shady, especially if the sampling step is greater then one pixel (r > 1). In this case, significant differences of depth will have dark halo. Despite of simplicity, the algorithm produces improper results in case if a single object of the scene is situated near from the camera, and all other objects are relatively far from the camera, one will be overshadowed too.

In case of using this algorithm as the edge detection filter, it will look okay, but for effect of ambient occlusion it produces improper results. To overcome this unnecessary dark halo we add so called bleach factor, which will make too dark color values lighter. Then output color looks the following way:

$$oc = dp * \left(2 - \frac{dp}{L}\right) \tag{5}$$

Here "dp" is a dot product of tangent and screen normal; "L" is the bleach factor which is in range [0..1]. For low values of "L", initially too dark zones will be lighter, and for high values of "L" dark zones stay dark. Then we simply accumulate these four samples of corrected depth and divide by total count of samples (which equals 4) to set the output color in correct range. For clarity of algorithm, we have wrote the source code of the fragment shader in Adobe Pixel Bender language (see the "Appendix A" section of this paper). Pixel Bender do not support loops, that's why all the four depths of neighbor points were sampled step-by-step by repeating key parts of a code. Note, that in the "Appendix A" we use depth map as an input image. For clarity, this depth map is encoded as grayscale texture, and its red, green and blue channels have the same color values in range [0..127] or [0..1] in Pixel Bender's notation. This depth map in less precise (8-bit), than the one, described above (our 32-

IOP Conf. Series: Materials Science and Engineering **240** (2017) 012016 doi:10.1088/1757-899X/240/1/012016

bit depth buffer). Pixel Bender's code illustrates the algorithm well, but is not applicable for real-time rendering, due to lack of performance. For practical use, we also have translated the code to Adobe Graphics Assembly Language (AGAL). This shader language is simple but yet powerful and consist of series of opcodes acting on registers, stored in GPU's memory. AGAL code is less human-readable, but times faster, than the same codebase written in Pixel Bender. We used almost the same approach to rendering of 3d meshes, as in previous work of our colleagues [6]. But in this case, we should render the scene twice:

- Render the scene with its whole appearance (mesh textures, lights, etc.) to the BitmapData. BitmapData is standard class of the Action Script 3.0 programming language. It stores color data of a pixels inside rectangular area.
- Render the packed 32-bit depth texture of the scene (no textures for meshes, no lights) to the BitmapData.
- Render the screen's rectangle, sample previously stored BitmapData objects of the scene and depth map, and perform the final post processing. Function "initBuffers" creates two triangles in GPU's memory to render screen's rectangle. Function "draw" draws its triangles with our post processing shader. This shader takes screen texture and depth texture (fs0, fs1) as input sources.

Final AGAL code has three modes:

- Render just a scene.
- Render the scene mixed with post-processing filter.
- Render just the filter.

In contrary of the "Appendix A", the AGAL version of shader is supplemented by new parameter "range". For most cases the range value 0.8 * camera.far produces the best result (where camera.far is a far clipping plane of a camera's frustum). One can change this parameter to make near-laying objects neater.

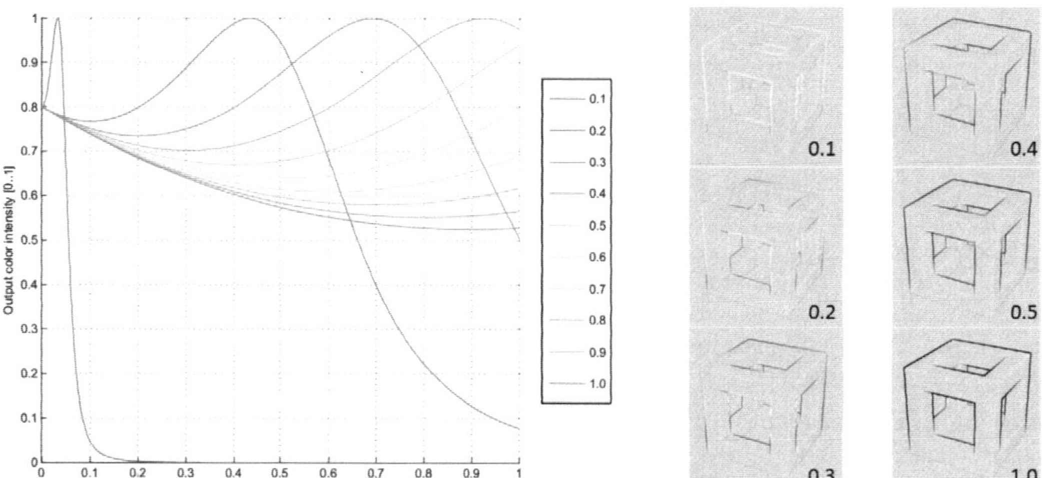


**Fig. 2.** Left: Output color depending on the angle between surface tangents and screen normal. Angle is specified in terms of dot product [and is in range [0..1]. Different rainbow colors correspond to the bleach factor. Right: Result of the shader's work. Each number at the right bottom corners of each frame points to the bleach factor.

## Conclusions

First, we checked the output of our shader written in Pixel Bender. We use a simple 8-bit grayscale depth map representing geometry of a cube without the inner central part. The result of a shader work is on the figure 3. On the left side is the depth map, which is used as the input texture of a shader. Central part of the Fig. 3 corresponds to the effect of ambient occlusion. The most occluded parts of the cube is darker. This image is produced with the following parameters: contrast=64 (of 100), r=7 (of 10), bleach=0.05 (of 1.0), brightness=0.1 (of 2.0). Right side of the Fig. 3 shows the result of work of the same shader, but with modified parameters: contrast=26, r=1, bleach=1, brightness=0. Now it works like the edge detection filter. Here are some notes about the meaning of these parameters:

1) In physical world, the contrast is the difference in color that makes an object (or its representation in an image or display) distinguishable. In our case, we adjust the resulting color by the following formula: $color=[\![color]\!]^{contrast}$.

"r" is the distance (in pixels) between currently processed pixel and its neighbors, where we sample the depth. Reasonable range of "r" is [1..10]. Low values of "r" produce narrow lines. High values of "r" produce thick lines.

The bleach factor makes too dark areas lighter and is well described in previous section of this paper.

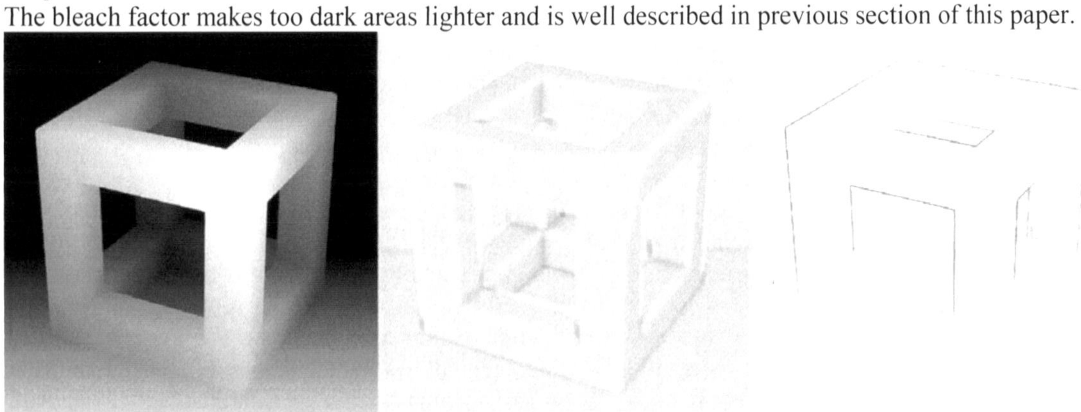

Fig. 3. From left to right: test 8-bit depth map; Pixel Bender's shader output working as ambient occlusion effect; Pixel Bender's shader output working as edge detection filter. The same codebase for all cases.

Now, it is time to see the results of our shader written in AGAL. For edge detection, AGAL version of our shader has slightly different parameters. The "range" parameter now controls the contrast of small differences of a depth of parts of three-dimensional model. If the parameter is within the distance from the camera to the object, then small details are better visible. If you want to display only the outer contour of the object, it is sufficient to set this parameter times greater. This is the one of advantages of our shader filter: it makes possible to display both the outline of a model and its "inner" edges.

For architectural and industrial rendering, it is often good to display not only the scene with ambient occlusion and lighting, but also to make accent on edges. For architectural and industrial rendering, it is often good to display not only the scene with ambient occlusion and lighting, but also to make accent on edges. Our shader can do these tasks simultaneously. Fig. 4 shows how both the effect of ambient occlusion and the edge detection filter look together. As in previous cases, we use the scene without any lighting (and no shadows). Nevertheless, final image looks good and has almost realistic lighting. This effect can be achieved with the same setting as for Fig. 4, but with low "r" (r=1.5) and low overall brightness.

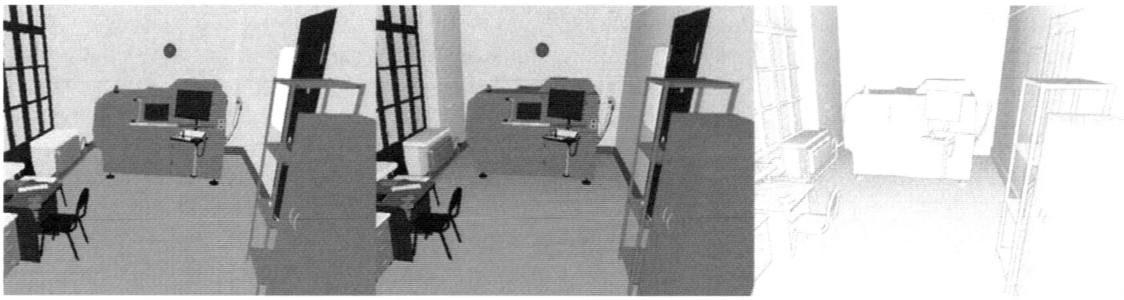

Fig. 4. AGAL shader output for ambient occlusion and for edge detection. Rendering of architectural scene (interior of a room). Left: scene texture; Center: scene texture and our filter; Right: raw output of our filter (without mixing with the scene texture).

Being fully screen space post-processing effect, the shader is almost independent on geometric complexity of a scene, so its performance is mostly specified by screen resolution. For screen resolution 1280x1024 we achieve 45-60 frames per second for scenes of two million polygons.

Brightness corresponds to uniform brightness of the resulting picture. In our case it's inverted brightness, so zero value points to bright picture and high values points to dark picture. We have presented real-time post-processing shader, which produces the effect of ambient occlusion and edge detection at the same time. The shader is written in two versions: in Pixel Bender 2.5 (for test purposes) and in AGAL (for real-time rendering). This effect is flexible, adjustable and useful for rendering of industrial design, architectural works and engineering. The algorithm is simple to implement and can be easily extended to other higher-level shader languages. It uses only depth texture without screenspace normal texture and without additional noise texture, which is key component of a rendering pipeline of most of known screenspace ambient occlusion algorithms. Our algorithm forces to render the scene twice: standard render pass + contours extraction pass. The shader is fast enough, but to increase the speed of processing, we plan to use new feature of AGAL: multiple render targets, so we can directly construct a real G-Buffer, rather than rendering the scene to separate textures.

## References

[1] Umbaugh S 2010 *Digital image processing and analysis: human and computer vision applications with CVIPtools (2nd ed. ed.)* Boca Raton, FL: CRC Press

[2] Langer M, Buelthoff H 2000 Depth discrimination from shading under diffuse lighting *Perception* 29 (6) pp 649–60

[3] Miller G 1994 Efficient algorithms for local and global accessibility shading *Proceedings of the 21st annual conference on Computer graphics and interactive techniques* pp 319–26

[4] Pharr M, Fernando R 2005 *GPU Gems 2: Programming Techniques for High-Performance Graphics and General-Purpose Computation (Gpu Gems)* Addison-Wesley Professional (Chapter 12)

[5] Scabia M Vertex and Fragment Shaders http://www.adobe.com/devnet/flashplayer/articles/vertex-fragment-shaders.html

[6] Gaisin R, Nikiforova A, Kashapov N, Kashapov R 2014 Development of the method of creation of the 3d model of topology of the surface by means of the optical microscope *IOP Conf. Ser.: Mater. Sci. Eng.* 69 **012007**

[7] Rusinkiewicz S 2004 Estimating Curvatures and Their Derivatives on Triangle Meshes *Second International Symposium on 3D Data Processing, Visualization, and Transmission (3DPVT)*

[8] Michael Burns, Janek Klawe, Szymon Rusinkiewicz, Adam Finkelstein, Doug DeCarlo 2005 *Line Drawings from Volume Data* In SIGGRAPH

[9] McGuire, Morgan and John F Hughes 2004 Hardware-Determined Feature Edges http://graphics.cs.williams.edu/papers/EdgesNPAR04/edges-NPAR04.pdf

# Formation of scattering indicatrix of retroreflective coatings with microspheres

**E R Galimov, Yu A Pryakhin, E E Tukbaev, N Ya Galimova and L R Fazlyev**

Kazan National Research Technical University named after A N Tupolev, Kazan, Russian Federation

kstu-material@mail.ru, morenko@imm.knc.ru

**Abstract.** The problems of calculation based on the third-order aberration and Fraunhofer approximation and in the ray approximation of reflection indicatrixes of retroreflective coatings containing the catadioptric system on the basis of glass microballoons are considered. The fundamental relations are formulated the experimental studies of indicatrix are carried out.

## 1. Introduction

Light reflectors of retroreflective type are widespread first of all at traffic management as coatings of road signs and vehicles numbers. They are also used in military and space technologies as the element of measuring systems [1]. Microspherical lens determining the light reflector optical properties is its main element. The light rays passing through the lens are reflected at its outer and inner surfaces and go out in directions including the opposite direction to the propagation of the incident beam [3]. A number of papers, for example [3, 4], are devoted to the research of retroreflective coatings based on the corner reflectors, but only a few works are devoted to the research of the microspherical retroreflective coatings [1-3].

## 2. Research part

The coating structure (variant) can be represented as a planar multilayer structure with microspheres situated in it (figure 1).

Peculiarities of retroreflective coatings reflection are determined by the optical properties of the radiation reflection by the microspheres which is characterized by the scattering indicatrix – angular dependence of the radiation flow of reflected (scattered) light. The light reflection from microspheres is well known in the phenomena of rainbow and halo – back reflection of light from water droplets [5]. When solving the plane wave diffraction problem, for example, on a homogeneous dielectric cylinder the expression for the scattered field can be written as a series over cylindrical waves – Debye series [5] with coefficients that take into account the (p-1) multiple internal reflections from the cylinder surface. If we consider only the first term of the Debye expansion (p=0) then it can be shown that in geometric optics a field is easy to find by multiplying the field of the corresponding beam on Fresnel reflection coefficient [5, 6].

Scattered field can be thought of as consisting of two parts – one of which is caused by the reflection and refraction on the sphere the other – by the diffraction of the wavefront at the external boundary of the scattering sphere. Interference between the various components leads to the appearance of fast oscillations of the intensity when changing the direction of observation. If the sphere radius r is much greater than the wavelength λ (in our case 2r/λ – 200) then the scattering field of such a sphere can be found using the geometrical optics approximation [6, p. 469].

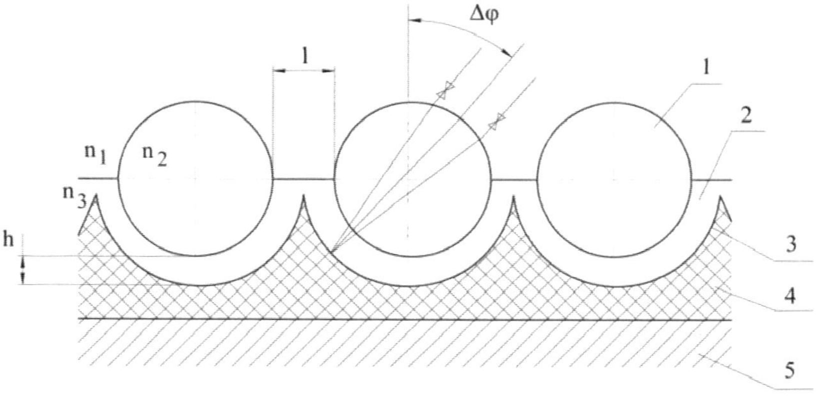


**Figure 1.** Scheme of retroreflective coating: 1 – ball, 2 – intermediate layer, 3 – reflecting layer, 4 – bonding layer, 5 – substrate; $n_i$ – refractive indexes of media, l –distance between the balls, h – balls indentation depth, Δφ – operating angle of retroreflective coating

The expression for the angles and the scattering indicatrix at diffraction on large spheres is given in [5].

Sphere as an optical lens system (figure 2) is given by the surface radius r, the refractive indexes of the environment and the sphere ($n_1$, $n_2$, $n_3$) and is also characterized by the paraxial parameters – focal length $f'$ and back-focal distance $S'_{F'}$. The back-focal distance of the sphere

$$f' = \frac{n_2 n_3 r_1 r_2}{n_2 r_2 (n_2 - n_1) + n_2 r_1 (n_3 - n_2) - d(n_2 - n_1)(n_3 - n_2)}, \tag{1}$$

$$S'_{F'} = f'\left\{1 - \frac{(n_2 - n_1)d}{n_2 n_1}\right\}, \tag{2}$$

where $n_i$, i = 1, 2, 3 – refractive indexes of environments, $r_1$, $r_2$ – radiuses of the front and rear surfaces of the lens.

Rear surface of the sphere along the rays in the catadioptric reflector has a mirror coating. The angle of the ray reflected in the opposite hemisphere (φ) can be found from the condition of rays' propagation at reflection in the microsphere at a given angle of incidence (ε)

$$\varphi(\varepsilon, n_1, n_2) = 4 \arcsin\left(n_1 \frac{\sin(\varepsilon)}{n_2}\right) - 2\varepsilon . \tag{3}$$

The sphere has catadioptric properties – a parallel beam of incident rays turns into a parallel beam of reflected rays – if its refractive index $n_2 = 2$ (at $n_1 = 1$, air).

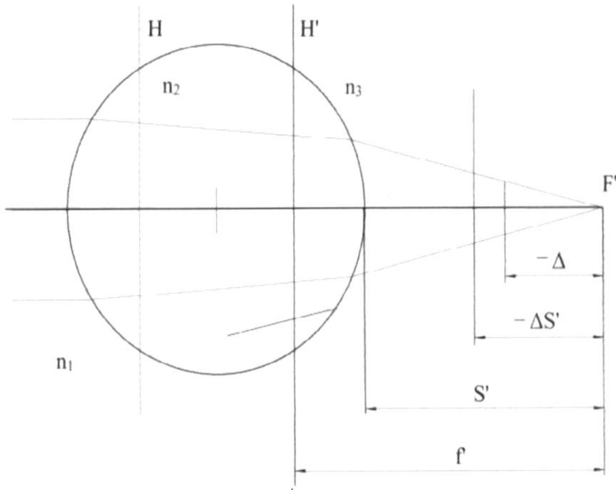


**Figure 2.** Paraxial parameters of the sphere: Δs' – longitudinal spherical aberration, S' – back-focal distance

In the studied scheme where an additional layer with a mirror surface is formed on the rear hemisphere of the ball the catadioptric reflector acquires catadioptric properties (figure 3).

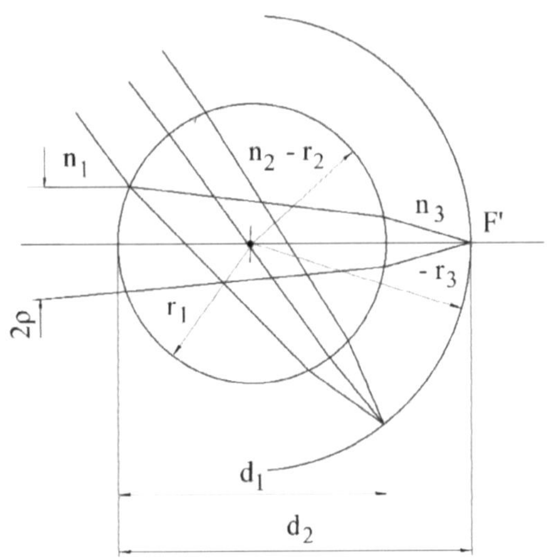


**Figure 3.** Optical scheme of catadioptric reflector: $n_1$, $n_2$ – refractive indexes of the environment and the sphere, $n_3$ – refractive indexes of the intermediate layer, 2ρ – aperture of the incident beam, $r_1$, $r_2$ – radiuses of the front and rear surfaces (for the ball $|r_1| = |r_2|$), $r_3$ – radius of the reflecting surface of the catadioptric reflector, F' – rear focal point of the sphere.

In order to determine the axial light intensity the spherical aberration of the third order was calculated, from the principal ray paths (normalization $\alpha_1 = 0$, $n_1 = 0$) [4] we find the longitudinal ($\Delta y_1$) and spherical aberration (ΔS), the focal length within the catadioptric reflector (F), Seidel sum $S_1$, the wave aberration as a function of change of $n_3$ and $d_2$L = L ($\Delta n_3$, $\Delta d_2$). The intensity distribution of the reflected radiation is described by the Fresnel-Kirchhoff integral in the Fraunhofer approximation for the exit pupil plane of the catadioptric reflector which is proportional to the Fourier transformation of the complex amplitude distribution of the reflected wave in the exit pupil plane. The maximum value of the indicatrix of reflection (scattering) is found by moving from the spatial frequencies in this plane to the spatial angular spectrum.

Measurements of the optical properties of different according to the technology retroreflective coatings were made some experimental scattering indicatrixes are presented in figure 4.

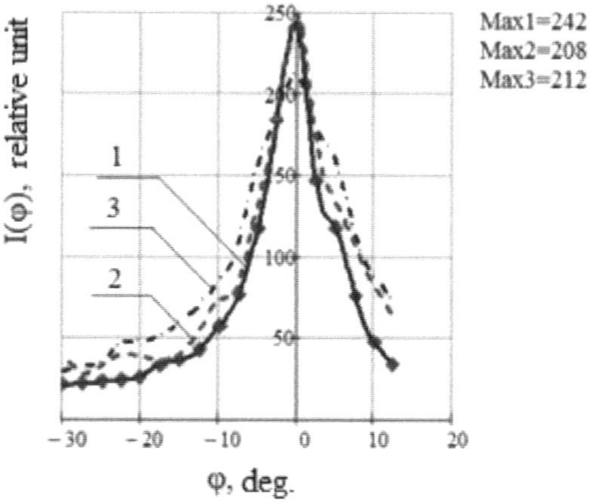


**Figure 4.** Scattering indicatrixes of different samples: I (φ) – intensity, φ – scattering angle

In Figure 4, curve 1 – the scattering indicatrix of the "diamond type" of the retroreflective film of "3M" company, curve 2 – the scattering indicatrix of film of company "3M" (white, engineer grade), curve 3 – the scattering indicatrix of coating, developed in KNRTU-KAI.

It should be noted that the half-width of the scattering indicatrix of different designs' reflective coatings are quite close to each other. Maximum reflection scattering phase of the developed coating (curve 3) is close to the maximum reflection of the «3M» film company (curve 2) at about the same viewing angles.

## 3. Conclusions

Thus, on the basis of the conducted research it was found that the main optical characteristics (the maximum intensity of the reflected signal, the width of the reflection chart) the developed light-return coverage correspond to foreign models film of the company "3M" engineering type.

The findings served as the basis for the development of new reflective coatings based on polymer powder compositions, filled with microsphere reflectors. As a result of comprehensive experimental and theoretical studies of the regularities of changes in optical coatings defined by the structure, the nature of the components, dispersion and uniformity of the microsphere reflectors and regime parameters of applying and the formation of functional coating layers were defined.

## References

[1] Starovoytov E I and D V Savchuk 2013 Research and optimization of application of corner reflectors for location of space objects *Space engineering and technology* number 1 pp 38-43
[2] Denisyuk G V and V I Korneev 1982 Fraunhofer diffraction from the corner reflector with metallized reflective facets (OMP) number 12 pp 1-2.
[3] Ivanov L A, Kizevetter D V, Kiselev N N, Malyugin V I and V A Slugin 2006 Measurement of retroreflection of glass microspheres and prediction of retroreflective coatings quality *Journal of optical technology* (vol 73) number 1 pp 35-40
[4] Tsvetkov A D 2011 Catadioptric retroreflector *Journal of optical technology* (78, 3)
[5] Solimeno S, Crosignani B and P DiPorto 1989 *Diffraction and waveguide propagation of optical radiation* (Moscow)
[6] Born M V and E Wolf E 1970 *Principles of Optics* (Moscow: Nauka)

# Mathematic simulation of high-efficiency process of grain cleaning

**Y V Galyautdinova[1], I A Gaysin[2], A D Samigullin[3], A R Samigullina[4] and R R Galyautdinov[5]**

[1]Naberezhnye Chelny Institute (branch), Kazan Federal University
Sjujumbike street 10a, Naberezhnye Chelny, 423800 Tatarstan, Russia

samigullin86@mail.ru

**Abstract:** The article presents the results of field experiment and the results of computer simulation of the grain cleaning process pneumosorting machine PSM-0,5. The results of the comparison of two methods of research are presented.

## 1. Introduction

Production and processing of grain form in economic system of the country a number of large sectors, such as grain production, the elevator industry, flour-grinding, cereal and feed milling production, which make a grain complex of the country. Value and a role of grain as goods in economy of the state are difficult to overestimate. These are goods which have stable, steady demand at all seasons of the year, in any region, so it is absolutely liquid.

In the conditions of market economy effectiveness of agricultural production is substantially defined by competitiveness of production. Considering these features, production of the majority of types of food products requires integration of several agro-industrial enterprises, which are different in a production orientation.

These circumstances cause relevance of in-depth studies of economic-mathematical models and methods of the analysis and assessment of economic efficiency of technologically integrated grain processing production systems.

Questions of application of mathematical models, methods and means of the informational software for the analysis and assessment of economic efficiency of grain processing production are poorly studied . Relevance of these problems defined the choice of a subject and statement of the purpose and research problems. The pneumosorting machine PSM-0,5 is chosen as an object of a research.

## 2. Research

At the moment there are two ways of carrying out a research of operation of the pneumosorting machine. The first way is a carrying out field tests which means manufacture of a test piece of the actual sizes or with a scale factor and carrying out numerous laboratory trials. Therefore this method is the most expensive, but at the same time it allows to receive the most precise and reliable results.

The second way - computer simulation which allows to reduce significantly the material inputs (there is no need of creation of actual model and laboratory installations) and to reduce time of

IOP Conf. Series: Materials Science and Engineering **240** (2017) 012018 doi:10.1088/1757-899X/240/1/012018

obtaining results. But at the same time it demands creation of the precise detailed computer PSM model with correctly given initial and calculated parameters. At the same time the most long period can be the time spent for creation of computer model, and directly calculation time. Accuracy of similar calculations depends on parameters of a grid of finite elements.

For further acceleration of engineering of new pneumosorting machines it was decided to make computer simulation of process of grain cleaning on the PSM-0,5 device with verification by results of field tests.

In the software package of STAR-CCM+ numerical simulation of grain cleaning in PSM was made. When carrying out a research the geometry of numerical simulation completely coincided with the original of installation. The same as during test boundary conditions were accepted. As a result of computer simulation the following results in a type of volume distribution of fields of vectors of speeds in the channel of the pneumosorting machine were received. They are given on fig. 1 and 2 for mass flow of 0,528 kg/s and on fig. 3 and 4 for mass flow of 0,368 kg/s.

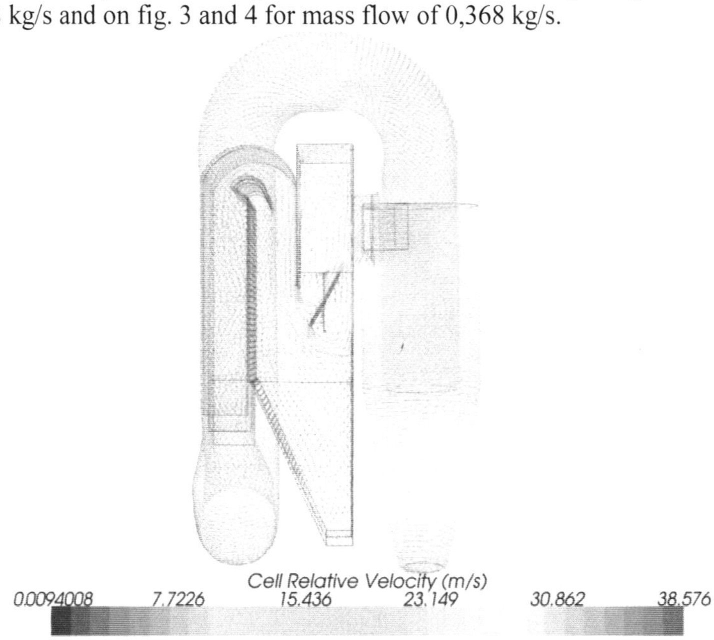


Fig. 1. Scene of distribution of vectors of speeds at the provision of a valve of speed control at the level of 10 of a graduated scale.

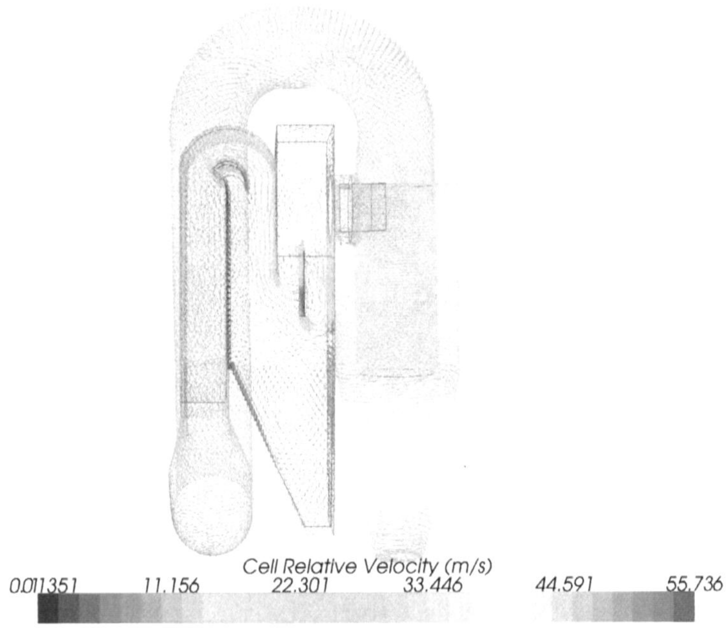


Fig. 2. Scene of distribution of vectors of speeds at the provision of a valve of speed control at the level of 5 of a graduated scale.

Analyzing the received results on fig. 1 and 2, it is possible to draw a conclusion that input speed to the pneumosorting canal is more than admissible one at which grain will be carried away to the pneumosorting canal. It will have an adverse effect on quality of operation of the pneumosorting machine.

In connection with the received results of computer simulation it is decided to conduct a research with smaller mass flow of air.

On the profiles of speeds given on fig. 3 and 4, we can see, that after replacement of the blower and the electric motor speed reaches desired values and becomes the same as in the existing installation. Thus, at less potent electric motor input speed to the pneumosorting canal remain the same and quality of cleaning remains the same.

IOP Conf. Series: Materials Science and Engineering **240** (2017) 012018 doi:10.1088/1757-899X/240/1/012018

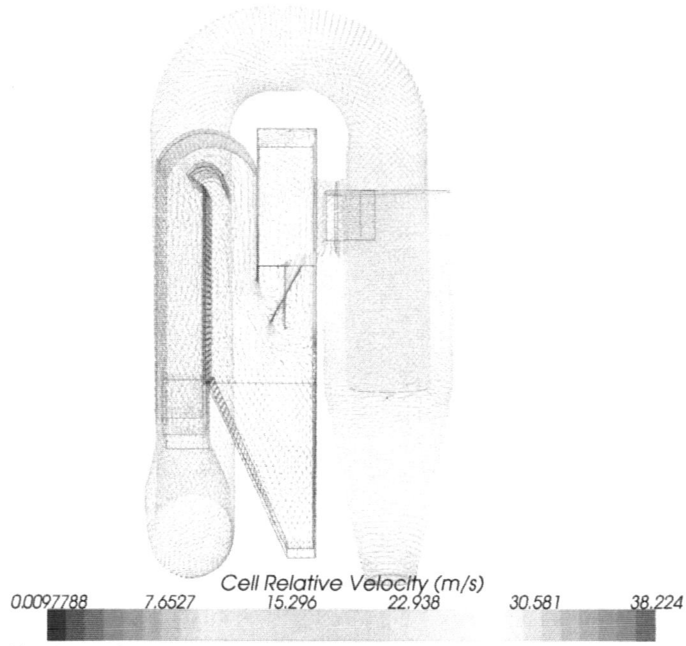


Fig. 3. Scene of distribution of vectors of speeds at the provision of a valve of speed control at the level of 10 of a graduated scale.

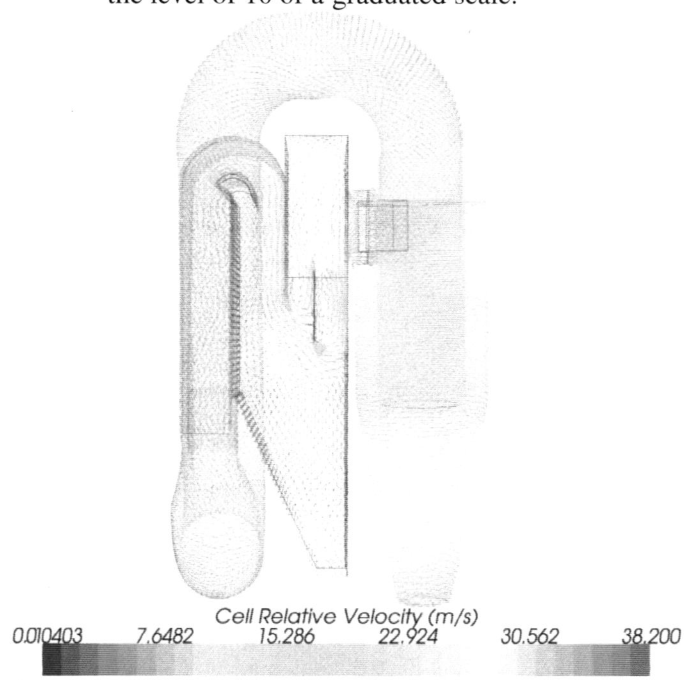


Fig. 4. Scene of distribution of vectors of speeds at the provision of a valve of speed control at the level of 5 of a graduated scale.

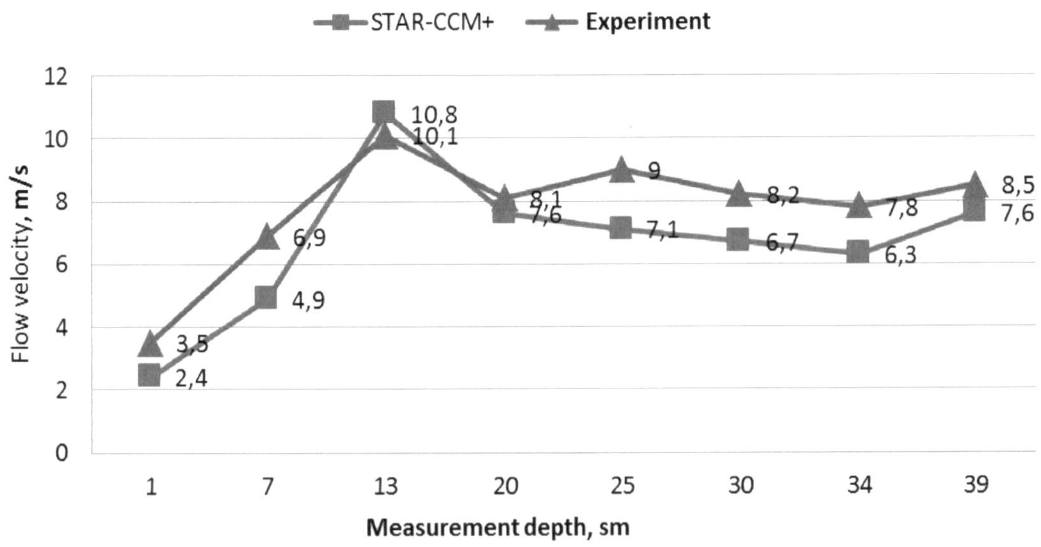


Fig. 5. The schedule of comparison of results of a natural experiment and data of computer simulation in the software package of Star CCM+.

Comparing results on fig. 5 it is possible to tell that the results received by a numerical experiment slightly differ from the values received by a physical experiment. Unessential mistakes of carrying out a physical experiment are a consequence of errors of devices. Also in a numerical experiment the ideal model is shown where the camera is absolutely tight, at a physical experiment it is difficult to reach it.

**3. Conclusion**

Thus, it is possible to draw a conclusion that the values received at computer simulation in the program complex STAR-CCM+ can be used at engineering of new pneumosorting machines.

**References**

[1] Gimadiyev A.M. Influence of technological and transport vehicles on a morphological and physiological maturity of seeds. 2009. 39 p.

[2] Gimadiyev A.M. Methodical grant on a subject "Pneumosorting machine PSM-6, PSM-6C" Technical description and operation manual: OOO "Naberezhnye Chelny printing house ", 2009,42 p.

[3] Gimadiyev A.M. Methodical grant on a subject " Pneumosorting machine PSM-25, PSM-25C, PSM-25M. PSM-25MC" Technical description and operation manual: OOO "Naberezhnye Chelny printing house", 2009, 52 p.

[4] Gimadiyev A.M. Paths of decrease in a traumatizing of grain and seeds when carrying out postharvest processing by rotary barriers on an example pneumosorting machine PSM. 2009, 127 p.

[5] Yerov YU.V., Nurullin E.G., Karimov KH.Z., Salakhiyev D.Z. Innovations in postharvest processing of grain and seeds.- Kazan: "Word", 2009 . 104 p.

[6] Loyko V. I. Management of economic efficiency of technologically integrated grain processing production systems. / Loyko Valeriy Ivanovich. Bogoslavskiy Stanislav Nikolayevich. Makarevich Oleg Aleksandrovich.// Scientific magazine of KUBGAU, №60(06), 2010 y.

[7] Gaysin I.A., Sarapulova J.V., Khabibullin S.S.and Saubanov R.R. Energy saving in the grain processing industry. *Informational and social and economic aspects of creation of the modern technologies*: The on-line online scientific and technical magazine No. 1(58) - N. Chelny, 2011. http://sets.ru.

IOP Conf. Series: Materials Science and Engineering **240** (2017) 012019 doi:10.1088/1757-899X/240/1/012019

# Development of expert systems for modeling of technological process of pressure casting on the basis of artificial intelligence

**K N Gavarieva [1], L A Simonova [1], D L Pankratov [1] and R V Gavariev [2]**

[1] Kazan Federal University, 18 Kremlyovskaya street, Kazan, 420008, Russian Federation

[2] Kazan National Research Technical University named by Tupolev – KAI

Soldatkinakn@mail.ru

**Abstract**. In article the main component of expert system of process of casting under pressure which consists of algorithms, united in logical models is considered. The characteristics of system showing data on a condition of an object of management are described. A number of logically interconnected steps allowing to increase quality of the received castings is developed

## 1. Introduction

In the field of casting under pressure in order to solve problems, which are targeted to achieving high-quality castings and exclusion of defects in the design phase, and the formation of optimal technological processes, automated information systems do not exist.

In recent years, the question of the application of expert systems (ES) in the production process in the simulation based on artificial intelligence increasingly began to be considered. [1]

At modeling phase of technological process an expert system must take into account the state of machinery during the operational phase, material, labor and financial resources of the enterprise [2].

The structural model of the system consists of realized elements (main line), modules and connections (dashed lines) as well as external plug projecting and modeling modules (dashed line) (Figure 1) This expert on the basis of the defect model system allows you to group and synthesize the information flow between the elements. The knowledge base of the expert system is based on a model of the casting considering verifying inference rules.

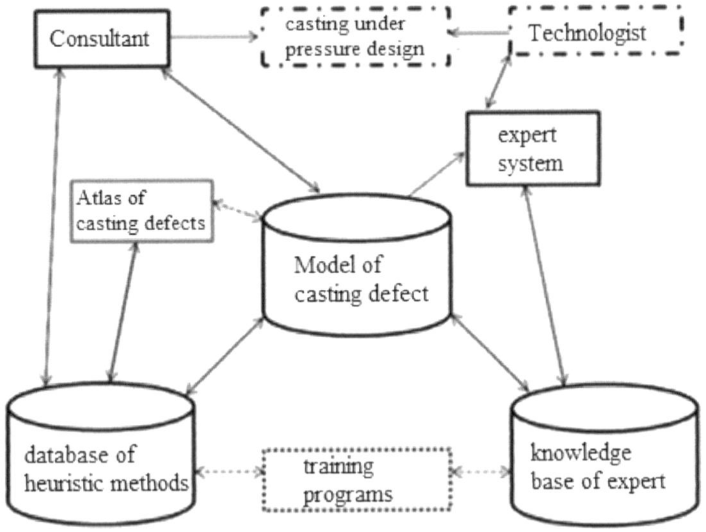


Figure 1 Structural model of expert system

## 2. Main part

The selection of knowledge representation model shall be based on predetermined objectives and peer review in the form of a set of heuristic techniques, interconnected logical chain.

The database of heuristic techniques takes an important place in the scheme, i.e. virtually on all of them, logical inference process is based. Defect model provides full description of heuristic methods and their relation to the withdrawal procedures.

The integrated expert system for solving problems of casting quality improving along with the proposed information model takes one of the central places.

The most rational for pressure casting process is the use of the productive expert systems with reverse output chain. This is due to the fact that the analysis or simulation of pressure casting process involves the use of physical laws describing the casting process parameters such as temperature distribution, pressure, fluid motion, the phase transition material. In this process, these components operate simultaneously on the final result, so the theoretical description and prediction of the influence of process parameters on the final result is a complex task requiring specific data of practical tests and experiments that make up a specific knowledge base of pressure casting process.Знания в экспертной системе прописываются в виде продукций (правил) вида «ЕСЛИ условие ТО вывод». The knowledge in the expert system are prescribed in the form of productions (rules) of the form "If the condition the conclusion ". So, one of the most important rules of pressure casting process is the following: if you increase the temperature of the melt, then received casting will have a reduced strength; if lower temperature of the melt, then received casting will have emptiness in its entirety. Components of the rules of "condition" and "output" are facts that can be recorded using the components of the "object-attribute-value".

As the object acts obtained casting or any particular parameter of the resulting casting, such as porosity or strength, and the attribute is some parameter of pressure casting, that are directly or indirectly affects the object, i.e. melt temperature, mold temperature, pressure, speed of pressing, the material of melt and the mold, the number of the forming mold nests, cooling and other kinds. On the basis of practical data the choice of specific attribute values allows to get the object with predetermined properties. Выбор поставленной задачи задается формулировкой цели консультации, которая решает определенную проблему какого-либо параметра качества реальной отливки. The choice of the task is given by the formulation of advice objective, which solves a specific problem of a quality parameter of real castings. To set goals of the consultation technologist selects a pair of «object - the attribute» from the list on the basis of information from the knowledge base of the expert system.

The logical conclusion is performed by the expert system after the user sets the goal of consultation. At this moment in the knowledge base there is a search for the first hypothesis included in the homogeneous group of hypotheses that corresponds consultation purpose.

However, it is necessary to take into account the correct installation of facts sequence to create a knowledge base [3]. In case of error there is a lengthening of the output chain and the emergence of the issues which are not resolved, that will eventually lead to a mismatch of problems of expert systems – a design of expert arguments. After determining the first hypothesis, the logical conclusion is performed in accordance with the algorithm of feedback chain of arguments to prove the current hypothesis. These steps are performed for all hypotheses designed to advice target as long as one of them has been proved. Proven hypothesis is defined as a recommendation. It is important to note that in the expert system one recommendation is equivalent to a group of recommendations (action algorithm) the subject area. If there is no evidence of any of the hypotheses the hypothesis with a large measure of posterior probability is accepted in the form of a recommendation. The expert system may include a large variety of different questions, which are set depending on the user's skill [4]. If the fact coefficient is less than the user's skill level, the expert system must identify its probability, based on experience. Otherwise, to fix the fact the expert system finds the relevant rule in the knowledge base and makes a calculation on the Bayes formula, inserting the values of the rules found in it.

The selection of knowledge representation model is characterized by a specified task and knowledge of experts [5]. Experts are presented as a set of heuristic techniques, interconnected among themselves with the logical chain. The choice is determined based on the used model of knowledge representation, the reality of the output of external calculation modules using the context file, as well as the availability of the source software modules for variety and adaptation expert system regarding the problems to be solved.

Допустим, Заданная цель – минимизировать количество брака при ЛПД. Учитывая, неясность формализации цели, необходимым является использование методов искусственного интеллекта. For example, a given target - to minimize the number of rejects at casting under pressure. Considering the uncertainty of formalization goals it is necessary to use methods of artificial intelligence. However, in the process of inference several subtasks appear which must be solved by standard computational methods [6, 7]. But for these tasks, there is already a clear algorithm and prepared programs.

In order to realize the expert system in the area of issues and logical inference it is necessary to adapt it to achieve the goal [8]. In this regard, it was approved additional types of questions, which allows more quickly and accurately retrieve data.

In addition to the generally established issue on the assessing the degree of statement factuality there reveals the possibility of excluding hypotheses, hard selection of hypotheses (e.g., selection of color defects), some hypotheses matching (for example, production can be two types of defect only). Sophisticated production rules are shown in the hierarchical graph. The graf has a clear structure of consideration of substantive rules (P) and recommendations (R) under the control and forecasting of the causes that lead to the appearance of defects in the casting and possible recommendations of their reduction and elimination (Figure 2). A fragment production rules is presented in Table 1.

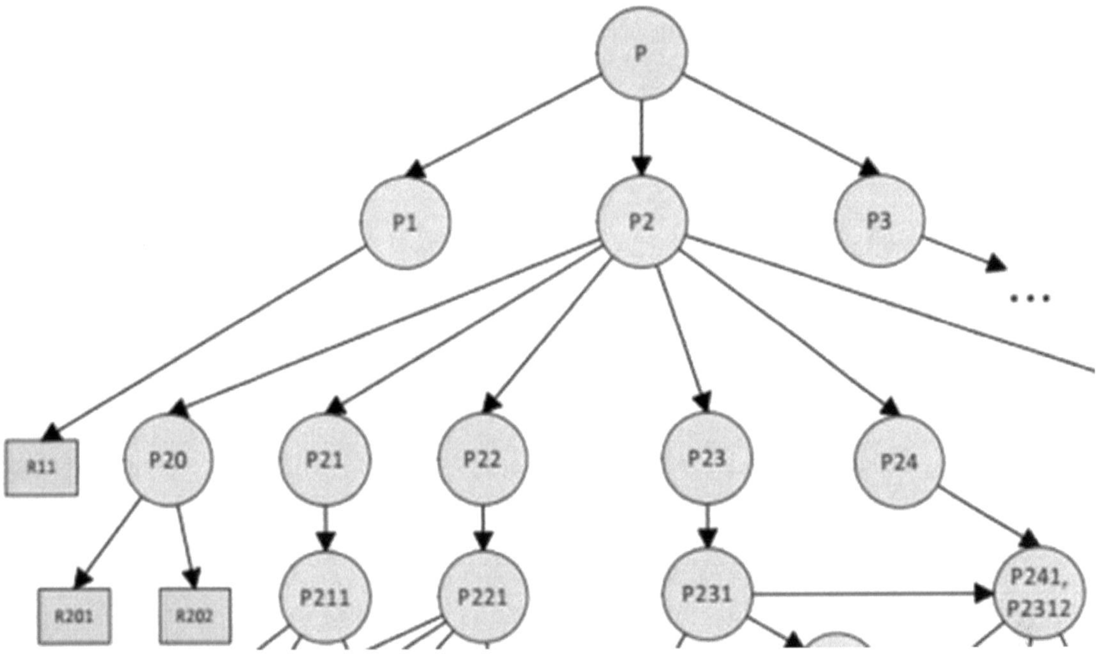


Figure 2 – Graf of production rules and recommendations (fragment)

Table 1 - production rules (fragment)

| Symbol | Имя правила Rule name | Content of rule | |
|---|---|---|---|
| | | Condition «If» | Action «then» |
| 1 | 2 | 3 | 4 |
| P2 | Cold seam | Too low injection rate. Too small wall thickness | Raise the temperature of the melt and form (P20), if necessary, change the location of the gate (P211) in order to improve flow conditions and to increase the injection rate (P22) and equalize the wall thickness (P23) and improve air removal from the mold (P24). |

## 3. Conclusion

Thus, the use of fuzzy logic in intellectual superstructure of the expert system allows to predict the quality of the castings, depending on the main parameters of die-casting process specified by the technologist and molds designer. Thus, in case of lack of experience, knowledge or proficiency of such specialists in any particular case, it is possible to carry out optimization of existing or new forecasting process by introducing linguistic variables, defining the main parameters of the production of castings, such as melt and mold temperature, locking piston pressure, pressing speed, the complexity of the mold and the casting configuration.

**References:**

[1] Korostelev V F 2004 *Theory, technology and automation* (M .: New Technology) P 224
[2] Sabirov D H 2005 Computer technology in the foundry industry (*works of the seventh congress of casters of Russia Novosibirsk*) Vol. II pp 208-16

[3] Balandin G F and Kashirtsev L P 1967 *Structural and mechanical properties of alloys AlLSi in the range of crystallization and analysis of their fluidity* (Heat exchange between the casting and the mold, Minsk Higher School) P 59

[4] Vasilyev V A, Levkin M I, Pavlenko A I 1988 Reconfiguration of the integrated management system with the help of an expert system *Problems of Cybernetics. Automation systems of moving objects* Vol 139 pp 90-97

[5] Raskazchikov A N 2008 Modeling of casting process with pressure superimposed on the crystallizing metal *Mathematical Methods in Engineering and Technology: Proceedings of the International Scientific Conference XXI Saratov* pp 41-44

[6] Pankratov D L, Gavariev R V, Gavarieva K N 2016 Influence of multilayer coatings on the operational stability of molds for injection molding *IOP Conference Series: Materials Science and Engineering* Vol 134 Iss 1 **012031**

[7] Gavariev R V, Savin I A and Leushin I O 2016 Impact of the functional coating on service durability of injection molds for the zinc alloys pressure casting *Tsvetnye Metally* No 1 pp 66-70

[8] Shaparev A, Savin I 2016 Calculation of the Amount of the Reduction Required for the Formation of Compound Layers during Cold Rolling of Bimetals *Materials Science Forum* Vol 870, pp 328-33

# Investigation of reliability indicators of information analysis systems based on Markov's absorbing chain model

**I R Gilmanshin[1,3] and A P Kirpichnikov[2]**

[1]Kazan Federal University, 18 Kremlyovskaya street, Kazan, 420008, Russian Federation
[2] Kazan National Research Technological University, 420015, Russia, Kazan, C. Marksa street, 68
[3]Kazan National Research Technical University named after A N Tupolev, Kazan, Russian Federation

Is-er@yandex.ru

**Abstract**. In the result of study of the algorithm of the functioning of the early detection module of excessive losses, it is proven the ability to model it by using absorbing Markov chains. The particular interest is in the study of probability characteristics of early detection module functioning algorithm of losses in order to identify the relationship of indicators of reliability of individual elements, or the probability of occurrence of certain events and the likelihood of transmission of reliable information. The identified relations during the analysis allow to set thresholds reliability characteristics of the system components.

An integral part of modern energy supply systems are energy consumption accounting systems. In most cases, their functionality is expanded by monitoring and monitoring functions. The monitoring and control function carries in deep analytical potential of revealing of extraordinary situations at power supply objects and detection of excessive losses in networks. The economic effect of timely, early detection of emergency situations and localization of areas of excess losses is hard to overestimate. Detailed problem-oriented monitoring and control of energy supply provides an increase in reliability of power supply. To this end, analytical systems for early detection of losses are integrated into the systems for commercial accounting of energy consumption. In the course of studying the algorithm for the functioning of the module for the early detection of excess losses, as well as individual characteristics of extraordinary situations, the question arises of assessing the degree of influence of the latter on the quality of the transmitted information. It is necessary to identify the relationship between the probabilities of occurrence of extraordinary events and the probability of transfer of full-fledged data. The study of signs of extraordinary situations allows us to assert the stochastic character of the latter.

A detailed investigation of the algorithm for the operation of the early loss detection module allows one to refer it to the Markov processes [1, 2, 3]. That, in turn, creates prerequisites for the application of existing methods of analysis of the algorithm as the Markov process [6, 8].

The analysis of the algorithm of functioning as the Markov process is reduced to three stages. At the first stage, the process is formalized in terms of the mathematical apparatus of the theory of Markov processes, a state graph and a transition probability matrix are constructed [1]. At the second stage, proceeding from a specific task, necessary mathematical operations are performed, the interrelation of process parameters is derived. At the third stage, the final calculation of the required parameters is carried out.

We represent the stages of the data analysis process by positions, and the probability of transition to the next stage is denoted by the corresponding variables. Thus, the probability of transition to the next stage will be uniquely determined by the value of the corresponding index. The investigated Markov chain will consist of the following positions:

$S_6$ - characterizes the receipt of data from the meter;
$S_5$ - investigation of data for the detection of energy carrier leakage;
$S_4$ - the stage of complex analysis of the readings of the meters for the detection of incorrect work;
$S_3$ - analysis of energy quality indicators for compliance with the standard;
$S_2$ - final conclusion on the successful passage of all stages of control;
$S_1$ - detection of the deviation of the investigated data from the set value.

And the states $S_6$, $S_5$, $S_4$, $S_3$ – are transient states that form a non-returnable set, and $S_2$ and $S_1$ – are absorbing and create an absorbing set. Consequently, the Markov chain, which simulates the algorithm for the functioning of the module for the early detection of excess losses, is an absorbing Markov chain.

By G.I. Novikov's definition, the absorbing Markov chain contains a state that the process has never reached before. A graphical representation of the Markov chain is the state graph depicted in Fig. 1.

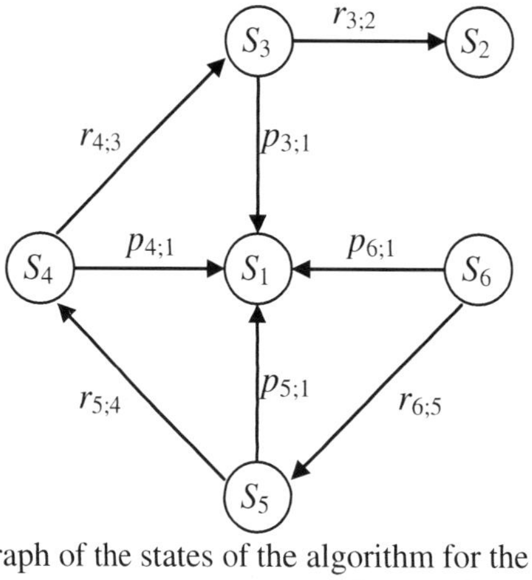


Где: $r_{6;5}+p_{6;1}=1$
$r_{5;4}+p_{5;1}=1$
$r_{4;3}+p_{4;1}=1$
$r_{3;2}+p_{3;1}=1$

Fig. 1. The graph of the states of the algorithm for the operation of the module for early detection of excess losses

When considering the transition matrix $P$, it is convenient to give the canonical form by combining all absorbing states into one group and all transient states to another.

Suppose there are s non-return states and $r$-$s$ absorbing states, then the canonical form of the matrix $P$ takes the following form:

$$P = \begin{pmatrix} S & 0 \\ R & Q \end{pmatrix}$$

Here the region "0" is composed entirely of zeros. The submatrix $Q$ (of dimension $s \times s$) describes the behavior of the process before the exit from the set of non-transient states, the submatrix $R$ (of dimension $s \times (r\text{-}s)$) corresponds to transitions from irrecoverable to ergodic states, and the matrix $S$ (dimension $(r\text{-}s) \times (r\text{-}s)$) refers to the process after reaching the ergodic set.

There exists a strict theorem [5], asserting that the powers of $Q$ tend to zero. Consequently, when the matrix $P$ is raised to higher degrees, all elements of the last $S$ columns will tend to zero. This is a matrix version of the corresponding theorem

From the definition of the absorbing Markov chain it follows that $S = I_{(r-s)\times(r-s)}$, i.e. $S$ Is an identity matrix of dimension $(r\text{-}s)\times(r\text{-}s)$. Therefore, the canonical form of the matrix $P$ has the form:

$$P = \begin{pmatrix} I & 0 \\ R & Q \end{pmatrix}$$

The matrix of transition probabilities $P$ takes the form:

$$P = \begin{array}{c|cc:cccc} & S_1 & S_2 & S_3 & S_4 & S_5 & S_6 \\ \hline S_1 & 1 & 0 & 0 & 0 & 0 & 0 \\ S_2 & 0 & 1 & 0 & 0 & 0 & 0 \\ \hdashline S_3 & p_{3;1} & r_{3;2} & 0 & 0 & 0 & 0 \\ S_4 & p_{4;1} & 0 & r_{4;3} & 0 & 0 & 0 \\ S_5 & p_{5;1} & 0 & 0 & r_{5;4} & 0 & 0 \\ S_6 & p_{6;1} & 0 & 0 & 0 & r_{6;5} & 0 \end{array}$$

This matrix already has a canonical form (ergodic and irrecoverable states are grouped together). The first two of its states are absorbing. We divide the matrix $P$ into blocks $S$, $R$ and $Q$, dimension $2 \times 2$, $4 \times 2$ and $4 \times 4$, respectively.

We obtain three submatrices of the following form:

$$S = \begin{pmatrix} 1 & 0 \\ 0 & 1 \end{pmatrix};\; R = \begin{pmatrix} p_{3;1} & r_{3;2} \\ p_{4;1} & 0 \\ p_{5;1} & 0 \\ p_{6;1} & 0 \end{pmatrix};\; Q = \begin{pmatrix} 0 & 0 & 0 & 0 \\ r_{4;3} & 0 & 0 & 0 \\ 0 & r_{5;4} & 0 & 0 \\ 0 & 0 & r_{6;5} & 0 \end{pmatrix}.$$

Moreover, the submatrix $Q$ (of dimension $s \times s$) describes the behavior of the process before the exit from the set of non-transient states, the submatrix $R$ (of dimension $s \times (r\text{-}s)$) corresponds to transitions from transient states to absorbing ones, and the submatrix $S$ (dimensions $(r\text{-}s) \times (r\text{-}s)$) Describes the process after reaching the absorbing set.

According to [5], by investigating the Markov chain, one can give a quantitative estimate of the average number of changes in the states of the process prior to absorption. In the simulated process, this value reflects the average number of cycles of the algorithm recalculation.

To estimate the probability of the end of the process when it is in the current state, we use the theorem: If $b_{i;j}$ the probability that the process emerging from the irrecoverable state of $S_i$ stops in the absorbing state $S_j$, then

$$\{b_{i;j}\} = B = N \cdot R, \qquad S_i \in T, \quad S_j \in \hat{T}.$$

Proof: leaving the state $S_i$, the process can be absorbed in $S_j$ in one or more steps. The capture probability at the first step is $p_{i;j}$.. If this does not happen, the process can get into either another absorbing state, or some kind of irrecoverable state of $S_k$. In the latter case, the wandering particle can be captured by the desired state with the probability $b_{k;j}$. Consequently,

$$b_{i;j} = p_{i;j} + \sum_{\substack{k=1 \\ (S_k \in T)}}^{n} p_{i;k} \cdot b_{k;j},$$

That in the matrix form takes the following form:

$$B = R + Q \cdot B.$$

Matrix *B* shows the probabilities of the end of the data processing in one or another absorbing state. In this example, the first column reveals the probability of transition to the state $S_1$ (detecting the deviation of the data under study from the set value) when it is in the current state of the graph (row of the matrix), and the second - the probability of transition to the state $S_2$ (the final conclusion about the successful passage of all stages of control ).

Thus, the set of obtained data establishes the probabilistic characteristics of the mathematical model of the algorithm for the functioning of the module for the early detection of excess losses. Practical application of the proposed algorithm of information analysis is possible on the basis of statistical data and expert estimates of the probability of occurrence of the corresponding events [4, 5, 7].

As a result of the investigation of the algorithm for the functioning of the module for the early detection of excess losses, it was proved that it can be modeled using absorbing Markov networks. Of particular interest is the investigation of the probabilistic characteristics of the algorithm for the functioning of the module for early detection of losses for the purpose of revealing the relationship between the reliability indicators of individual elements or the probability of occurrence of individual events and the probability of reliable information transmission.

The dependencies revealed during the analysis make it possible to establish threshold values of the reliability characteristics of the nodes of the system. For example, by choosing a certain probability of the transmission of reliable information, it is possible to impose restrictions on the threshold values of the probabilities of occurrence of individual events of the algorithm, which, in turn, imposes limitations on the reliability characteristics of the system hardware. In addition, the results obtained make it possible to assess the influence of reliability characteristics of individual system nodes on the probability of reliable data transmission. In addition to the values themselves, formulas for calculating the variance for all data groups are obtained. The developed algorithms have considerable practical potential both in designing and in evaluating the work of already existing systems.

## References

[1] Kirpichnikov A P 2007 *Applied Queuing Theory: Textbook. Allowance* Kazan: Kazan Mathematical Society 150 p
[2] Basyrov I R 2001 Expansion of Petri nets, problem-oriented to modeling multi-assortment production systems *Uchenye zapiski KFEI: Sat. Sci. Works* Issue 16 pp 177-82
[3] Kemeni J, Snell J 1970 *Finite chains of Markov* Moscow: Nauka 272 p
[4] Gilmanshin I R, Ferenets A V 2009 *Energy saving in housing and communal services: construction of a complex of a centralized automated system for collecting, monitoring and analyzing household energy consumption* News of higher educational institutions. Problems of energy No 9-10 pp 82-88
[5] Gilmanshin I R, Ferenets A V 2007 *Automated accounting of energy resources consumption as a condition for effective functioning of the housing and communal services* Bulletin of the Kazan State Technical University A.N. Tupolev No 3 pp 18-20
[6] Maiorov S A 1978 *Fundamentals of the theory of computer systems: Textbook. Allowance* Moscow: Visshaya Schkola 408 p
[7] Larionov A M, Mayorov S A, Novikov G I 1987 *Computing systems, systems and networks: Textbook. For higher educational institutions* L.: Energoatomizdat 288 p
[8] Gilmanshin I R 2009 Mathematical model of the module for early detection of losses of the energy consumption control system *IV Tinchurin readings: materials of the international. Sci* T 1 pp 66-68
[9] Ventsel E S, Ovcharov L A 1991 *The theory of random processes and its engineering applications* Moscow: Nauka 384 p

# Instrumental research method of qualitative composition of landfill gas in the surface layer of landfills

**I R Gilmanshin[1], N F Kashapov[1], S I Gilmanshina[1], A I Galeeva[2]**
[1]Kazan Federal University, 18 Kremlyovskaya street, Kazan, 420008, Russian Federation
[2]Kazan State Power Engineering University, Kazan, 420066, Russian Federation

Is-er@yandex.ru

**Abstract.** The article analyzes the practice of waste management in Russia. The system of target indicators of the efficient landfills functioning is formalized. The method of instrumental analysis of concentration and qualitative composition of landfill gas in the surface layer of the Samosyrovo landfill is presented.

As of 2014, in the Russian Federation there are 5 168.3 million tones of . wastes have been generated.[1] Out of the total volume of annually generated waste about 60 million tons are SDW, from which the result of the life activity of the population is about 50 million tons and 10 million tons of industrial waste. The annual increase in waste volumes is 3-4%. [2]. In addition, according to statistics, specific volumes of waste generation in Russia are 350 kg / person in a year that is significantly lower than the average for Europe (503 kg / person per year), and even than in the EU-12 (420 kg / person), whose income level is close to Russian. [3]

According to the established practice, almost all the volume of solid waste is stored on landfills and dumps, and only 4-5% of the waste is involved in recycling. A low percentage of SDW involvement in processing is due to the lack of a sufficient number of enterprises for recycling of waste and waste disposal facilities. In total, 1092 specially arranged places for waste disposal (solid waste landfills) are registered in Russia, which is several times less than even authorized landfills, which are about 10 thousand. According to Rosprirodnadzor data, in 45% of the constituent entities of the Russian Federation the project capacity of landfills is lower than the established annual waste accumulation rate . All above it determines the prerequisites for reforming the industry of handling, processing, utilization and disposal of waste in the Russian Federation. The existing system of waste management needs to be promptly adjusted taking into account the global world trends.

Along with systematical upgrade of the waste management system, it is necessary to establish the following targets:

- percentage of waste to the disposal from the total volume of generated waste,
- damage to the environment associated with the removal of land from use, pollution of the environment,
- energy intensity of the burial process and subsequent reclamation of the landfill.

The condition for optimizing these criteria comprehensively determines the effectiveness of the waste management system.

IOP Conf. Series: Materials Science and Engineering **240** (2017) 012021 doi:10.1088/1757-899X/240/1/012021

Consider the selected criteria.

1. The proportion of waste to the disposal in the total volume of waste generated. The formation of waste is inextricably linked with the process of human life. Moreover, the growth of welfare and economic development only increase the volume of waste generated. Therefore, a system of total disposal of waste with full volume inevitably leads to increased costs associated with organizations of landfills, their content and reclamation. At the same time, land plots are withdrawn from circulation directly by landfills and the scope of use of adjacent sites is limited. The existing optimization mechanisms presuppose preliminary segmentation (including separate collection), mechanical modification of the state and form of waste (pressing), and problem-oriented methods of influence that determine the reduction in the volumes and degree of aggressiveness of wastes sent for disposal (thermal, chemical, microbiological, etc.).

2. Damage to the environment. Morphological composition of waste along with the burial system determines the intensive, continuous negative impact on the environment. Filtrates, the emission of harmful substances and gases into the soil and the atmosphere require monitoring and a set of compensatory measures. Moreover, the high effectiveness of protective measures can be achieved only when environmental safety requirements are taken into account at all stages of the life cycle of the landfill.

3. Energy intensity of the burial process and subsequent reclamation of the landfill. High material and energy costs associated with the implementation of comprehensive measures for the effective management of landfills are holding back their (measures) full implementation. At the same time, these activities are costly only to the first sight.

Certainly, a separate evaluation of the effectiveness of individual measures as a rule does not reveal their investment attractiveness. However, in the complex, systemic consideration of the effect, taking into account indirect positive factors, the economic efficiency of the project is fundamentally changing.

The Samosyrovo solid waste landfill was selected as the object of research. The polygon consists of 4 burial maps. The first map is an experimental site - the burial site is full and is subject to reclamation and degassing. The estimated timeframe for filling the first waste storage card is 6 years. Years of the beginning and end of operation of the first card - 2004 - 2009. Design capacity of the first card: 937 872 tons. (5 345 780 $m^3$) The area of the first polygon map is 4.44 ha. The height of the polygon (m), the height of the prism of the burial of the first card is 26 meters from the surface of the earth. Method of disposal of waste: clay sandy quarry - fill with uniform stowage, quarry fill - fill. Morphological composition of wastes on the first map: consumption waste of 562 713.68 $m^3$ (98 721.7 tons), industrial waste 2 138 349 $m^3$ (375 149 tons). The hazard class of the stored waste: mainly 3 to 4 hazard classes. Conditions and method of waste disposal degree - fill with uniform waste disposal, compaction, layer-by-layer repopulation of soil with a layer pitch of 2 meters in height. The thickness of the layer is 250 mm.

Based on the analysis of the initial data on the landfill, a preliminary plan was drawn up for the assessment of the energy potential of the first SDW landfill map:

1 Instrumental study of qualitative composition of landfill gas in the surface layer 1 of the SDW landfill map

2 Determination of quantitative parameters of landfill gas emissions from the surface of the landfill

3 Investigation of quantitative and qualitative characteristics of the landfill gas withdrawn through the arrangement of wells. The study of the actual morphological composition of waste in the body of the test site at the reference points.

4 Modeling the process of methanogenesis in the body of the test site and assessing the energy potential of the first SDW landfill map

Instrumental study of qualitative composition of landfill gas in the surface layer 1 of the SDW landfill map.

Instrumental investigation of the qualitative composition of the landfill gas in the surface layer 1 of the SDW landfill map was carried out using an analytical sampling probe for sampling the qualitative composition of the landfill gas in the surface layer of the polygon body from a depth of 500 mm.

As a gas analyzer, a complex analysis instrument for the qualitative composition of Geotech GA2000 gaseous from Keison Products, England was used.

Based on the set radius of the sampling point coverage of 20 meters, 23 samples were analyzed.

The results of the analysis are presented in Tables 1-4.

Table 1 Methane content in the landfill, in%

| Row 6 | Row 5 | Row 4 | Row 3 | Row 2 | Row 1 |
|---|---|---|---|---|---|
| - | 3,8 | 2,5 | 12,4 | 20 | 16,8 |
| 0,9 | 1,3 | 0,6 | 19,3 | 2,1 | 0,3 |
| 0,7 | 0,6 | 0,4 | 0,3 | 0,6 | 0,3 |
| 0,4 | 1,5 | 1,4 | 0,3 | 0,7 | 0,3 |

Table 2 Carbon dioxide content at the landfill, in%

| Row 6 | Row 5 | Row 4 | Row 3 | Row 2 | Row 1 |
|---|---|---|---|---|---|
| - | 21,5 | 0,4 | 18 | 30,8 | 25,2 |
| 4,4 | 0,8 | 1,4 | 30,5 | 2,3 | 5,9 |
| 4,8 | 0,4 | 0,2 | 13,9 | 10,4 | 3 |
| 10,7 | 0,8 | 15,1 | 10 | 7,2 | 21,3 |

Table 3 Oxygen content in the landfill, in%

| Row 6 | Row 5 | Row 4 | Row 3 | Row 2 | Row 1 |
|---|---|---|---|---|---|
| - | 2,6 | 12,3 | 7,1 | 2 | 1,8 |
| 12,4 | 13,9 | 14,4 | 1,7 | 12,8 | 10,5 |
| 12,1 | 15 | 15 | 5,8 | 8,4 | 12,5 |
| 8,7 | 13,6 | 5,7 | 8,2 | 10,8 | 1,9 |

Table 4 Carbon dioxide content in the landfill, in ppm

| Row 6 | Row 5 | Row 4 | Row 3 | Row 2 | Row 1 |
|---|---|---|---|---|---|
| - | 31 | 55 | 96 | 134 | 52 |
| 23 | 24 | 40 | 112 | 60 | 21 |
| 29 | 25 | 33 | 19 | 34 | 22 |
| 21 | 31 | 75 | 16 | 27 | 27 |

During the analysis of the results of the study of the qualitative composition of the landfill gas in the surface layer 1 of the SDW landfill map, the increase (up to 20%) of methane concentration in the northeastern part of the map of the 1 landfill is established; significant uneven distribution of methane concentration in the surface layer of the top area of the first polygon map; presence of local areas (bubbles / cavities) of increased (up to 60%) concentration of methane in the body of the first map of the landfill; the surface layer of the first polygon map has an uneven layer of overspill, at a number of sites the thickness of the overpacking layer exceeds 500 mm; the actual stage of the life cycle of organic inclusions in the body of the first polygon map allows us to continue the studies of quantitative emission of landfill gas into the atmosphere; the actual stage of the life cycle of organic inclusions in the body of the first map of the landfill allows us to positively consider the feasibility of building an energy complex for landfill gas, but the data obtained are not sufficient for a comprehensive assessment of the energy potential.

**References**

IOP Conf. Series: Materials Science and Engineering **240** (2017) 012021 doi:10.1088/1757-899X/240/1/012021

[1] Brown K A, Maunder D H 1994 Using landfill gas: A UK perspective *Renewable Energy* 5 (5-8) pp 774-81
[2] Machado S L, Carvalho M F, Gourc J P, Vilar O M, Nascimento J C F 2009 Methane generation in tropical landfills: Simplified methods and field results *Waste Management* 29 (1) pp 153-61
[3] *Global methane emissions and opportunities for its reduction* Global Methane Initiative (GMI) URL: https://www.globalmethane.org/documents/analysis_fs_eng.pdf
[4] Thompson S, Sawyer J, Bonam R, Valdivia J E 2009 Building a better methane generation model: Validating models with methane recovery rates from 35 Canadian landfills Waste Management 29 (7) pp 2085-91
[5] Themelis N J, Ulloa P A 2007 Methane generation in landfills *Renewable Energy* Volume 32 Issue 7 pp 1243-57
[6] Gilmanshin I R 2014 Bioorganic waste of the Republic of Tatarstan: features and prospects for utilization taking into account the regional aspect *Modern problems of globalization of the world economy and socio-cultural development of man: materials of the reports of the final scientific-practical conference* pp 74-76
[7] Azimov Y I, Galeeva A I, Gilmanshin I R, Gilmashina S I, Ferenets A V 2014 Innovative ways of processing production and consumption wastes with obtaining high-performance products *Innovative machine-building technologies, equipment and materials* pp 180-84
[8] *Multriwell durable gas ecstraction* URL: http://interactivepdf.uniflip.com/2/74963/290615/pub/
[9] Kashapov N F, Gil'manshin I R, Konahina I A 2014 System analysis of the energy complex of engineering enterprise as a basic tool of effective energy *management IOP Conference Series: Materials Science and Engineering* Vol 69 (1) **012024**
[10] I.R.Gil'manshin, N.F.Kashapov. Energy service contracts in regional engineering center for small and medium businesses // IOP Conference Series: Materials Science and Engineering, Bristol-UK. Vol.69, Conf.1, December 2014.
[11] Asiya Galeeva, Nafisa Mingazova, Iskander Gilmanshin. Sustainable Urban Development: Urban Green Spaces and Water Bodies in the City of Kazan, Russia // Mediterranean Journal of Social Sciences MCSER Publishing, Rome-Italy. Vol. 5, No. 24, November 2014. p. 356-360.
[12] Sh.M.Valitov, I.R. Gilmanshin Financial control of the implementation of programs to improve the energy efficiency of educational institutions // Kazan economic bulletin. - Kazan: "Izdatelstvo of KFU", 2014 No. 3 (11). p. 44-51.

IOP Conf. Series: Materials Science and Engineering **240** (2017) 012022 doi:10.1088/1757-899X/240/1/012022

# Competence formation of engineering directions students in the field of energy saving as a way to create new generation technologies

**I R Gilmanshin**[1,2], **S I Gilmanshina**[1]
[1] Kazan Federal University, 18 Kremlyovskaya street, Kazan, 420008, Russian Federation
[2] Kazan National Research Technical University named after A N Tupolev, Kazan, Russian Federation

Is-er@yandex.ru

**Abstract.** The urgency of the formation of competence in the field of energy saving in the process of studying engineering and technical disciplines at the university is substantiated. The author's definition of the competence in the field of energy saving is given, allowing to consider the necessity of its formation among students - future engineers as a way to create technologies of a new generation. The essence of this competence is revealed. The system of work, pedagogical conditions and technologies of its formation in the conditions of the federal university is substantiated.

Economic and political changes in the world community dictate new requirements for bachelors and masters of technical directions - the formation of professional engineering thinking and the knowledge that allows to get adaption in the modern world of energy-saving technologies. The question how ready are graduates of engineering directions of the federal university for innovative professional activity, taking into account the requirements of the country's development in conditions of energy saving and energy efficiency is arising [1, 2]. And it reflects the dynamics of changes in socio-cultural values, ideals, goals of social development. As a consequence, the advancing development of engineering education causes the need for the formation of engineering competencies in the field of energy saving among students - future bachelors and masters.

Under the competence in the field of energy saving, we mean the competence related to the ability to see, to understand, to sustain energy resources and environment as a source of energy resources; to take in a account environmental and engineering knowledge as values; to choose valuable, targetable and meaningful settings for their engineering and technical actions with innovative underground, independently identify technical and environmental contradictions and make decisions on resource-saving. This definition makes it possible to consider the necessity for future students of the studied competence to form a new generation of technologies.

Since all kinds of competencies are literally built into activities, it becomes necessary to re-examine the specifics of engineering and technical activities. At the same time, we will rely on the view taken in science of the specialist's activities [3, 4] and the professional development of the personality [5-8].

The activity, according to [3], is a form of active, purposeful interaction with the environment (including other people) that responds to the interaction that caused this interaction in something. The need is the prerequisite of activity. The main characteristics of the activity are its objectivity, purposefulness, mediation, motivation and productivity. The means of activity are scientific-theoretical and practical knowledge. Activity is a system of actions united in a single whole by a motive motivating it and aimed at achieving the set goal. The meaning of activity is determined by the motive. The motive is connected with the satisfaction of needs. The result of engineering and technical activities are transformations, both in engineering and technology, and in the world around them. This is the specificity of a person's professional engineering and technical activities.

To highlight the levels of productivity of activity of the educator of University on formation of competence in the field of energy saving among the students of engineering we use the classification N.V.Kuzmina [9], which is established according to five levels. The first level is reproductive, when the educator is able to convey what he knows. The second adaptive level in which the educator is able to adapt its message to the specifics of the audience. On the third level of productivity – locally-modeling the knowledge of students the educator owns the learning strategies, knowledge, and skills for the individual sections of the course that allows you to shape the purpose of education, to set goals, develop an algorithm for their solutions and to use pedagogical tools for the inclusion of students in educational-cognitive activity. The fourth level of the system-modeling knowledge of students in which the educator owns strategy of forming the desired system of knowledge, abilities and skills in the discipline as a whole. The fifth level of the system-modeling the activities and behavior of students in which the educator owns strategy of turning their discipline into a means of formation of personality of a future bachelor of engineering possessing competence in the field of energy saving. Taking this classification as a basis, it should be stated that educators of engineering disciplines should not be limited to the first two or three levels of efficiency of pedagogical activity. It is important for them to correspond to the fifth level, which is not possible if the educator does not own it by himself, in this case, competence in the field of energy saving.

The formation of students ' competence in energy saving, we suggest, is connected with training of their active mastering of humanistic methodology, the creative transformation of the world, the harmonization of relations in system "man – machinery, technology – nature – society" based on laws of protection and efficiency. There is no doubt that the most important role in this process is given to sustainable engineering education. This issue has attracted the attention of specialists [10]. since the environment can be studied with its attributes as a physical, chemical, biological, etc., which, in turn, is associated with the advanced understanding of ecology. However, training complex, contributing to the formation of competence in the field of energy efficiency for the future bachelors of engineering in the process of studying of profile disciplines on the basis of the principle of humanization, currently in the literature.

In the light of the discussing problem, we clarify that the main emphasis in activities to develop the students ' competence in energy saving should be on building a clear understanding of the role and place of modern scientific and technical knowledge and innovative technologies in the system of ecological concepts and energy saving measures. It is necessary to form practical skills to identify, to explain and to analyze the scientific basis of certain ecological situations and events, energy saving measures, and also to predict them on the basis of the effects induced by anthropogenic activities in transformed ecosystems, the biosphere as a whole and extravagance in relation to energy. (Features of scientific explanation and prediction considered in [11]).

Based on the above requirements, we systematized the work on competence formation of students in the field of energy saving in the process of studying the natural sciences and engineering disciplines. This system consists of the following components: a) the formation of an environmentally directed engineering thinking (a new mentality associated with the views on environmentally safe and sustainable development of society); b) the development of independence thinking (ability to independence in the knowledge of new energy-saving materials, technologies, techniques and activities, the ability to identify the contradictions in solving complex educational problems of environmental and resource orientation); c) the armament of modern information technologies, means and methods of innovation activities in the field of energy saving.

To accomplish the objectives of the training information should be purposefully selected. The pedagogical conditions of formation of students' competence in the field of energy saving in the process of engineering education are: the development and application of special tasks for the formation of independent thinking based on the complex educational problems of environmental and resource-saving orientation; the application of the method of environmental projects; organization of training of technical disciplines on the basis of modern interactive and information technologies.

Under the technologies of formation of competences in the field of energy saving, we understand the set of methods, a number of forms, methods and means which are used in the educational process that lead to the goal in accordance with the needs of society. In our research we think that in the holistic educational process of modern learning theories should be in harmony. The necessary emphasis on technology, providing formation of competences in the field of energy efficiency through the mastering of the content of the main types of future professional activity, allowing to master the social experience and the development of mobility and adaptation, ability to work in team, ability to take responsibility, how to respond to comments and criticism.

Nowadays in the conditions of the Kazan Federal University it is a large-scale transition to modern tools and active methods and forms of learning [12]. Among the all the methods of active learning, the most flexible structure has a didactic game is a complex phenomenon for which there is no consensus in the pedagogical science. We take in a account the following meaning: didactic game is a form of recreation subject and the social content of certain activities, systems modeling of relations characteristic of this type of practice (e.g. engineering). In the didactic game there are elements of various forms and methods of training accumulated. Its educational and training value is due to the problem of the game, which is the core and source of their roles. In the preparation of didactic games with the assistance of special teaching material on energy efficiency (e.g. alternative energy) must: 1) to choose a theme and analyze its relationship with the learning material; 2) to formulate the purpose and objectives (taking into account the relationship of the chosen topic with academic material); 3) to determine the structure of the game (including theme, goals, objectives, participants); 4) to identify qualitative indicators (to find out the claims of each student participant to determine the optimal partners); 5) to establish objective indicators (to estimate the time and conditions of the game, the necessary theoretical and experimental material, equipment, safety).

In conclusion, we note that there is no hard part of the implementation of gaming technologies in the system of formation of students ' competence in the field of energy saving in the process of studying of special disciplines, such as "energy Saving technology", "Technology and organization of production and services" with the use of stands, simulators and other modern training tools. It is a creative process that characterize the personality of the educator.

## References

[1] Gilmanshin I R, Kashapov N F 2014 Energy Service Contracts in the Regional Engineering Center for Small and Medium Enterprises *IOP Conference Series: Materials Science and Engineering* Vol 69 69 (1) **012010**

[2] Gilmanshin I R, Ferenets A V, Azimov Yu I, Galeeva A I, Gilmanshina S I 2015 Innovative technologies of waste recycling with the production of high performance products IOP *Conference Series: Materials Science and Engineering* Vol 86 **012014**

[3] Zimnia I A 1997 *Pedagogical psychology* Rostov n / a: Denix 480 p

[4] Rubinshtein S L 1989 *Fundamentals of general psychology*: in 2 volumes M .: Pedagogika 328 p

[5] Zeer E F 2006 *Psychology of professions* M .: Academic Project; The Mir Foundation 336 p

[6] Klimov E A 2004 Psychology of professional self-determination. Moscow: Academy 302 p.

IOP Conf. Series: Materials Science and Engineering **240** (2017) 012022 doi:10.1088/1757-899X/240/1/012022

[7] Gilmanshina S I, Sagitova R N and others 2015 Professional Thinking Formation Features of Prospective Natural Science Teachers Relying on the Competence-Based Approach *Review of European Studies* Vol 7 No pp 341-49
[8] Gilmanshin I R, Gilmanshina S I 2016 The formation of students' engineering thinking as a way to create new techniques, technologies, materials *IOP Conf. Ser .: Mater. Sci. Eng* 134 **012006**
[9] Kuzmina N V 1990 *The professionalism of the teacher's personality and the master of industrial training* M .: Higher School 117 p
[10] Gilmanshina S I, Gilmanshin I R 2015 Building of axiological competence of graduate students by means of project-based learning *IOP Conference Series: Materials Science and Engineering* Volume 86 Number 1 **012029**
[11] Gilmanshina S I, Gilmanshin I R, Sagitova R N and Galeeva A I 2016 The Feature of Scientific Explanation in the Teaching of Chemistry in the Environment of the New Information of School Students' Developmental Education *International Journal of Environmental and Science Education* Volume 11 Issue 4 pp 349-358
[12] Gilmanshina S I, Sagitova R N, Gilmanshin I R 2016 Innovative Component of Preparation of Bachelors and Masters in the System of University of Natural Science Education *The European Proceedings of Social & Behavioral Sciences EpSBS* Volume XII pp 55-59

# Teaching of students technology early professional orientation of schoolchildren

**S I Gilmanshina[1], R N Sagitova[1], I R Gilmanshin[1,3] and A R Kamaleeva[2]**
[1] Kazan Federal University, 18 Kremlyovskaya street, Kazan, 420008, Russian Federation
[2]Institute of pedagogics, psychology and social problems, 420039, Kazan, Isaev St., 12
[3] Kazan National Research Technical University named after A N Tupolev, Kazan, Russian Federation

gilmanshina@yandex.ru

**Abstract.** The necessity of early professional orientation of schoolchildren on the engineering profession and a new type of teacher was proved. Theoretically substantiated and experimentally tested pedagogical conditions of training of students - future teachers of technology early professional orientation of schoolchildren in the system of university preparation of teacher of a new type. This development of courses of special disciplines, aimed at developing of future teachers of readiness for early career guidance activities; development of interactive group projects for schoolchildren of different age groups (including primary school), expanding their understanding of the world of professions; practical testing of career guidance projects dealing with children's audience.

## 1. Introduction

Today, in a rapidly changing world, the demands of the society to form a conscious motivation for their future professional activity as a result of the students' early professional orientation increase. Engineers and teachers are especially in demand and are under the close attention of the state. To popularize engineering and teacher education, much is being done at the federal level. These are various specialized programs: All-Russian competitions "Engineer of the Year", "Teacher of the Year", grant and scholarship programs for engineering students and pedagogical departments. Nevertheless, the question of the early professional orientation of schoolchildren to the profession of an engineer or a teacher of a new type is very relevant and requires starting to form a readiness for professional self-determination already in grades 1-4. At the same time it is necessary to rely on the view taken in science on the activities of the teacher [1 - 3] and the professional development of the individual [4, 5].

At the same time, if there are a lot of methodological developments for high school students using various career guidance methods and techniques [6], then for junior level of such developments it is clearly not enough.

The problem of research: what are the pedagogical conditions for teaching students of pedagogical departments the technologies of early vocational guidance of schoolchildren in the system of university training of a new type of teacher.

The purpose of the research: to develop pedagogical conditions for teaching students - future teachers the technologies of early vocational guidance of schoolchildren in the system of university teacher training of a new type.

Methodological basis of the research: competency-activity approach, the approach of innovativeness of educational activity.

## 2. The basic part

The Primary School plays a significant role in the overall educational system. In these years there is a general development of the child's personality, its inclusion in the system of social relations, the beginning of the formation of a culture of activity and behavior, the formation of its intelligence. It is during this period that the cognitive activity of a person becomes the leading one and determines its development. Students at the primary level are curious, respond quickly to new information, which is conducive to expanding their understanding of the world of professions. Later on, based on the received material, the student will be able to analyze the professions himself more sensibly. And the more professions a child will know, the fewer mistakes he will make later in the process of forming his professional trajectory.

In addition, the introduction of the Federal Education Standard of the new generation implies the implementation of the program for the formation of universal educational activities (UAL) in children of primary school age. Otherwise, in elementary school it is necessary to form the ability to learn, i.e. Extract information, extract from it the necessary knowledge, the ability to correctly navigate the world of professions, consciously approach the choice of their future profession. At the same time, knowledge and consideration of the psychological characteristics of children of primary school age allow us to correctly structure the structure and content of vocational guidance in primary classes. For example, children of this age should be gradually introduced into the sphere of professions, learn to navigate in this area and remember that in grades 1-4, only the brightest, most effectively presented information is remembered.

Thus, in modern socio-economic conditions for universities is the task of training a new type of teacher capable of performing early career guidance work among junior schoolchildren, taking into account the factors determining the formation of professional intentions of the individual and the sources of their identification.

In Kazan (Privolzhsky) federal university, in 2011 an innovative model for the training of subject teachers in profile institutes is being implemented in Russia [7]. For example, teachers of chemistry, respectively, are trained at the Chemical Institute. A.M. Butlerova. Scientific and pedagogical activity in the training of teachers of chemistry at the Chemical Institute made it possible to distinguish three pedagogical conditions for teaching students the technologies of the early vocational guidance of schoolchildren.

The first condition is the development of courses of special disciplines aimed at forming the future teachers' readiness for early career guidance. In accordance with this condition, new disciplines are included in the curriculum for the training of teachers of chemistry of a new type in the direction of "Pedagogical Education, Profile Chemistry" in addition to the traditional disciplines "Theory of Teaching Chemistry" (1st year) and "The Methodology of Chemistry" (2, 3 courses) : "Innovations in chemical education" (for courses 1 and 3), "Didactic games in teaching chemistry" (for 1 course), "International education systems" and "Education system of Tatarstan" (for 4 courses), the material of which allows students to acquaint with modern technology and career guidance activities. As a result of mastering these disciplines, the students form important competences for their successful career guidance activity in accordance with the Federal Education Standard of the new generation. We mention some of them: awareness of the social importance of their future profession; Ability to implement education, upbringing and development, taking into account social, age, psychophysical and individual characteristics, including special educational needs of students; readiness for pedagogical support psycho-educational process; the ability to design individual educational routes of students; readiness for the formation of students' abilities to professional self-determination.

The second condition is the development by the future teachers of group interactive projects [8] for students of different age groups (including primary school) expanding their understanding of the world of professions. Its essence is that since by the beginning of schooling a verbal-logical thinking begins to form in a person, including the ability to operate with words, understand the logic of judgments, be able to argue one's own point of view, all career-oriented projects must assume interactivity and a logical explanation of the phenomena and processes that are demonstrated. Of course, the age characteristics of students in grades 1-4 have some differences, both among themselves and in comparison with middle and senior students. For example, long stories, conversations about professions at this age are not effective. When developing projects for primary school, it is necessary to take into account the specificity of students' cognitive interests (what they are really interested in): play, listen and discuss fairy tales, draw, solve puzzles and puzzles and the like.

The third condition is the practical approbation of vocational guidance projects in communication with the children's audience. For example, students of 2-4 courses - future chemistry teachers have developed and tested a number of projects on early vocational guidance. So in the Children's City of the professions "KidSpace" (Kazan) interactive chemical expositions for children of the 1st and 2nd grades were held, in the Technical Creativity Houses for Children and Youth (in the cities of Kazan and Ulyanovsk) - an interactive for pupils of 3-4 classes, in the museum Kazan chemical school - for students of grades 5-6.

*Conclusion.* The pedagogical conditions for teaching students-future teachers the technologies of early vocational guidance of schoolchildren in the system of university teacher training of a new type are theoretically grounded and experimentally tested. This is the development of courses of special disciplines aimed at forming the future teachers' readiness for early career guidance; The development of group interactive projects for students of different age groups (including primary school), expanding their understanding of the world of professions; Practical approbation of career-oriented projects in communication with the children's audience.

## 3. References

[1] Zimnyaya I A 1997 *Pedagogical psychology* Rostov n/D: Deniks 480 p

[2] Gilmanshina S I, Sagitova R N and et. 2015 Professional Thinking Formation Features of Prospective Natural Science Teachers Relying on the Competence-Based Approach *Review of European Studies* Vol 7 No 3 pp 341-49

[3] Gilmanshina S I, Gilmanshin I R, Sagitova R N and Galeeva A I 2016 The Feature of Scientific Explanation in the Teaching of Chemistry in the Environment of New Information of School Students' Developmental Education *International Journal of Environmental and Science Education* Volume 11 Issue 4 pp 349-58

[4] Klimov E A 2004 *Psychology of professional self-determination* Moscow: Academy 302 p

[5] Samigullina G S, Gilmanchina S I, Gaisin I T, Gilmanshin I R, Akchurina I R 2015 *Professional and Creative Development of Natural Geographic Course Teachers within the Process of Professional Retraining* International Education Studies 8(4)

[6] Sibgatova K I, Gilmanshina S I, Khalikova F D, Gilmanshin I R et al. 2015 Peculiarities of Pupils and Vocational College Students' Career Guidance Modeling in the Integrated System "School – College – Enterprise" *Asian Social Science* 11(1) pp 386-91

[7] Gilmanshina S I, Sagitova R N, Gilmanshin I R 2016 Innovative Component of Preparation of Bachelors and Masters in The System of University Natural Science Education *The European Proceedings of Social & Behavioural Sciences EpSBS* Volume XII, pp 55-59

[8] Gilmanshina S I, Gilmanshin I R 2015 Building axiological competence of graduate students by means of project-based learning *IOP Conference Series: Materials Science and Engineering* Volume 86 **012029**

IOP Conf. Series: Materials Science and Engineering **240** (2017) 012024 doi:10.1088/1757-899X/240/1/012024

# Training of chemistry teachers for sustainable development

**S I Gilmanshina[1], R N Sagitova[1], G F Melnikov[1] and R R Fedotova[2]**
[1]Kazan Federal University, 18 Kremlyovskaya street, Kazan, 420008, Russian Federation
[2]Kazan National Research Technical University named after A.N.Tupolev, 420111, Kazan, Russia

gilmanshina@yandex.ru

**Abstract**. Proposed and piloted teacher training plan containing elements of the concept of sustainable development. teacher training plan includes the development of general and specialized courses in chemical disciplines, organization of activities, taking into account the principles of Green Chemistry.

## 1. Introduction

The concept of sustainable development of society in recent decades has become a fundamental principle of human development. Sustainable development understands how meeting the needs of the present does not undermine the ability of future generations to meet their own needs [1]. An important role in the development of this concept belongs to the quality of education and the formation of civil responsibility in the younger generation, so that in the future, based on the knowledge gained, they could form a society functioning on a sustainable basis [2]. The solution of this task is impossible without teachers, and, therefore, the training of teachers committed to the idea of sustainable development is becoming particularly relevant.

## 2. The basic part

Training of teachers of chemistry at the Kazan Federal University is carried out at the Kh. A.M. Butlerova. Particular attention is paid to the innovative component in the preparation of bachelors and masters [3]. A purposeful work is being carried out to teach the scientific explanation of natural phenomena [4], the formation of professional thinking in students [5], the value-semantic competence [6], the skills of career-oriented work [7]. However, since the concept of sustainable development involves obtaining new energy-efficient materials [8] and optimizing waste utilization [9, 10], these aspects should be reflected in curricula and the organization of training sessions in the process of preparing students of the university.

Consequently, the success of forming future chemistry teachers' understanding of the concept of sustainable development is related to the need to carry out work in three areas. First, to develop a curriculum that allows you to form a solid knowledge of classical chemistry, the methodology of its teaching and the concept of "green" chemistry. Secondly, when carrying out laboratory and practical exercises, regularly demonstrate the importance of waste recycling and non-waste production, and third, systematically carry out work on the formation of the ecological and chemical culture of students.

Let us briefly review the activities of the Department of Chemical Education for each of the indicated areas.

The curriculum developed by the Department of Chemical Education includes classical chemical disciplines such as inorganic chemistry, organic chemistry, analytical chemistry, physical chemistry, colloid chemistry, chemistry of macromolecular compounds, and a number of methodological disciplines on the theory and methodology of teaching chemistry at school. Disciplines "Applied chemistry", "Methods of synthesis of substances" are offered for understanding of the features of technological processes. A modern chemistry teacher should well represent environmental problems and ways to solve them. Such disciplines as "Chemical phenomena in the environment", "Chemistry of soils", "Chemical bases of ecology" form the knowledge of students about the peculiarities of the behavior of chemicals in the environment, the discipline "Problems of environmental and analytical monitoring of environmental pollution", "Fundamentals of chemical safety "Complement the courses of classical chemistry, introducing the properties and methods of determining pollutants in nature. The course "Modern problems of chemistry", which not only reveals and refines the problems of chemistry, but also reveals the features of a new chemical philosophy - "green" chemistry, is conclusive and generalizing. All of the above courses are accompanied by a laboratory workshop in addition to lectures.

Organization of training sessions for future teachers of chemistry is aimed at the formation of professional competencies, education of their ecological and chemical culture, ecological literacy, a sense of responsibility for their actions when dealing with chemicals. The formation of theoretical knowledge among students is supported by the development of practical skills in working with chemicals, the need to follow the basic principles of "green" chemistry in the planning and implementation of a chemical experiment. The importance of working with small quantities of substances in a chemical experiment is suggested and explained, and the need for proper disposal of chemical waste is indicated.

The plan of laboratory works is composed in such a way that some of the substances synthesized by students are used in related disciplines. For example, in a laboratory workshop on the chemistry of coordination compounds, students are asked to obtain sodium hexanitrocobaltate (III), which is then used in laboratory work on analytical chemistry as a reagent for potassium cations. Similarly, in laboratory classes on organic chemistry, when studying the chemical properties of compounds, students receive substances of new classes, which in turn are used in the study of the relevant topic. For example, when studying the chemical properties of alcohols, the corresponding aldehydes and ketones (oxidation reactions) are obtained, which are used in the study of the corresponding classes of organic substances. This approach allows to partially or completely reduce the needs of the department in reagents and follow the philosophy of "green" chemistry. Students - future teachers, performing a laboratory workshop, involuntarily accept the philosophy of both lean chemistry and "green" chemistry. Later, working in school, teachers will be able to convey the principles of "green" chemistry and sustainable development of society to their students.

Educational work with students, as can be seen from the examples given, is conducted continuously throughout the entire educational process. However, the concept of sustainable development covers the most diverse aspects of human and social life and can be viewed from three perspectives: economic (consumption restriction, production optimization), social (harmonization of relations in society) and environmental (philosophy of "green" chemistry). In order to broaden the understanding of sustainable development concepts, students are encouraged to participate in environmental events organized at the university. During pedagogical practice, students develop skills of conducting chemistry lessons, hold conversations with schoolchildren on various topics, touching on issues of sustainable development. As the entire population of the country passes through the school, there is a hope that over time the issues of ecological and chemical culture will be resolved.

*Conclusion.* The specifics of training teachers of chemistry for the sustainable development of society are revealed. It is related to the activities of teachers in three areas - developing a curriculum, organizing training sessions, educational work with students - future teachers. It is established that the curriculum should include, along with classical chemical and methodological disciplines, environmental and chemical disciplines and disciplines that form and develop a philosophy of "green" chemistry. The organization of the laboratory workshop should contribute to the formation not only of narrowly professional competencies, but also the education of students in the ecological and chemical culture of the individual.

**References**

[1] Yevteev S A and Perelet R A 1989 *Our Common Future: Report of the World Comission on Environment and Development (WCED)* Oxford University Press, UK

[2] Collins T J 2017 Review of the twenty-three year evolution of the first university course in green chemistry: teaching future leaders how to create sustainable societies *Journal of Cleaner Production* Volume 140 pp 93-110

[3] Gilmanshina S I, Sagitova R N, Gilmanshin I R 2016 Innovative Component of Preparation of Bachelors and Masters in The System of University Natural Science Education *The European Proceedings of Social &Behavioural Sciences EpSBS* Volume XII pp 55-59

[4] Gilmanshina S I, Gilmanshin I R, Sagitova R N and Galeeva A I 2016 The Feature of Scientific Explanation in the Teaching of Chemistry in the Environment of New Information of School Students' Developmental Education *International Journal of Environmental and Science Education* Volume 11 Issue 4 pp 349-58

[5] Gilmanshina S I, Sagitova R N and et. 2015 Professional Thinking Formation Features of Prospective Natural Science Teachers Relying on the Competence-Based Approach *Review of European Studies* Vol 7 No 3 pp 341-49

[6] Gilmanshina S I, Gilmanshin I R 2015 Building axiological competence of graduate students by means of project-based learning *IOP Conference Series: Materials Science and Engineering* Volume 86 Number 1 **012029**

[7] Sibgatova K I, Gilmanshina S I, Khalikova F D, Gilmanshin I R et al 2015 Peculiarities of Pupils and Vocational College Students' Career Guidance Modeling in the Integrated System "School – College – Enterprise" *Asian Social Science* 11(1) pp 386-91

[8] Gilmanshin I R, Ferenets A V, Azimov Yu I, Galeeva A I, Gilmanshina S I 2015 Innovative technologies of waste recycling with production of high performance products *IOP Conference Series: Materials Science and Engineering* Vol 86 pp12014-16

[9] Gilmanshin I R, Kashapov N F, Azimov Y I, Gimanshina S I, Ganeeva D A, Valishev R D 2015 Utilization of biogas solid waste landfills by building power complex on the basis of alternative energy installations *Kazanskii economicheskii vestnik* Vol 2 (16) pp 41-44

[10] Gilmanshin I R, Kashapov N F, Gilmanshina S I, Galeeva A I 2016 Landfill energy complex based on the renewable energy installations *IOP Conference Series: Materials Science and Engineering* Volume 134, Number 1 **012007**

IOP Conf. Series: Materials Science and Engineering **240** (2017) 012025 doi:10.1088/1757-899X/240/1/012025

# Thermal and hydraulic efficiency of the corridor tube bundle in conditions of pulsating flow of fluid

**V K Ilyin[1], L S Sabitov[2], A I Haibullina[1], A R Hayrullin[1]**
[1]Kazan State Power Engineering University, Kazan, 420066, Russian Federation
[2]Kazan Federal University, 18 Kremlyovskaya street, Kazan, 420008, Russian Federation

sabitov-kgasu@mail.ru

**Abstract.** The method of determining the thermal and hydraulic efficiency η of tube bundles with a pulsating flow is given in the paper. Based on numerical simulation, the effect of the Reynolds number Re of the dimensionless pulsation amplitude β and the Strouhal number Sh on the thermal and hydraulic efficiency η of the corridor tube bundles was investigated. The optimum pulsation regimes corresponding to the maximum η are found.

Tube bundles are the main elements of heat transfer devices which are widely used in various industries. Therefore, nowadays, there are many works devoted to the study of heat transfer and hydrodynamics in tube bundles in order to improve the efficiency of heat transfer equipment. [1-10]

In work [2], the heat transfer of an oscillating cylinder in a fluid flow was studied numerically. The dimensionless maximum frequency of the oscillation velocity of the cylinder lies in the range $0{,}5 \leq V_m \leq 1$, where $V_m = v_m/u_0$, here $v_m$ and $u_0$ maximum speed of oscillation of the cylinder, m/sec and velocity of fluid flow, m/sec. the range of Reynolds numbers was $100 \leq \text{Re} \leq 500$. It is noted that as the $V_m$ increases, the heat transfer intensification increases, and with increasing Re decreases. When Re = 200, $V_m = 0{,}25$, $V_m = 0{,}5$, $V_m = 1$ the increase in heat transfer were 10,2 %, 13,3 %, and 21,9 % . When $V_m = 1$, Re = 100, Re = 200 and Re = 500 the increase in heat transfer were 17 %, 21,9 %, и 28,7 %. The maximum local increase in heat transfer along the perimeter of the cylinder was observed at the bottom of the cylinder.

In [3], an experimental method was used to study heat transfer in the case of a transverse flow past a cylinder in a pulsating air flow. The Strouhal numbers Sh were in the range $0{,}18 \leq \text{Sh} \leq 2{,}8$, Reynolds numbers Re $205 \leq \text{Re} \leq 822$, amplitude of pressures $40 \leq p \leq 276$, Pa. An increase in the heat transfer coefficient by 2.1 times was registered. It is established that with increasing numbers of Re and Sh, the intensification of heat transfer decreases, and with increasing pressure amplitudes, p increases in intensification.

The authors of [4] used numerical methods to study the heat transfer of a semicircular cylinder placed in a horizontal channel under forced fluctuations in the flow of liquid. The pulsations were sinusoidal, water with Prandtl number Pr = 7 was used as the working fluid, Reynolds number Re lay in the range $10 \leq \text{Re} \leq 100$, Strouhal number $0 \leq \text{Sh} \leq 2$, amplitude of vibration $0 \leq A \leq 0{,}6$. The maximum increase in the dimensionless coefficient of drag of the form (total drag coefficient) was

22%, Nusselt number was 10%. It is noted that the maximum increase in heat transfer irrespective of Re is observed at Sh = 1.

In [5, 6], an experimental method was used to study heat transfer in a corridor bundle of tubes under conditions of a pulsating flow of a liquid. Flow pulsations were asymmetrical. As a working fluid, water Pr □ 5.5 was used. Re lie in the range $100 \le \text{Re} \le 500$, the ripple frequency is $0{,}125 \le f \le 0{,}5$ Hz, the amplitude of the oscillation is $1{,}25 \le A/D \le 4{,}5$. The maximum increase in heat transfer by 90% was observed at Re = 500, $f = 0.5$ Hz, $A / D = 4.5$. It is shown that with increasing Re, the heat transfer decreases, and with increasing $f$ and $A / D$, heat transfer increases.

In pulsating currents, the effect of pulsations on heat transfer is mainly considered or the effect of pulsations on the shape resistance coefficient is investigated, but the thermal and hydraulic efficiency of heat exchange elements is not considered. Although the efficiency of the heat exchange equipment depends on the values of the hydraulic resistance and the heat transfer coefficients of the heat exchange elements, which in turn affect the final metal capacity of the heat exchanger and the necessary power of the pumps required for pumping the heat carriers. In this paper, we propose a technique for determining the thermal and hydraulic efficiency of heat exchange elements in pulsating flows.

To assess the effectiveness of methods of intensification in stationary flows, the coefficient of thermal and hydraulic efficiency of Kirpichev [7]

$$E = q / N, \tag{1}$$

where $q$ – specific heat sink, W/m$^2$; $N$ – specific coolant flow rate, W/m$^2$

$$N = \Delta p \cdot v = \xi \frac{\rho \cdot v^2}{2} \cdot v = \xi \frac{\rho \cdot v^3}{2},$$

Here $\Delta p$ – pressure drop in the tube bundles, Pa; $v$ – Average fluid velocity in the bundle, m/sec; $\rho$ – Fluid density, kg/m$^3$; $\xi$ – coefficient of hydraulic resistance.

The power for pumping a coolant in a bundle of tubes with a pulsating current is averaged over a period of pulsation $T_p$, c

$$N_p = \frac{1}{T_p} \int_0^{T_p} \Delta p_{p(t)} v_{p(t)} dt. \tag{2}$$

Here $\Delta p_{p(t)}$ – instantaneous values of the pressure drop during pulsating flow in a bundle of tubes, Pa; $v_{p(t)}$ – instantaneous velocities with a pulsating flow, m/sec; $t$ – time, sec.

The effectiveness of the method of intensification can also be estimated with the help of the specific coefficient of thermal and hydraulic efficiency under the condition $\text{Re}_p = \text{Re}_{st}$ (Reynolds analogy factor) [1] in the form

$$\eta = E_p / E_{st} = (\text{Nu}_p / \text{Nu}_{st}) / (\xi_p / \xi_{st}), \tag{3}$$

where $\text{Re}_{st}$, $\text{E}_{st}$, $\text{Nu}_{st}$, $\xi_{st}$ – Reynolds number, Kirpichev efficiency coefficient, Nusselt number and hydraulic resistance in a steady-state channel, $\text{Re}_p$, $\text{E}_p$, $\text{Nu}_p$, $\xi_p$ – the average for the period of pulsations of the Reynolds number, the Kirpichev efficiency coefficient, the Nusselt number and the equivalent hydraulic resistance in the channel for pulsating flow.

Reynolds for steady-state $\text{Re}_{st}$ and pulsed flow is calculated as follows $\text{Re}_p$:

$$\mathrm{Re}_{st} = \frac{v_{st} D}{\nu}; \tag{4}$$

$$\mathrm{Re}_{p} = \frac{v_{p} D}{\nu}, \tag{5}$$

where $D$ – diameter of tube bundle, m; $\nu$ – Kinematic viscosity, m$^2$/sec; $v_{st}$ – The average flow velocity for stationary flow along the narrowest section in the bundle, m/sec; $v_p$ – The velocity during pulsed flow was averaged over the period of pulsation $T_p$

$$v_p = \frac{1}{T_p} \int_0^{T_p} v_{p(t)} dt .$$

For steady flow in a tube bundle, the coefficient of hydraulic resistance was determined as follows

$$\xi_{st} = \frac{\Delta p_{st} \cdot 2}{\rho v_{st}^2}. \tag{6}$$

where $v_{st}$ – average flow velocity for stationary flow, m/sec.

For a pulsating flow, the equivalent coefficient of hydraulic resistance $\xi_p$ (coefficient of hydraulic resistance of the channel with equivalent energy (power) expenditure for pumping the coolant with superposition of pulsations) was determined as follows

$$\xi_p = \frac{N_p \cdot 2}{\rho v_p^3}, \tag{7}$$

According to the authors of [8, 9], it is more rational from a practical point of view to evaluate the effectiveness of methods of heat exchange intensification from the value of heat removal from a unit surface at equal specific powers spent on overcoming the hydraulic losses $N_{st} = N_p$

$$E_N = \left.\frac{E_p}{E_{st}}\right|_{N_{st}=N_p} = \left.\frac{\mathrm{Nu}_{st}}{\mathrm{Nu}_p}\right|_{N_{\text{ст}}=N_p} = \left.\frac{\mathrm{Nu}_p / \mathrm{Nu}_{st}}{\left(\xi_p / \xi_{st}\right)^{\frac{m}{3}}}\right|_{\mathrm{Re}_{st}} . \tag{8}$$

Where $m$ – the exponent with the number Re in the criterial equation for the calculation of heat transfer in pulsating flow (equation (10)). Thus, the expressions for the efficiency coefficient at equal specific powers $E_N$ will be calculated as follows

$$E_N = \frac{\mathrm{Nu}_p / \mathrm{Nu}_{st}}{\left(\xi_p / \xi_{st}\right)^{\frac{m}{3}}} \tag{9}$$

Heat transfer in a corridor bundle of tubes with pulsating flow can be found from the following relationship [10]

$$\mathrm{Nu}_p = 3{,}05 \cdot \mathrm{Re}_{st}^{0{,}42} \cdot (\beta \mathrm{Sh})^{0{,}2} . \tag{10}$$

Here Strouhal number

$$\text{Sh} = \frac{fD}{v_{st}},$$

where the ripple frequency $f$

$$f = \frac{1}{T_p}, \text{ Hz; } T_p = T_1 + T_2, \text{ c},$$

where $T_p$ – the period of pulsation, which consists of the sum of two half-cycles $T_1$ – half-period of the pulse delivery into the shell space of the tube bundle, $T_1 = 0{,}5$ (const) и $T_2$ – half-life pressure relief, $T_2$ – is set depending on $f$. The dimensionless amplitude of pulsations $\beta = A/D$, here $A$ return of fluid in a tube bundle, m.

Equation (10) is valid for the following conditions $100 \le \text{Re} \le 1000$, $0{,}026 \le (\beta\text{Sh}) \le 2{,}6$, $2{,}6 \le \text{Re}(\beta\text{Sh}) \le 260$, $1{,}25 \le \beta \le 4{,}5$, At the relative longitudinal and transverse step of the tubes bundle $s_1/D = 1{,}3$, $s_2/D = 1{,}3$ and Pr $\approx 5{,}5$. The coefficient of determination is ($R^2 = 0{,}84$).

On Fig. 1–8 there are shown the dependencies Re, β, Sh, $f$ from η and $E_N$ calculated by according to equations (3) and (9), respectively, for data obtained in numerical simulation. [10] On Fig. 1–3 there are shown Re from η and $E_N$ for different β and $f$. On Fig. 1–3 it could be seen that as the numbers Re increase, the coefficient η grows independently of β and $f$, and the different influence of Re on $E_N$. When $f = 0.5$ Hz with an increase in Re from 100 to 300, an increase in the index $E_N$ from 1.09 to 1.32 is observed, with a further increase in Re, a decrease $E_N$ to 1.21 occurs. At $f = 0.25$ Hz and 0.1666 Hz, the maximum $E_N$ for various β is observed in the Re range from 500 to 900.

In Fig. 4 shows the dependences of $\text{Nu}_p/\text{Nu}_{st}$ and $\xi_p/\xi_{st}$ on βSh, which show that with increasing product βSh, heat transfer and hydraulic resistance increase in non-stationary flow as compared to stationary flow. In the entire range of βSh and Re numbers, the growth of $\xi_p/\xi_{st}$ exceeds $\text{Nu}_p/\text{Nu}_{st}$ (Fig. 5). The maximum value of the thermohydraulic efficiency η = 0,62 is observed at minimum Re = 100 and βSh = 0.026.

β and $f$ differently affect the indicator $E_N$ (Fig. 6, 7). If the decrease in the pulsation amplitude β (Fig. 6) $E_N$ occurs, then an increase in the ripple frequency $f$ (Fig. 7) leads to an increase $E_N$. A similar trend is observed for all the Re numbers studied (Fig. 8). The maximum value $E_N$ = 1.32 is observed for Re = 300 and Sh/β = 0.15

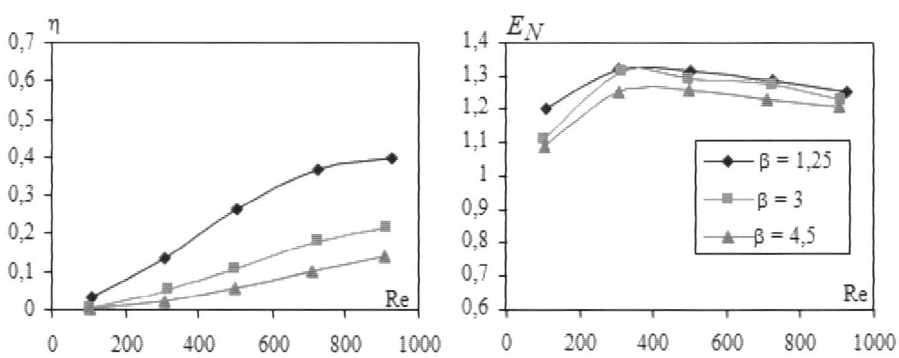


Fig. 1. Dependence Re from η and $E_N$ for $f = 0{,}5$ Hz

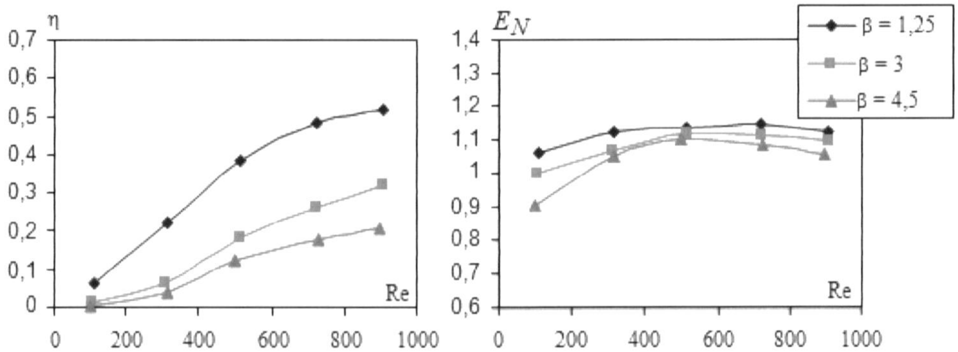


Fig. 2. Dependence Re from η and $E_N$ for $f$ = 0,25 Hz

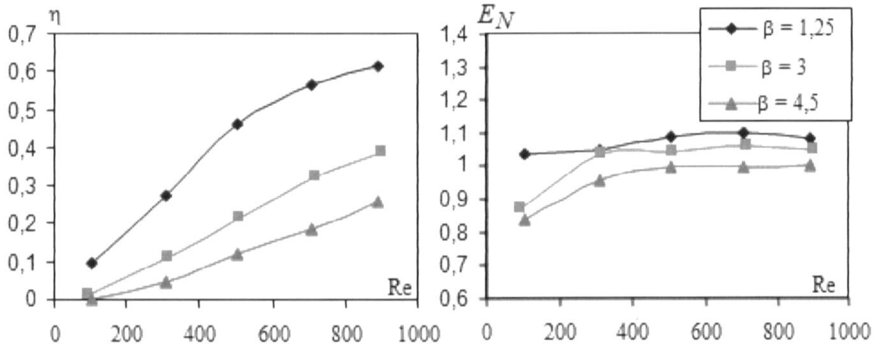


Fig. 3. Dependence Re from η and $E_N$ for $f$ = 0,1666 Hz

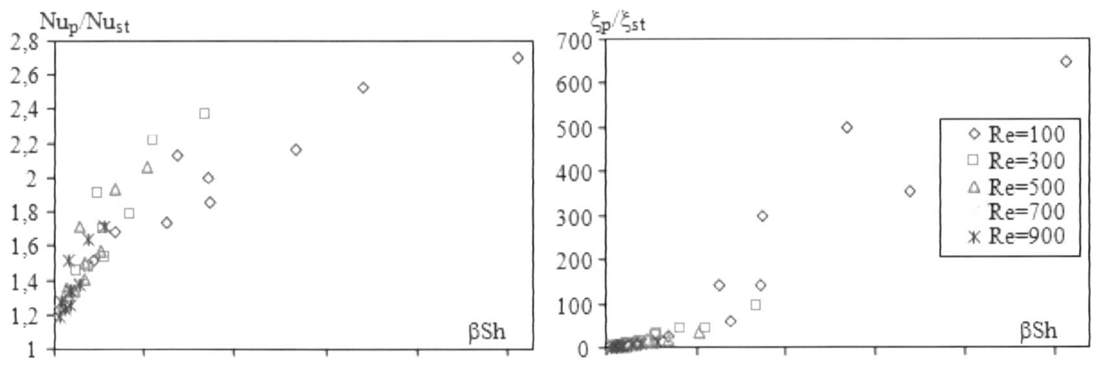


Fig. 4. Dependence $Nu_p/Nu_{st}$ and $\xi_p/\xi_{st}$ from βSh

IOP Conf. Series: Materials Science and Engineering **240** (2017) 012025 doi:10.1088/1757-899X/240/1/012025

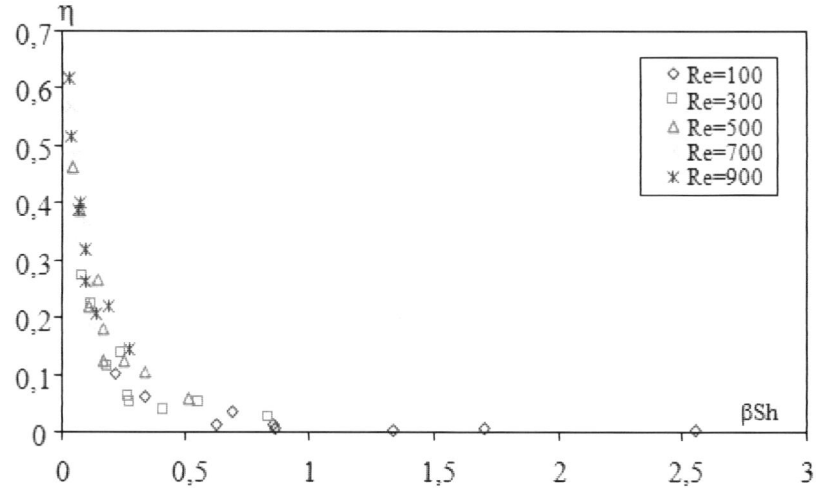


Fig. 5. Dependence η from βSh

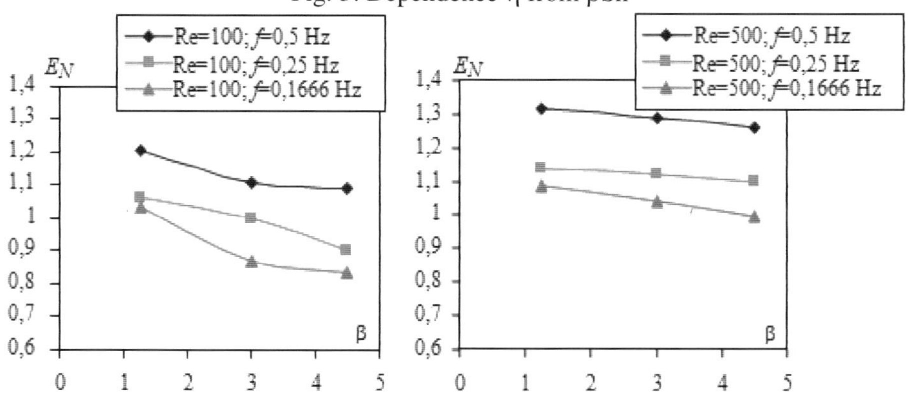


Fig. 6. Dependence $E_N$ from β for Re = 100 and 500

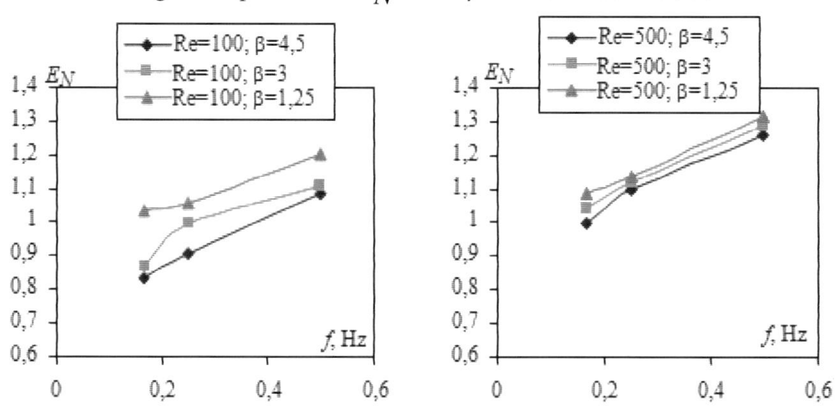


Fig. 7.Dependence $E_N$ from *f* for Re = 100 and 500

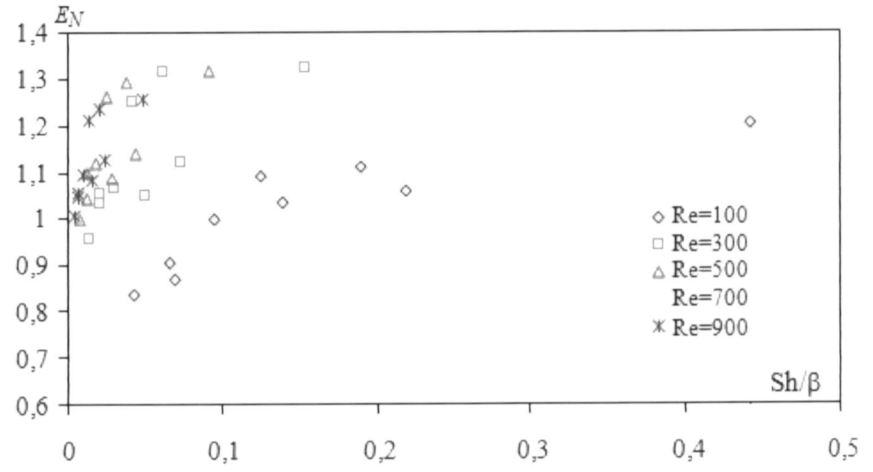


Fig. 8. Dependence $E_N$ from Sh/β

**Conclusion**

A technique is proposed for evaluating the effectiveness of pulsations in order to increase the heat transfer of heat exchange elements.

The parameters of the specific coefficient of thermal $\mathrm{Re}_p = \mathrm{Re}_{st}$ η and hydraulic efficiency are given under the condition and the efficiency factor at equal specific powers spent for pumping the heat carrier $E_N$ for external heat transfer in the corridor bundle in pulsating currents.

It is shown that with an increase in Re there is an increase in η, for $E_N$ maximum regimes are observed at Re about 300 and 500.

With an increase in the frequency Sh and the amplitude β of pulsations, a decrease η occurs. The increase in Sh and β influences differently on $E_N$, the increase in Sh leads to an increase $E_N$, and the increase β, on the contrary, to a decrease $E_N$.

The maximum value is η = 0.62 for Re = 100 and βSh = 0.026. The maximum value $E_N$ = 1.32 is observed at Re = 300 and Sh / β = 0.15.

**References**

[1] Popov I A, Yakovlev A B, Shelkov A V, Rigkov D V, Obuhova L A 2011 Topical methods of heat transfer intensification for heat transfer equipment of heat power engineering *Power engineering of tatarstan* no 5 pp 25-29

[2] Wu-Shung Fu, Tong Bao-Hong 2002 Numerical investigation of heat transfer from a heated oscillating cylinder in a cross flow International *Journal of Heat and Mass Transfer* V 45 pp 3033–43

[3] Guoneng Li, Youqu Zheng, Guilin Hu, Zhiguo Zhang, Yousheng Xu 2016 Experimental Study of the Heat Transfer Enhancement from a Circular Cylinder in Laminar Pulsating Cross-flows *Heat Transfer Engineering* V 37(6) pp 535–44

[4] Neelesh B, Amit K 2017 Pulsating flow and heat transfer analysis around a heated semi-circular cylinder at low and moderate Reynolds numbers *Journal of the Brazilian Society of Mechanical Sciences and Engineering Mar* Vol 54

[5] Khaibullina A I, Khairullin A R, Sinyavin A A, Ilin V K 2014 Heat transfer at in-line tube bank under low-frequency asymmetrical impulses impact on fluid flow *European physical journal* No 76 pp 010041-43
[6] Khaibullina A I, Ilyin V K 2014 Experimental investigation of external heat transfer from tube bundle under condition of low-frequency asymmetrical impulses of fluid cross-flow with reynolds numbers ≤ 500 *Energy Problems* No 1-2 pp 11–19
[7] Popov I A, Mahyanov H M 2009 *Physical basics and manufacture use of heat transfer intensification. Heat transfer intensification* Kazan: Center of innovative technologies 560 p
[8] Migai V K 1987 *Modeling of heat-exchange power equipment* L.: Energoatomizdat 264 p
[9] Nazmeev Yu G, Konakhina I A 2001 *Organization of energy technology complexes in the petrochemical industry* M.: Publishing house MPEI 364 p
[10] Khaybullina A I, Khayrullin A R, Ilin V K 2016 Heat transfer in the flow channel in tube bundle corridor type under imposed on the flow liquid of upstream low-frequency asymmetrical pulsations *Energy Problems* No 11-12 pp 64–75

IOP Conf. Series: Materials Science and Engineering **240** (2017) 012026 doi:10.1088/1757-899X/240/1/012026

# Factors influencing on the thermal flow with the cross-section of the corridor tube bundle in low-frequency non-symmetric pulsations

**V K Ilyin[1], L S Sabitov[2], A I Haibullina[1], A R Hayrullin[1]**
[1]Kazan State Power Engineering University, Kazan, 420066, Russian Federation
[2]Kazan Federal University, 18 Kremlyovskaya street, Kazan, 420008, Russian Federation

sabitov-kgasu@mail.ru

**Abstract.** Factors influencing on the thermal flow of cylinder of corridor tube bundle in pulsating flow of fluid was analyzed numerically. The range of Reynolds numbers was Re <1000 pulsation frequency f = 0.5, Hz, relative amplitude A / D = 1.25. It is shown that the increase in the thermal flow in tube bundle when pulsations are applied to the flow of fluid is mainly due to the convective component. The change in the temperature pressure and turbulence contributes less to the increase in the thermal flow.

Nowadays the study of heat transfer in pulsating flows in the elements of heat exchange equipment and in other industrial equipment is of great practical importance [1-3]. In past studies [4-6], an experimental and numerical method was used to study the heat transfer in the case of low-frequency, nonsymmetric pulsations of the fluid flow in a corridor tube bundle. It is established that when external pulsations are applied to the fluid flow, an increase in the external heat transfer of the corridor tube bundle takes place.

In this paper, by means of numerical simulation, it is planned to analyze the factors influencing the change in the thermal flow in the low-frequency asymmetric pulsations of the fluid flow in the corridor tube bundle for the Reynolds number range Re <1000 of the pulsation frequency f = 0.5, Hz, relative amplitude A/D = 1, 25.

**Mathematical model**

A numerical method was used to describe the heat transfer in a tube bundle during pulsating flows. The computational region of the model (Figure 1) consisted of a two-dimensional bundle of tubes arranged in a corridor order. The diameter of the tubes was D = 0.01 m, the pitch of the tubes was horizontal and vertical $S_{1,2}$ = 0.013 m. The number of tubes in the direction of the liquid was z = 10.

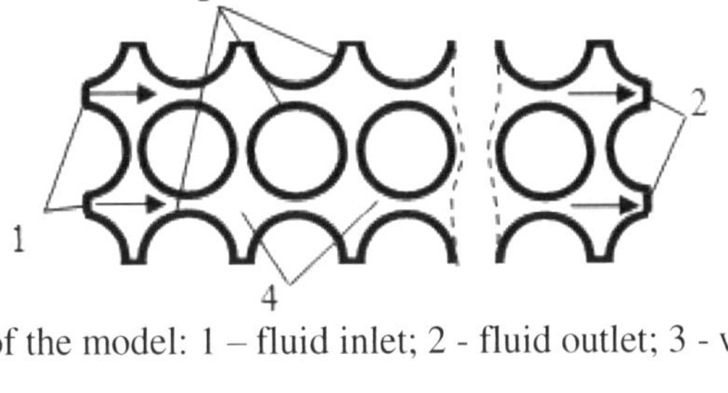


Fig. 1. Calculated region of the model: 1 – fluid inlet; 2 - fluid outlet; 3 - wall; 4 - annular space

IOP Conf. Series: Materials Science and Engineering **240** (2017) 012026 doi:10.1088/1757-899X/240/1/012026

The incompressible fluid flow was described by the system of Navier-Stokes equations (Reynolds averaged Navier-Stokes RANS)

$$\frac{\partial \bar{u}_j}{\partial x_j} = 0 \quad (1),$$

$$\frac{\partial \rho \bar{u}_i}{\partial t} + \frac{\partial \rho \bar{u}_j \bar{u}_i}{\partial x_j} = -\frac{\partial \bar{p}}{\partial x_i} + \frac{\partial}{\partial x_i}\left((\mu + \mu_t)\left(\frac{\partial \bar{u}_i}{\partial x_j} + \frac{\partial \bar{u}_j}{\partial x_i}\right)\right), \quad (2)$$

where $\bar{u}_i$, $\bar{u}_j$ – averaged speed components; $\rho$ – fluid density; $\mu$ – dynamic viscosity; $\bar{p}$ – pressure; $\mu_t$ – turbulent viscosity; ($i$ = 1, 2 $j$, = 1, 2).

Heat transfer was described by the equation (Fourier-Kirchhoff)

$$c_p\left(\frac{\partial}{\partial t}\rho T + \frac{\partial}{\partial x_j}(\rho \bar{u}_i T)\right) = \frac{\partial}{\partial x_j}\left[(\lambda + \lambda_t)\frac{\partial \bar{T}}{\partial x_j}\right], \quad (3)$$

where $c_p$ – heat capacity of fluid; $\lambda_t = c_p \mu_t / \mathrm{Pr}_t$ – turbulent thermal conductivity; $\mathrm{Pr}_t = 0{,}85$ – turbulent Prandtl number; $\bar{T}$ – temperature.

Turbulent viscosity was calculated using the Spalart-Allmaras turbulence model (SARC) [7]

For stationary flow at the inlet in the calculated region (Fig. 1), the velocity $u_n$ depending on the desired Reynolds numbers $\mathrm{Re} = Du_n / \nu$ (4), where $\nu$ – kinematic viscosity of fluid, m$^2$/sec; where $u_n$ – the velocity of the fluid averaged over the narrowest cross-section of the tube shell space, m/sec; $D$ – outer diameter of the tube in the bundle, m. To create a nonstationary flow in the tube at the input, a velocity profile was defined as a function of time $u_n(t)$ With the necessary frequency $f$ and relative amplitude $A/D$ pulsations, where $A$ – displacement of fluid particle backward, m in the narrowest section of the annular space of the bundle. The type of the velocity profile was set in a manner similar to the profile obtained experimentally in the work. [4] In Fig. 2, as an example, the velocity profile for the conditions $f$ = 0.5 Hz, A/D = 1.25, Re = 100 is given. The averaged velocity $\langle u_n \rangle$ over the pulsation period $T_p$ corresponds to the stationary flow velocity, i.e. $\langle u_n \rangle = u_n$.

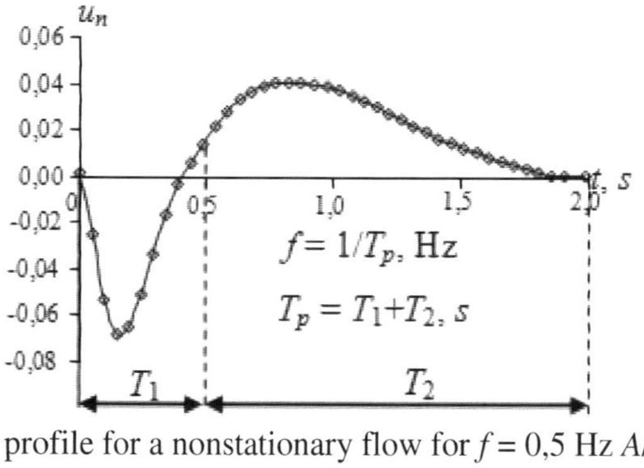


Fig. 2. The velocity profile for a nonstationary flow for $f$ = 0,5 Hz $A/D$=1,25; Re = 100

On the walls of the tubes the following boundary conditions were set: fluid temperature $T_{wall}$ = 315 K; Speed $u$ = 0, m / sec; pressure $\partial P / \partial n = 0$ and absolute roughness $\Delta$ = 0, m.

On the area between the tube halves, from above and below, the constancy of speed $\partial u / \partial n = 0$, pressure $\partial P / \partial n = 0$, and temperature $\partial T / \partial n = 0$ was determined. The pressure at the exit from the bundle was constant P = 101325 Pa.

ANSYS Fluent 14.0 was used to solve the system of equations (1-3). The system of equations was solved by the finite volume method (FVM) for which the calculation area was divided into 64856 reference volumes. The average volume size was $2\times1^{-4}$ m from the condition for the dimensionless coordinate in the boundary layer $y^+ = 1$. [7] The time step was 0.01 sec.

The thermal flow $q$ was averaged over the wall surface of the fifth cylinder in the bundle and in time (for r = D / 2, depending on the angular mark - φ (Figure 3))

$$q_{r,\varphi} = \frac{\int_{\varphi_1}^{\varphi_2} q d\varphi}{\pi D/16} \tag{4}$$

The mean $q_{r,\varphi}$ on time was carried out for one period of pulsations $T_p$

$$q_{t,r,\varphi} = \frac{\int_{0}^{T_{\Pi}} q_{r,\varphi} dt}{T_{\Pi}}. \tag{5}$$

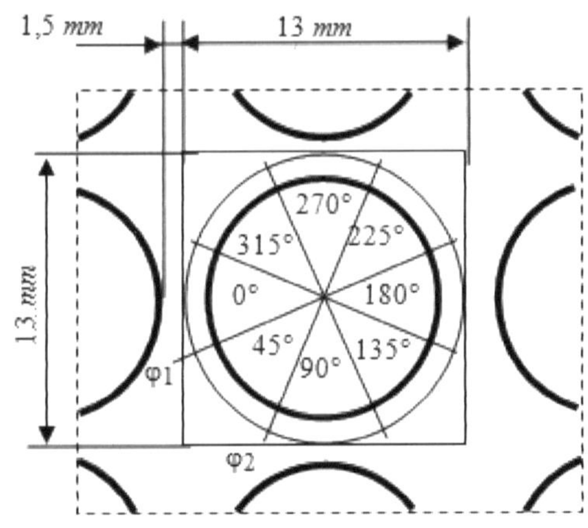


Fig. 3. The layout of the control cylinder and the angle mark φ

**Table 1**. The angle mark

| | | | | | | | | |
|---|---|---|---|---|---|---|---|---|
| ( | 0° | 4<br>5° | 90<br>° | 13<br>5° | 18<br>0° | 22<br>5° | 27<br>0° | 31<br>5° |
| (<br>2 | 3<br>37,5 | 2<br>2,5° | 67,<br>5° | 11<br>2,5° | 15<br>7,5° | 20<br>2,5° | 24<br>7,5° | 29<br>2,5 |
| (<br>1 | 2<br>2,5° | 6<br>7,5° | 11<br>2,5° | 15<br>7,5° | 20<br>2,5° | 24<br>7,5° | 29<br>2,5° | 33<br>7,5° |

The temperature of the fluid was averaged over the region around the cylinder for $x = [0, 2L + D]$, $y = [0, 2L + D]$ (Fig. 3)

$$T = \frac{\int_{y_1}^{y_2}\int_{x_1}^{x_2} T dx dy}{(2L + D)^2 - \pi D^2 / 4}. \quad (6)$$

To averaging over time T, we used the following expression

$$T_t = \frac{\int_0^{T_\Pi} T dt}{T_\Pi}. \quad (7)$$

The fluid velocity module $\bar{u}$, $\bar{u}_i, i = 1, 2$ was as follows,

$$U = |\bar{u}| = \sqrt{\sum_{i=1}^{2} \bar{u}_i^2}. \quad (8)$$

The velocity module averaged over time $U_t$, corresponded to a period of time $t = [0, T_p]$.

The velocity module was also averaged over the region around the fifth cylinder in the beam for $r = [D/2, D/2+L]$ and was according to the formula

$$U_{r,\varphi} = \frac{\int_{r_1}^{r_2}\int_{\varphi_1}^{\varphi_2} U d\varphi dr}{S(\varphi, r)}. \quad (9)$$

Here $S(\varphi, r)$ area of averaging, φ – «angle mark».

To averaging over time $U_{r,\varphi}$, we used the following expression

$$U_{t,r,\varphi} = \frac{\int_0^{T_p} U_{r,\varphi} dt}{T_p} \quad (10)$$

Turbelent heat transfer $\lambda_t$ averaged over the region around the cylinder at $r = [D/2, D/2+L]$

$$\lambda_{t_{r,\varphi}} = \frac{\int_{r_1}^{r_2}\int_{\varphi_1}^{\varphi_2} \lambda_t d\varphi dr}{S(\varphi, r)}. \quad (11)$$

Here $S(\varphi, r)$ area of averaging, φ – « angle mark ».

To averaging over time $\lambda_{t_{r,\varphi}}$, we used the following expression

IOP Conf. Series: Materials Science and Engineering **240** (2017) 012026 doi:10.1088/1757-899X/240/1/012026

$$\lambda_{t_{t,r,\varphi}} = \frac{\int_0^{T_p} \lambda_{t_{r,\varphi}} dt}{T_p}. \tag{12}$$

In order to better understand the processes taking place in the pulsating current, and also to identify the causes that lead to an increase or decrease in the thermal flow, in a nonstationary flow, in comparison with the stationary one, the change in a number of forces (quantities) that contribute to the change in the thermal flow has been studied. This is the contribution of the convective component, turbulent, the change in the thermal flow due to the thermal conductivity and the average driving force of the process - the temperature pressure.

The convective component was estimated by the velocity of the flow taken modulo $U_{r,\varphi}$.

The turbulent component and the change in the thermal flow due to thermal conductivity were estimated by the effective thermal conductivity

$$\lambda_{e_{r,\varphi}} = \lambda + \lambda_{t_{r,\varphi}}, \tag{13}$$

$$\delta\lambda_{e_{r,\varphi}} = \delta\lambda + \delta\lambda_{t_{r,\varphi}}. \tag{14}$$

Since the fluctuations in thermal conductivity $\delta\lambda$ were of the order of 0.01%, then $\delta\lambda_{e_{r,\varphi}} \approx \delta\lambda_{t_{r,\varphi}}$, in fact, the turbulent component $\delta\lambda_{e_t}$ was actually estimated as a gain. The temperature pressure $\Delta T$ was estimated as follows

$$\Delta T = T_{wall} - T \tag{15}$$

The dimensions of the area for determining $U_{r,\varphi}$, $\lambda_{e_{r,\varphi}}$,, as well as the location of their local zones along the perimeter of the cylinder at different angle marks φ are shown in Fig. 3. The values $U_{r,\varphi}$, $\lambda_{e_{r,\varphi}}$ were averaged over the entire area around the cylinder (from 0° to 360°, and also by one eighth inner region of the ring, depending on the angle mark φ, which took values from 0° to 315° (Table 1). φ= 0°, the frontal part of the cylinder is assumed to be perpendicular to the flow of liquid circulating through the bundle. The increase $\delta\Theta$ in the sought value in the pulsating flow $\Theta_p$ compared with the stationary flow $\Theta_{st}$ was estimated as follows

$$\delta\Theta = \left(\Theta_p / \Theta_{st} - 1\right)\cdot 100 \text{ \%,}. \tag{16}$$

Further, for the sake of convenience, we omit the subscript *r* in (5), (6), (10) - (15), omit the subscript φ for the averaged quantities over the entire region around the cylinder in the same formulas.

When observing the change in the above-listed quantities $\lambda_e$, $\Delta T$, $U$, together with the thermal flow density q in the nonstationary flow as compared to the stationary one, one can indirectly answer the question at the expense of which value the increase or decrease in the thermal flow in the liquid flow occurs when pulsations are applied, and also trace how pulsations affect $\lambda_e$, $\Delta T$, $U$, while it remains unknown what contribution each of these quantities makes to the change in the thermal flow.

The contribution of each of the above values to the change in the thermal flow is estimated on the basis of the Newton-Richman law and the Peakle Pe criterion.

IOP Conf. Series: Materials Science and Engineering **240** (2017) 012026 doi:10.1088/1757-899X/240/1/012026

From the Newton-Richmann law it is known that $q = \alpha\Delta T$ (17). Here heat transfer $\alpha$ can be represented as the product of heat transfer due to convection $\alpha_c$ and the sum of heat transfer due to turbulent thermal diffusivity $a_t = \lambda_t / (c_p\rho)$ (18) and the usual temperature conductivity *a*

$$\alpha = \alpha_c \cdot (a_t + a). \quad (19)$$

Where sum $a_t + a = a_e = \lambda_e / (c_p\rho)$ (20) – effective thermal conductivity.

The contribution ratio $\alpha_c$ and $a_e$ can be expressed in terms of the effective Peakle number $\mathrm{Pe}_e$

$$\alpha_c / a_e \approx \mathrm{Pe}_e = U \cdot L / a_e, \quad (21)$$

where $L$ =1,5 – ring width, mm around the cylinder (Fig.3). Let's find $a_e$ from (22)

$$a_e = \alpha_c / \mathrm{Pe}_e, \quad (22)$$

taking this into account, we rewrite equation (20) in the following form

$$\alpha = (\alpha_c \cdot \alpha_c) / \mathrm{Pe}_e = 2\alpha_c / \mathrm{Pe}_e, \quad (23)$$

Find $\alpha_c = \sqrt{\alpha \cdot \mathrm{Pe}_e}$. (24)

The contribution of each quantity $\alpha_c$, $a_e$, and $\Delta T$ in the change in the heat flux was estimated among themselves as a percentage, where the sum of these three values was 100%, for this purpose the rule of logarithms

$$\ln(q) = \ln(\alpha_c \cdot a_e \cdot \Delta T) = \ln(\alpha_c) + \ln(a_e) + \ln(\Delta T); \quad (25)$$

$$\ln(\alpha_c)/\ln(q) \cdot 100 + \ln(a_e)/\ln(q) \cdot 100 + \ln(\Delta T)/\ln(q) \cdot 100 = 100, \%. \quad (26)$$

**Results and its discussion**

Below are the results of numerical simulation for $f$ = 0,5 Hz, $A/D$ = 1,25 and Re < 1000 (Fig. 4).

From Fig. 4 it can be judged how changes in the averaged temperature averaged over the region around the cylinder and in time $\delta\Delta T_t$ convective $\delta U_t$ and turbulent component $\delta\lambda_{e_t}$ can affect the change in the thermal flow $\delta q_t$ at different numbers Re. On Fig.4 it is shown that $\delta q_t$ depends more on $\delta U_t$ and $\delta\Delta T_t$. Increase $\delta q_t$ grows from the minimum value 50 % to maximum 125 %, in the range of numbers Re from 100 to 300, while an increase of the growth of temperature pressure is observed $\delta\Delta T_t$. Further in the range of numbers Re from 300 to 900 occurs the decline $\delta q_t$ together with $\delta\Delta T_t$ and $\delta U_t$. The value of the increase in effective thermal conductivity $\delta\lambda_{e_t}$ varies the opposite of the speed increase $\delta U_t$. If $\delta\lambda_{e_t}$ grows with increase of number Re then $\delta U_t$ declines.

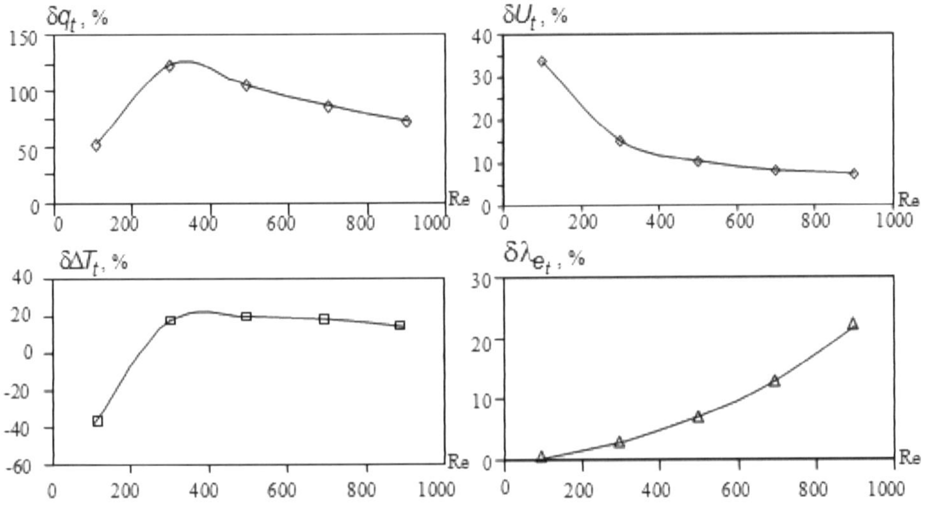


Fig. 4. Relations $\delta q_t$, $\delta U_t$, $\delta \Delta T_t$, $\delta \lambda_{e_t}$ from Re for $f$ = 0,5 Hz, $A/D$ = 1,25

Below are the factors contributing to the change in thermal flow as a percentage, with a total of 100% (see formula 26). According to Fig. 5 that in both steady-state and pulsating flows, a significant contribution to q is exerted by the convective component $\alpha_c$. The part of $\alpha_c$ among the factors , influencing the change in the thermal flow lies in the range $\alpha_c$ = [59÷69] %, part of $\Delta T$ = [16÷23] %, $a_e$= [11÷20] %. In stationary flow, with increasing of numbers Re the part of $\alpha_c$ increases, but $a_e$ decreases, with an almost unchanged part of $\Delta T$ . In the pulsating flow with increasing numbers Re to 500, the fractions $a_e$ and $\alpha_c$ decrease, but $\Delta T$ increases, with an increase to 900, the proportion of shares remains the same.

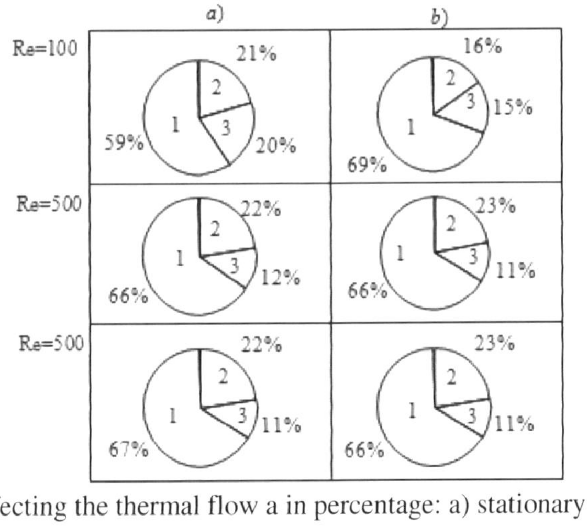


Fig. 5. Factors affecting the thermal flow a in percentage: a) stationary flow; b) pulsating current,1 – $\alpha_c$ ;2 – $\Delta T$; 3 – $a_e$

Consider how the values of local increments vary $\delta q_{t,\varphi}$, $\delta U_{t,\varphi}$, $\delta \lambda_{e_{t,\varphi}}$ (Fig. 6) on the perimeter of the cylinder, depending on the numbers Re. Maximum $\delta q_{t,\varphi}$ and $\delta U_{t,\varphi}$ in the entire range of Re observed in the frontal and aft part of the cylinder, at φ = 0°, and 180°, which is quite obvious, since it is in these zones in the steady flow that stagnant zones with poor circulation of the fluid arise. Minimum values $\delta q_{t,\varphi}$ and $\delta U_{t,\varphi}$ fixed at $\varphi = 90°$. Minimal value of increase $\delta\lambda_{e_{t,\varphi}}$ occurs at $\varphi = 0°$ and 180°, in the range of numbers Re from 100 to 500 growth $\delta\lambda_{e_{t,\varphi}}$ was negative. Maximum value $\delta\lambda_{e_{t,\varphi}}$ fixed at $\varphi = 45°$. The change of numbers Re differently influence on the growth of $\delta q_{t,\varphi}$, $\delta U_{t,\varphi}$ and $\delta\lambda_{e_{t,\varphi}}$. The increase of values $\delta q_{t,\varphi}$ is observed throughout the perimeter of the cylinder with increasing Re from 100 to 300, then there is a decline up to Re ≈ 900. The growth of $\delta U_{t,\varphi}$ decreases with increasing Re by all perimeter of cylinder. The growth of $\delta\lambda_{e_{t,\varphi}}$ increases in all range of numbers Re and all $\varphi$.

By Fig. 6 it could be done the conclusion that when numbers Re from 100 to 900 and frequency $f$ = 0,5 Hz, $A/D$ = 1,25 large growth of $\delta q_{t,\varphi}$ when $\varphi = 0°$ and 180° connected with large growth of $\delta U_{t,\varphi}$ at these ranges.

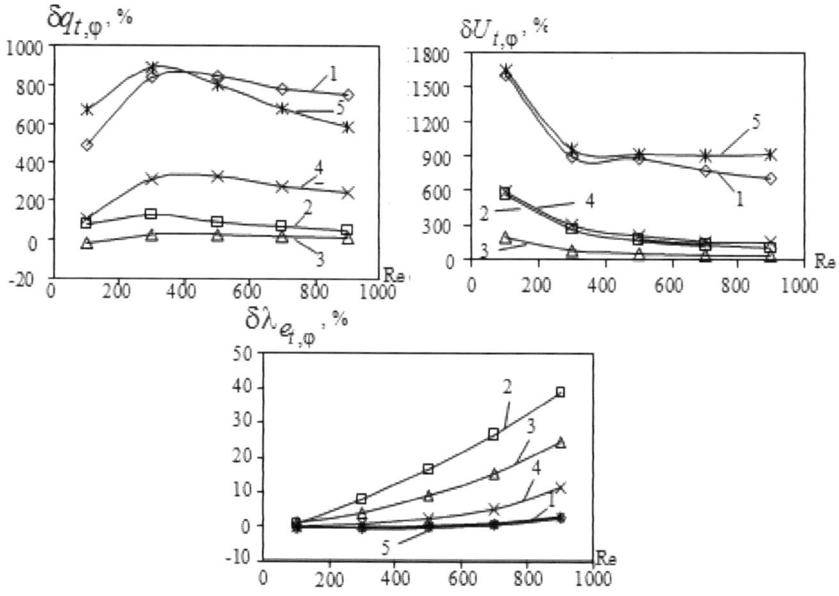


Fig. 6. The growth of $\delta q_{t,\varphi}$, $\delta\lambda_{e_{t,\varphi}}$ and $\delta U_{t,\varphi}$ by perimeter of cylinder, $f$ = 0,5 Hz, $A/D$ = 1,25: 1 – $\varphi = 0°$; 2 – $\varphi = 45°$; 3 – $\varphi = 90°$; 4 – $\varphi = 135°$; 5 – $\varphi = 180°$

In Fig. 7, 8 there are shown the dependences of the increments $\delta q$, $\delta U$, $\delta\Delta T$, $\delta\lambda_e$ from Re for one period of pulsations $T_p$. Growth $\delta U$ is observed at the initial moment of time from 0 to 0,1sec, then for some time the values $\delta U$ fluctuate up to the time $t$ = 1 sec, after which a decrease to $t$ = 2 sec occurs. The maximum increase for the entire period of time Tp is fixed for Re≈100. The maximum increase in the increase in the effective thermal conductivity $\delta\lambda_e$ is observed in the time t from 0.9 to 1.2 sec. At Re numbers of about 100 and 500, the increase in the temperature pressure $\delta\Delta T$ sags in a period of time from t = 0.1 sec to 0.9 sec, at Re about 900 from t = 0.1 sec to 0.7 sec, then the values $\delta\Delta T$ for all Re rise slightly higher than the corresponding instant t = 0 sec, after which the curves decrease to 2 sec. At Re numbers of about 500 and 900, there are small periods of time of not more than 0.1 sec, at which $\delta\Delta T$ have negative values. When Re ≈ 100 the gain $\delta\Delta T$

was negative for the whole period of pulsation $T_p$. $\delta q$ begin to grow from $t$ = 0.1 sec to 0.3 sec, at Re numbers of about 500 and 900, then it starts to decrease down to $t$ =2 sec.

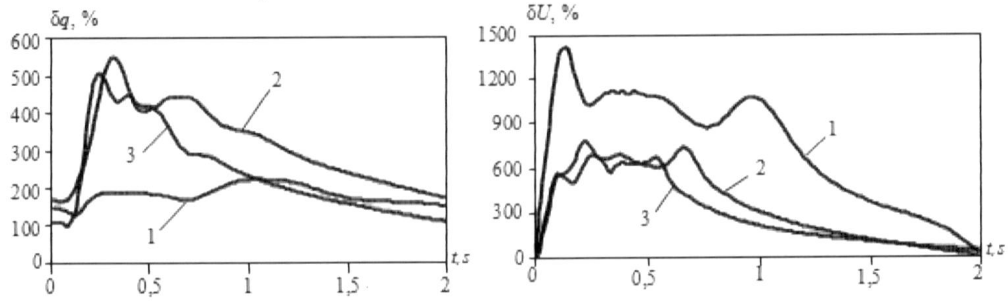


Fig. 7. Dependencies $\delta q$ and $\delta U$ from $t$ at $f$ = 0,5 Hz, $A/D$ = 1,25
1 – Re = 100; 2 – Re = 500; 3 – Re = 900

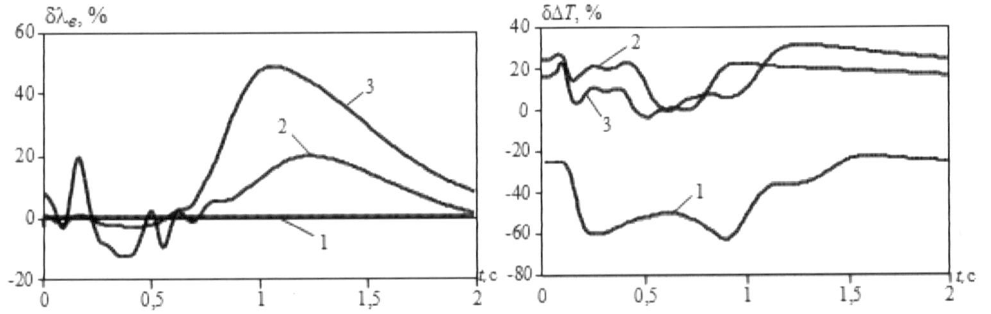


Fig. 8. Dependencies $\delta\lambda_e$ and $\delta\Delta T$ from $t$ at $f$ = 0,5 Hz, $A/D$ = 1,25:
1 – Re = 100; 2 – Re = 500; 3 – Re = 900

**Conclusion**

The results of a numerical experiment show that the change in the thermal flow in tube bundle during pulsating flows in tube bundles is mainly due to the convective component $U$. The temperature pressure $\Delta T$ and the turbulent component $\lambda_e$ as a whole contribute less to the change in the thermal flow. In this case, there are pulsation regimes for which a decrease in the temperature pressure $\Delta T$ and the turbulent component $\lambda_e$ are observed in comparison with the stationary flow. An increase in the Reynolds number Re leads to a decrease in the convective component $U$, in comparison with the stationary flow. This can be attributed to the fact that as the Re numbers increase, the ratio of the pulsation velocity to the stationary one decreases, which is expressed by the ratio $fA/u_n$, where $u_n$ is the velocity in the stationary flow, which increases with the increase in the numbers Re. $\lambda_e$ with increasing Reynolds numbers Re increases, while the temperature pressure $\Delta T$ can both decrease and increase with the Reynolds number Re.

An increase in the local values of the convective component $U_\varphi$ along the perimeter of the cylinder is mainly observed in the frontal and aft parts of the tube. This is due to the fact that in these areas with stationary flow there are stagnant zones with poor circulation of the fluid.

An increase in the local turbulent component $\lambda_{e_\varphi}$ along the perimeter of the cylinder during pulsating currents is observed in the places where the jet collides with the surface of the cylinder, i.e. in the lower part of the cylinder and the area between the front and bottom. Values $\lambda_{e,\varphi}$ are significantly lower in the frontal and aft part of the cylinder. Such a distribution $\lambda_{e,\varphi}$ along the perimeter of the cylinder in the nonstationary flow is probably due to the fact that for stationary flow in these regions the flow has a lower turbulence, with smaller separation zones than in the aft part of the cylinder, where a pronounced vortex zone is formed.

Considering the change in the convective component in time $U$, we can distinguish two lengths of time at which growth $U$ occurs. The first is associated with the acceleration of the flow, with the reverse flow of the fluid flow in the bundle of tubes, the second is associated with the acceleration of the fluid when it moves in the forward direction. For instantaneous values $\Delta T$ of the temperature pressure in the time interval between the segments at which growth $U$ occurs, a decrease $\Delta T$ is observed. The periods of time at which the growth of the instantaneous values $\lambda_e$ of the turbulent component arise, can vary significantly, depending on Re. In this case, the increase in the thermal flow during the pulsation period is mainly observed, for the same periods of time at which the convective component $U$ increases.

When considering the factors influencing the thermal flow in percentage ratio as a result of mathematical modeling, it was found that a greater contribution to the change in the thermal flow for tubes bundle is about 59-69% for both a steady flow and a non-steady flow of fluid in tube bundle has a convective component, the temperature pressure $dT$ in 16-23% and about 11-20% is the contribution of the effective thermal conductivity $a_e$.

**References**

[1] Amiri S, Taher R, Mongeau L 2017 Quantitative visualization of temperature field and measurement of local heat transfer coefficient over heat exchanger elements in sinusoidal oscillating flow *Experimental Thermal and Fluid Science* Vol 633 pp 1724–30

[2] Anup K, Atul K, Rajneesh B, Mark C 2017 The response of an elastic splitter plate attached to a cylinder to laminar pulsatile flow *International Journal of Heat and Mass Transfer* Vol 68 pp 423–43

[3] Musaeva D, Ilin V, Baake E, Geza V 2016 Influence of low-frequency pulsed Lorentz force on the melt flow and the structure of solidifying material *Heat Processing: Research & Development* No 1 pp 61-66

[4] Khaybullina A I, Khayrullin A R, Ilin V K 2016 Heat transfer in the flow channel in tube bundle corridor type under imposed on the flow liquid of upstream low-frequency asymmetrical pulsations *Energy Problems*,No 11-12 pp 64-75

[5] Khaibullina A I, Ilyin V K 2014 Experimental investigation of external heat transfer from tube bundle under condition of low-frequency asymmetrical impulses of fluid cross-flow with reynolds numbers ≤ 500 *Energy Problems* No 1-2 pp 11–19

[6] Khaibullina A I, Khairullin A R, Sinyavin A A, Ilin V K 2014 Heat transfer at in-line tube bank under low-frequency asymmetrical impulses impact on fluid flow *European physical journal* No 76 pp 010041-43

[7] Khaybullina A I, Khayrullin A R, Sinyavin A A, Ilin V K 2014 Modeling the turbulence of a pulsating flow of a heat carrier in a corridor bundle of tubes *Thermophysical foundations of energy technologies* pp 368-72

IOP Conf. Series: Materials Science and Engineering **240** (2017) 012027 doi:10.1088/1757-899X/240/1/012027

# External heat transfer in corridor and staggered tube bundles of different configuration under the application of low-frequency pulsations

**V K Ilyin[1], L S Sabitov[2], A I Haibullina[1], A R Hayrullin[1]**
[1]Kazan State Power Engineering University, Kazan, 420066, Russian Federation
[2]Kazan Federal University, 18 Kremlyovskaya street, Kazan, 420008, Russian Federation

sabitov-kgasu@mail.ru

**Abstract.** In this paper, the external heat transfer coefficient for cross flow around a tube bundle by a pulsating flow was studied by numerical method. ANSYS Fluent 14.0 program is used for the mathematical modeling. The Reynolds numbers lie in the range $100 \leq Re \leq 1000$, the Prandtl numbers $215 \leq Pr \leq 363$, the frequency and the dimensionless relative amplitude of the pulsations were in the range $0{,}2 \leq f \leq 0{,}5$, Hz, $15 \leq \beta \leq 35$, the pulsation ratio $0{,}25 \leq \psi \leq 0{,}5$. The external heat transfer for low-frequency pulsations was studied for 9 configurations of staggered tube bundles and 3 corridor tube bundles. Based on the modeling results, a criteria equation is obtained that allows calculating the external heat transfer in pulsating flows in tube bundles of different configurations. It is established that an increase in Re, Pr, $\psi$ leads to a decrease in heat transfer, and an increase in $f$ and $\beta$ to an increase in heat transfer.

## Introduction

Heat transfer in pulsating flows are being studied for several decades. At the moment, there is a huge amount of work devoted to the study of external and internal heat transfer in pulsating flows, specifically the flow of fluid in circular tubes [1], in channels with depressions [2], flowing around a single cylinder [ 3, 4], a tandem of cylinders [5], various protrusions, blocks and other obstacles [6, 7].

Experimental studies related to the flow structure and heat transfer in the case of a transverse flow past a single cylinder with superimposed pulsations on the air flow were performed by the authors of the paper. [3] The dimensionless pulsation frequency was in the range St = 0-1.76, relative ripple amplitude $\beta = 0$–$0{,}85$. In the studied range of parameters, the maximum increase in heat transfer was 15%.

In research [5], the heat transfer of a tandem of cylinders in pulsating flow was investigated by an experimental method. Flow pulsations had a symmetrical character. The dimensionless frequency parameter lay in the range of $58.1 < \beta < 451.7$, the Kelegane-Carpenter number was in the range $3.14 < KC < 15.7$. It is established that an increase in the parameter KC leads to an increase in heat transfer.

In research [8], a numerical and experimental method was used to study the heat transfer in the case of a transverse flow past a cylinder with low-frequency pulsations of sinusoidal character in the range $0{,}2 \leq f \leq 1{,}1$, Hz and high relative pulsations $10 \leq A/D \leq 38$, and the pulsation number Reynolds was in the 40 < $Re_{os}$ < 810. It is established that with increasing f and A / D it has been occurring an increase in heat transfer.

In most works, flow pulsations have a symmetric sinusoidal character, and in the presence of more papers devoted to flow past a cylinder, studies devoted to heat transfer in pulsating flows in bundles of tubes are extremely small. In this paper, it is planned to investigate heat exchange in tube bundles of various configurations, with symmetrical and asymmetrical pulsations. The heat exchange of a single cylinder in pulsations with different ratio $\psi$ was studied numerically in [9], tube bundles in paper [10]. It is shown that the pulsation ratio $\psi$ influences the heat transfer at constant Re and frequencies $f$..

**Mathematical model**

The calculated region of the model (Fig. 1) is shown in Fig. 1.

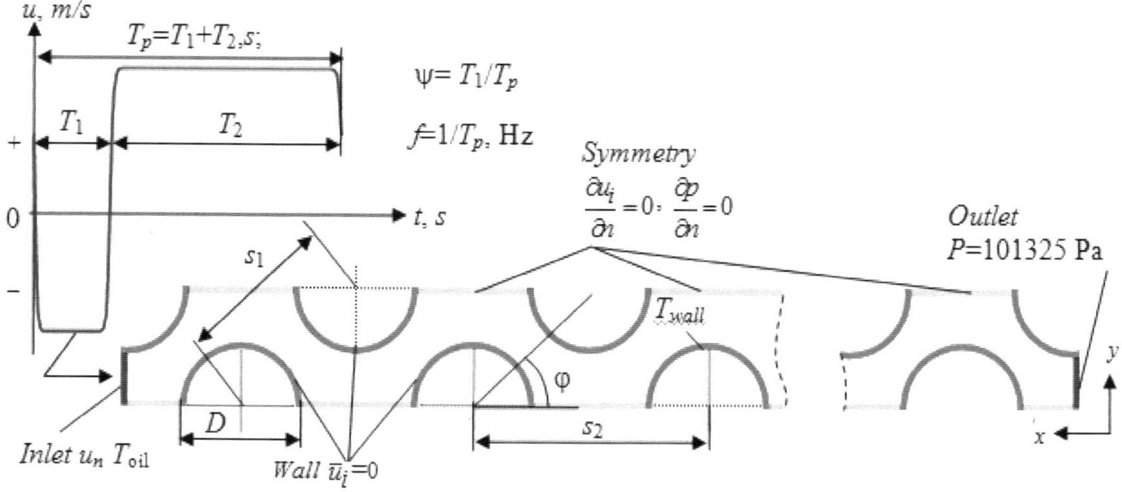


Fig. 1. The estimated area of the model

The flow of an incompressible fluid is described by the system of Navier-Stokes equations (Reynolds averaged Navier–Stokes RANS) averaged by the Reynolds method, which consists of the continuity equation [11]

$$\frac{\partial \bar{u}_j}{\partial x_j} = 0, \tag{1}$$

and equations of transfer for the mean values of quantities randomly pulsating in a turbulent flow using the theory of turbulent viscosity proposed by J. Boussinesque

$$\frac{\partial \rho \bar{u}_i}{\partial t} + \frac{\partial \rho \bar{u}_j \bar{u}_i}{\partial x_j} = -\frac{\partial \bar{p}}{\partial x_i} + \frac{\partial}{\partial x_i}\left((\mu + \mu_t)\left(\frac{\partial \bar{u}_i}{\partial x_j} + \frac{\partial \bar{u}_j}{\partial x_i}\right)\right), \tag{2}$$

where $\bar{u}_i$, $\bar{u}_j$ – averaged speed components; $\rho$ – fluid density; $\mu$ – dynamic viscosity; $\bar{p}$ – pressure; $\mu_t$ – turbulent viscosity; ($i$ = 1, 2 $j$, = 1, 2).

Heat transfer is described by the equation of convective heat transfer (Fourier-Kirchhoff)

$$c_p\left(\frac{\partial}{\partial t}\rho T + \frac{\partial}{\partial x_j}(\rho u_i T)\right) = \frac{\partial}{\partial x_j}\left[(\lambda + \lambda_t)\frac{\partial T}{\partial x_j}\right], \tag{3}$$

where $c_p$ – heat capacity of a liquid; $\lambda_t = c_p \mu_t / \Pr_t$ – turbulent thermal conductivity; $\Pr_t = 0{,}85$ – turbulent Prandtl number; $\bar{T}$ – temperature.

For turbulence simulating on the basis of the analysis [12, 13], the Spalart-Allmaras model (SA) was chosen in the modified SARC model with correction for curvature of the current line (RC).

$$\frac{\partial \rho \tilde{\nu}}{\partial t} + \frac{\partial \rho \bar{u}_j \tilde{\nu}}{\partial x_j} = \frac{\partial}{\partial x_j}\left(\frac{\rho(\nu + \tilde{\nu})}{\sigma_\nu}\frac{\partial \tilde{\nu}}{\partial x_j}\right) + \frac{C_{b2}\rho}{\sigma_\nu}\frac{\partial \tilde{\nu}}{\partial x_j}\frac{\partial \tilde{\nu}}{\partial x_j} + \rho P_\nu - \rho \varepsilon_\nu, \quad (4)$$

where $P_\nu$ – the velocity of turbulent viscosity generation, $\varepsilon_\nu$ – the velocity of its dissipation, $\tilde{\nu}$ - quantity that coincides with turbulent viscosity everywhere except near-wall regions.

The pulsating currents were modeled with the help of the velocity profile (dependence the velocity from time u (t)) corresponding to the necessary relative amplitudes β, frequencies *f*, Re and ratio ψ of pulsations, which was specified at the input by a tube bundle as the boundary condition (Fig.1). The relative amplitude was calculated as β = A / D, where *A* –the displacement of the liquid particle back, m in the narrowest section of the annular space of the beam; D- the diameter of the beam tube. The numbers Re were calculated as

$$\mathrm{Re} = \frac{uD}{\nu}, \quad (5)$$

where ν –the kinematic viscosity, $m^2$ /sec; $u_p$ –fluid velocity, m /sec for pulsating flow averaged over the pulsation period $T_p$ was equal to the velocity of stationary flow of $u_{st}$

$$u_{st} = u_p = \frac{1}{T_p}\int_0^{T_p} u_{p(t)} dt.$$

The velocity profile is obtained using the mathematical model of the hydraulic system of the pulsator-heat transfer. [14] In this article, numerical simulation was carried out to find out the real velocity profile in the bundles of tubes of heat transfer equipment during the generation of pulsations by means of a pulsating device.

The working fluid was oil. The temperature of $T_{oil}$ was set at the entrance to the bundle of tubes, depending on the Prandtl numbers Pr = ν/*a*, where the thermal diffusivity was $m^2$ /sec. The thermophysical properties of the oil corresponded to the oil "Turbine oil-T22". On the wall of the central tube in the beam (Fig. 1) the temperature $T_{wall} = T_{oil} - 1$ was set.

Numerical simulation was carried out for Prandtl range $215 \le \Pr \le 363$, ratio of pulsations $0{,}25 \le \psi \le 0{,}5$, while Re numbers and frequencies $f$ were in the range $100 \le \mathrm{Re} \le 1000$, $0{,}2 \le f \le 0{,}5$ Hz, relative amplitude $15 \le \beta \le 35$. The influence of different bundle configurations on heat transfer during pulsating flows have been studied for 9 configurations of staggered and 3 corridor bundles (Table 1).

Variants of bundles correspond to widely use in heat transfer equipment. [15]

Table 1

Geometrical characteristics of bundles

| № | φ | $s_1/D$ | *D*, мм |
|---|---|---|---|
| 1 | 30° | 1,25 | 16 |
| 2 | 30° | 1,5 | 16 |
| 3 | 30° | 1,75 | 16 |
| 4 | 45° | 1,25 | 16 |

| | | | |
|---|---|---|---|
| 5 | 45° | 1,5 | 16 |
| 6 | 45° | 1,75 | 16 |
| 7 | 60° | 1,25 | 16 |
| 8 | 60° | 1,5 | 16 |
| 9 | 60° | 1,75 | 16 |
| 10 | 90° | 1,25 | 16 |
| 11 | 90° | 1,5 | 16 |
| 12 | 90° | 1,75 | 16 |

Numerical simulation was performed in the ANSYS Fluent 14.0 program using the finite volume method (FVM). The optimal grid was calculated according to the procedure given in [13]. When evaluating the simulation results, the Nusselt numbers were calculated as follows

$$\mathrm{Nu} = \alpha \cdot D / \lambda, \tag{6}$$

where $\lambda$ – the thermal conductivity, W/(m · K); $\alpha = q/(T_{oil} - T_{wall})$ – heat transfer. Here $q$ –the heat flux, W / m$^2$ averaged over the surface of the central tube in the bundle and during the pulsation period. $T_{oil}$ also averaged over the entire design area and during the pulsation period.

**Results and discussion**

In Fig. 2 it is shown the dependence $\mathrm{Nu}_p/\mathrm{Nu}_{st}$ from Re with different $s_1/D$ for $\varphi = 90°$. With the increase of Re in the range $100 \leq \mathrm{Re} \leq 600$, $\mathrm{Nu}_p/\mathrm{Nu}_{st}$ drastically decreases, regardless of $s_1/D$, compared to the $600 \leq \mathrm{Re} \leq 1000$ range, where the ratio $\mathrm{Nu}_p/\mathrm{Nu}_{st}$ varies insignificantly. An increase in $s_1/D$ results in an increase in $\mathrm{Nu}_p/\mathrm{Nu}_{st}$. Considering the influence of $\psi$ on heat transfer, it can be seen that the intensification is mainly observed in regimes with a smaller $\psi$.

In Fig. 3 it is shown the dependence $\mathrm{Nu}_p/\mathrm{Nu}_{st}$ from Re for Pr = 293, $\beta$ = 15, $f$ = 0.5 for different $\varphi$ and $s_1/D$. When $\varphi = 30°$ $s_1/D = 1,25$ (Fig. 3, a), the increase in the heat transfer intensity over the entire range of $100 \leq \mathrm{Re} \leq 600$ decreases with increasing Re. When $s_1/D = 1,5$ and 1.75, the decrease of $\mathrm{Nu}_p/\mathrm{Nu}_{st}$ slows down after Re = 600. The effect $\psi$ on $\mathrm{Nu}_p/\mathrm{Nu}_{st}$ is more appeared for Re = 1000, and $\mathrm{Nu}_p/\mathrm{Nu}_{st}$ is higher at $\psi = 0.25$.

When $\varphi = 45°$ (Fig. 3, b), the effect of Re numbers on $\mathrm{Nu}_p/\mathrm{Nu}_{st}$ is, in general, similar to the tube bundle layout for $\varphi = 45°$ (Fig. 3, a).

At $\varphi = 60°$, $s_1/D = 1,25$ (Fig.3, b), with an increase in Re in the range $100 \leq \mathrm{Re} \leq 600$, $\mathrm{Nu}_p/\mathrm{Nu}_{st}$ decreases at first, with a further increase in Re = 1000, a slight increase in $\mathrm{Nu}_p/\mathrm{Nu}_{st}$ occurs. Effect $\psi$ on $\mathrm{Nu}_p/\mathrm{Nu}_{st}$ for bundle $\varphi = 60°$is less compared with bundles at $\varphi = 30°$ and 45°.

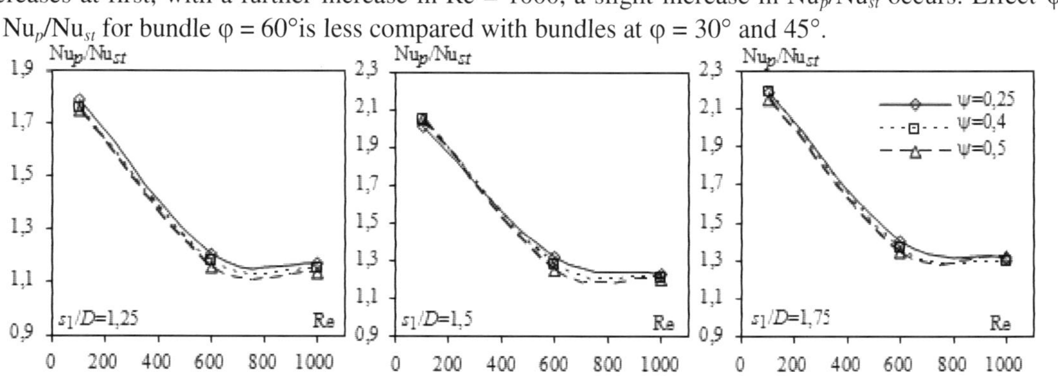


Fig. 2. Dependence $\mathrm{Nu}_p/\mathrm{Nu}_{st}$ from Re for Pr = 215, $\beta$ = 15, $f$ = 0,5 Hz, $\varphi = 90°$

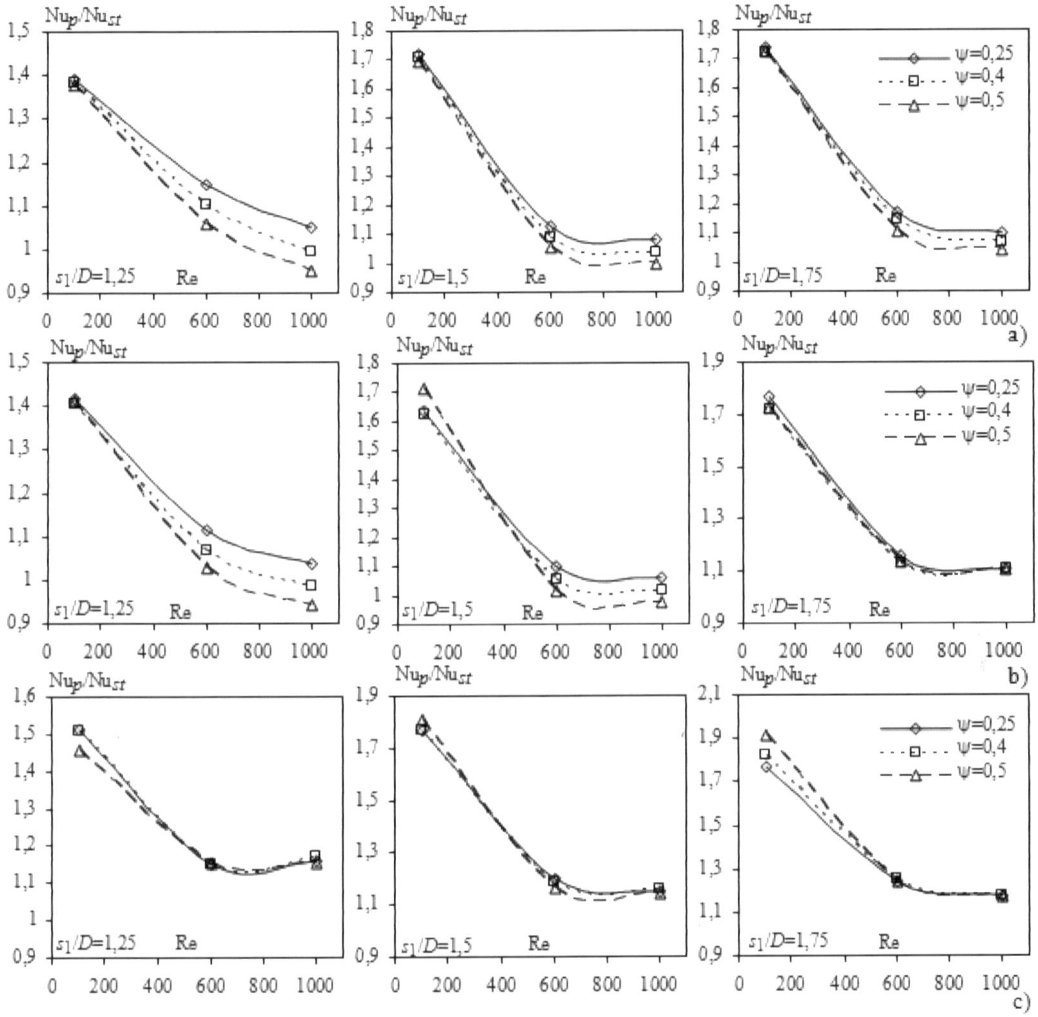


Fig. 3. Dependence $Nu_p/Nu_{st}$ from Re for Pr = 293, β = 15, $f$ = 0,5:
a) φ = 30°; b) φ = 45°; c) φ = 60°;

In Fig. 4, 5 it is shown the dependence of $Nu_p/Nu_{st}$ from Pr for different $s_1/D$, at φ = 90° and 60°, by which it is seen that an increase in the Pr number as well as an increase in Re leads to a decrease in heat transfer in the pulsating flow as compared with the stationary one, regardless of $s_1/D$.

In Fig. 6 it is shown the dependence of $Nu_p/Nu_{st}$ from β at φ = 90° from which it follows that an increase in β and $f$ leads to an increase in heat transfer. When Re = 600, Pr = 293, $f$ = 0.2 (Fig. 6, a), it is noticeable that with increasing $s_1/D$ the influence of ψ decreases, i.e. the curves ψ=0,25, 0,4 and 0,5 converge.

In pulsation modes with a lower intensity, i.e. at minimum values of β and $f$, it is observed $Nu_p/Nu_{st} < 1$ , which is mainly occurred when approaching the symmetric nature of pulsations ψ = 0.5 and with decreasing $s_1/D$.

In Fig. 7 it is shown the dependence of $Nu_p/Nu_{st}$ from $f$ for different combinations of staggered bundles at Re = 600, Pr = 215, β = 35, by which it is seen that an increase in $f$ leads to an intensification of heat transfer, regardless of the arrangement of the bundles.

The maximum increase in heat transfer is observed for a bundle with φ = 60° and $s_1/D$ = 1,75.

In Fig. 8 it is shown the dependence of $Nu_p/Nu_{st}$ from β/(Re·Pr·Fo) at ψ=0,25 for different φ and $s_1/D$. Here Fo - the Fourier number which is calculated as follows Fo=$a/(f \cdot D^2)$, where $a$ – the

IOP Conf. Series: Materials Science and Engineering **240** (2017) 012027 doi:10.1088/1757-899X/240/1/012027

thermal diffusivity of the oil, $m^2$ / sec. The dimensionless complex $\beta/(Re \cdot Pr \cdot Fo) = Af/u_{st}$ characterizes the ratio of the pulsation velocity to the stationary one, the increase of which leads to the growth of $Nu_p/Nu_{st}$ irrespective of the bundle arrangement and the ratio pulsation $\psi$.

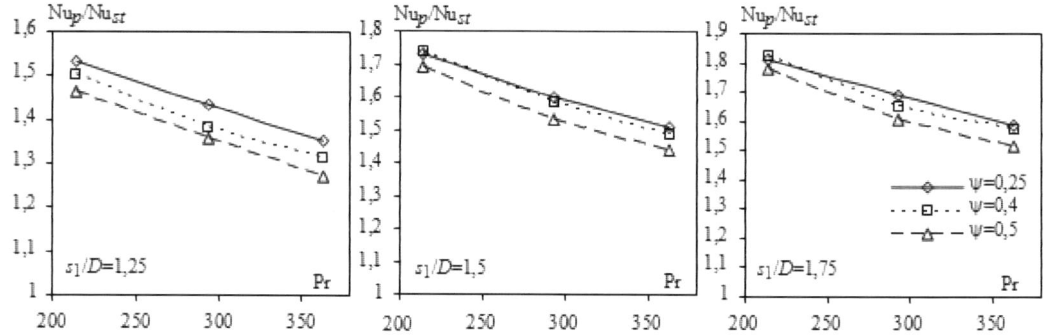


Fig. 4. Dependence of $Nu_p/Nu_{st}$ from Pr at Re = 100, $\beta$ = 25, $f$ = 0,2 Hz, $\varphi$ = 90°

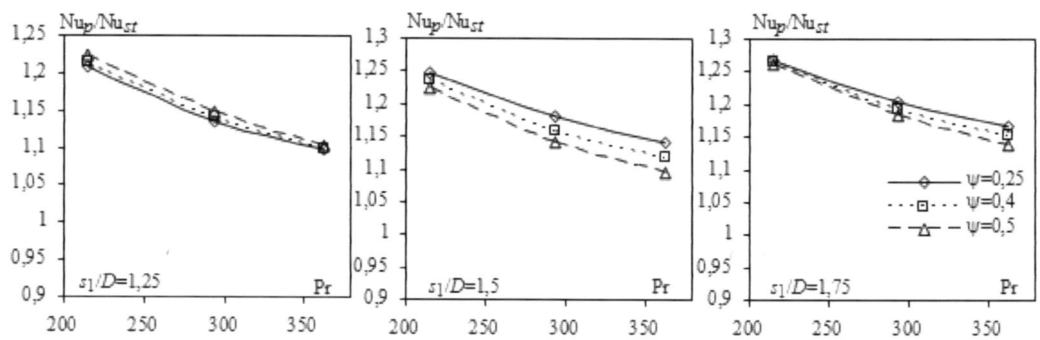


Fig. 5. Dependence of $Nu_p/Nu_{st}$ from Pr at Re = 600, $\beta$ = 25, $f$ = 0,312 Hz

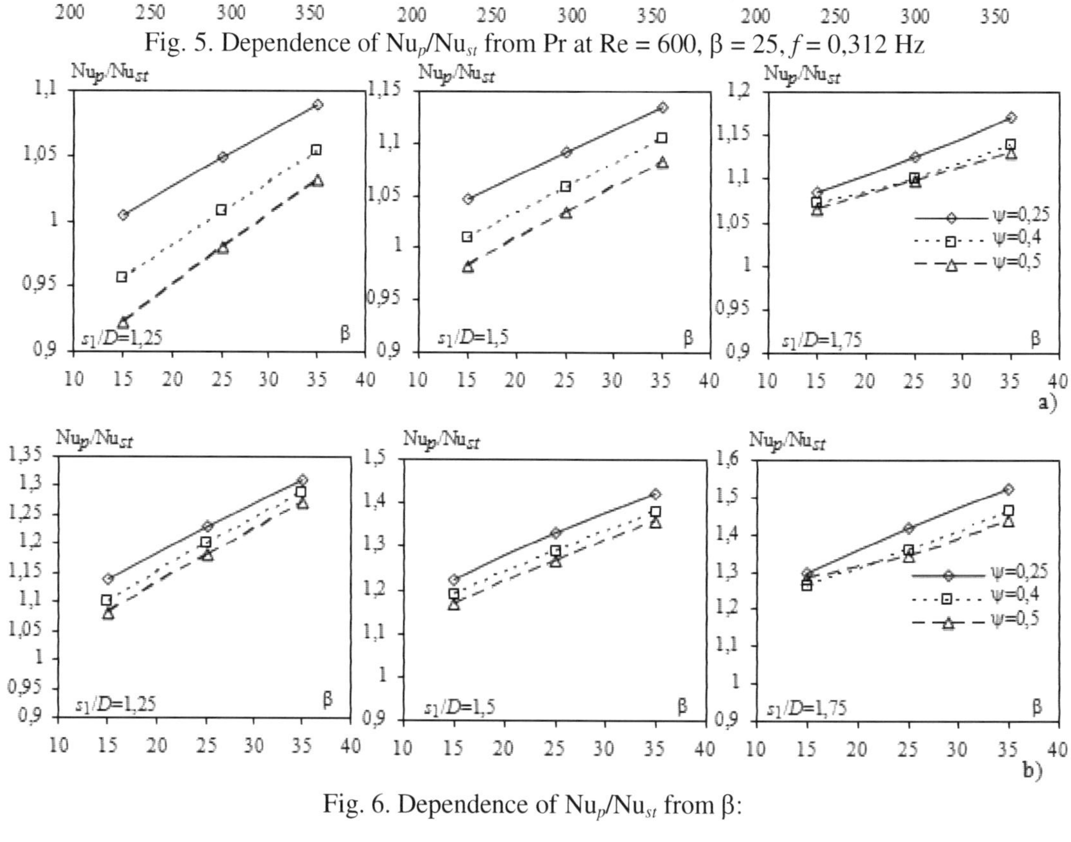


Fig. 6. Dependence of $Nu_p/Nu_{st}$ from $\beta$:

a) Re = 600, Pr = 293, $f$ = 0,2, Hz, φ = 90°; b) Re = 600, Pr = 293, $f$ = 0,5, φ = 90°;

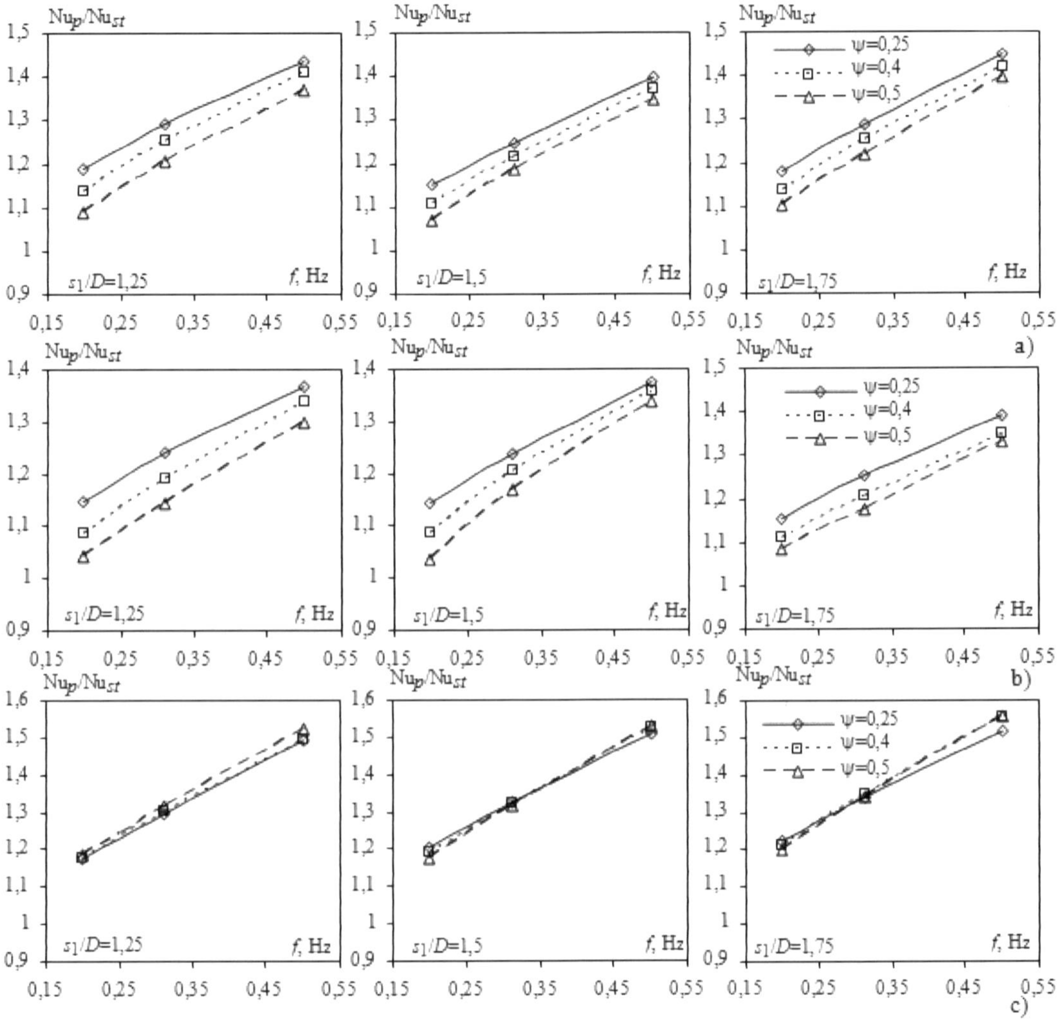


Fig. 7. Dependence of $Nu_p/Nu_{st}$ from $f$ at Re = 600, Pr = 215, β = 35:
a) φ = 30°; b) φ = 45°; c) φ = 60°;

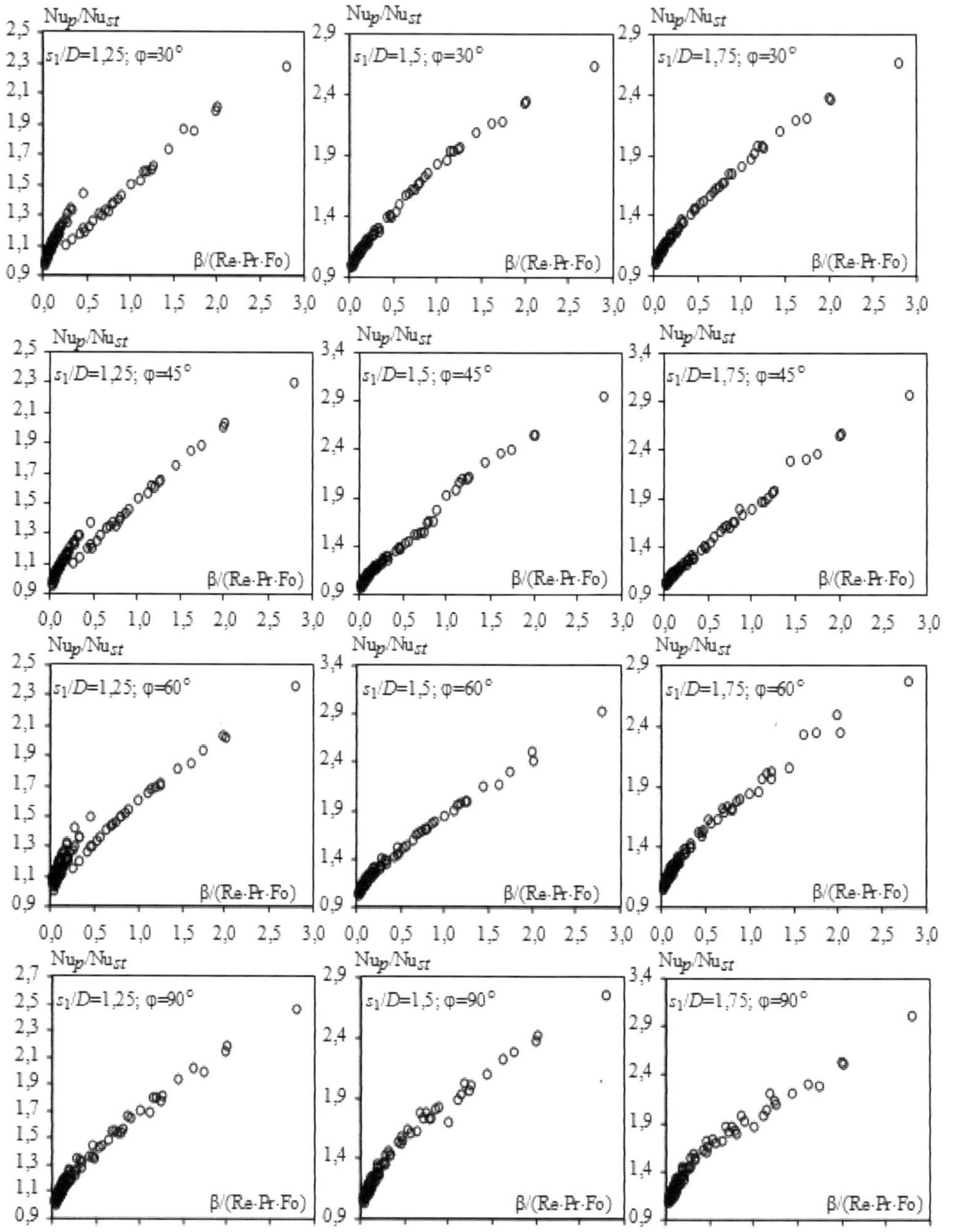


Fig. 8. Dependence of $Nu_p/Nu_{st}$ from $\beta/(Re \cdot Pr \cdot Fo)$ at $\psi$=0,25

As a result of generalization of numerical modeling data, a criterial equation of the following type

$$\frac{Nu_p}{Nu_{st}} = 0{,}954 Re^{-0{,}201} \cdot Pr^{-0{,}211} \cdot \beta^{0{,}184} \cdot Fo^{-0{,}230} \cdot \psi^{-0{,}053} \cdot \varphi^{0{,}085} \cdot s_1 / D^{0{,}287} \tag{7}$$

where $Nu_{st}$ – the Nusselt number for a stationary flow could be found by the criterion equation

$$Nu_{st} = 0{,}354 \cdot Re^{0{,}6} Pr^{0{,}33} \varphi^{-0{,}1} \cdot (s_1 / D)^{-0{,}45} \cdot (\mu_{ж} / \mu_{ст})^{0{,}14}, \tag{8}$$

Equation (7) is obtained for the ranges $215 \le Pr \le 363$, $0{,}25 \le \psi \le 0{,}5$, $100 \le Re \le 1000$, $15 \le \beta \le 35$, $5{,}81 \cdot 10^{-4} \le Fo \le 14{,}53 \cdot 10^{-4}$. Coefficient of determination $R^2 = 0{,}906$, maximum error of approximation is not more than $\delta max = 35{,}7$ %, mean deviation is not more than $\bar{\delta} = 5{,}5$ %.

**Conclusion**

A numerical method was used to investigate heat transfer in the case of cross flow around tubes bundles with the imposition of low-frequency pulsations on the flow of liquid. It is shown that, depending on the parameters of pulsations, the heat transfer of $Nu_p/Nu_{st}$ can be increased up to 3 times.

It is established that an increase in β and *f* lead to an increase in $Nu_p/Nu_{st}$, and an increase in Re, Pr and ψ lead to a decrease in $Nu_p/Nu_{st}$. It is noted that with an increase in geometric parameters φ and $s_1/D$, the $Nu_p/Nu_{st}$ ratio increases.

A criterial equation is obtained for the calculation of heat transfer in tube bundles in pulsating flows. In terms of exponential parameters, it has been found that the minimum effect on $Nu_p/Nu_{st}$ has ψ parameter value of the degree ψ (-0,053), the maximum effect on $Nu_p/Nu_{st}$ has the $s_1/D$ parameter value of the degree (0.287). Comparing the influence of β and Fo on $Nu_p/Nu_{st}$ , it is noticeable that Fo has a greater influence (the degree of Fo is greater than β by 25%).

**References**

[1] Li Hua, Zhong Yingjie, Zhang Xuemei, Deng Kai, Lin Haihao, Cai Luyin 2007. Experimental Study of Convective Heat Transfer in Pulsating Air Flow inside Circular Pipe *International Conference on Power Engineering-2007* pp 880-85

[2] Jin D X, Lee Y P, Lee D Y 2007 Effects of the pulsating flow agitation on the heat transfer in a triangular grooved channel *International Journal of Heat and Mass Transfer* Vol 50 pp 3062–71

[3] Mikheev N I, Molochnikov V M, Mikheev A N, Dushina O A 2017 Hydrodynamics and heat transfer of pulsating flow around a cylinder *International Journal of Heat and Mass Transfer* Vol 109 pp 254–65

[4] Armin W, Wolfgang P 2017 Dynamics of unsteady heat transfer in pulsating flow across a cylinder *International Journal of Heat and Mass Transfer* Vol 109 pp 1111–31

[5] Amiri S, Taher R, Mongeau L 2017 Quantitative visualization of temperature field and measurement of local heat transfer coefficient over heat exchanger elements in sinusoidal oscillating flow *Experimental Thermal and Fluid Science* Vol 633 pp 1724–30

[6] Khalil K, Bader A, Awadh A, Ioan P 2008 Mixed convection analysis of laminar pulsating flow and heat transfer over a backward-facing *step International Journal of Heat and Mass Transfer* Vol 51 pp 5785–93

[7] Moon Jeong Woo, Kim Seo Young, Cho Hyung Hee 2005 Frequency-dependent heat transfer enhancement from rectangular heated block array in a pulsating channel flow *International Journal of Heat and Mass Transfer* Vol 48 pp 4904–13

[8] Hiroshi Iwai, Tomoyuki Mambo, Naoki Yamamoto, Kenjiro Suzuki 2004 Laminar convective heat transfer from a circular cylinder exposed to a low frequency zero-mean velocity oscillating flow *International Journal of Heat and Mass Transfer* Vol 47 pp 4659–72

[9] Musaeva D A, Sinyavin A A, Gurianov A I 2012 Mathematical modeling of heat transfer processes under conditions of low-frequency non-symmetric pulsations of fluid flow *Energy problems* no 7–8 pp 19–27

[10] Musaeva D A, Gurianov A I, Sinyavin A A 2012 Investigation of influence of low-frequency impulses on heat transfer processes with cross-flow in corridor sheave of tubes *Modern science: researches, ideas, results, technologies* no 2 pp 183-86

[11] Wilcox David C 2006 *Turbulence Modeling for CFD. Third edition copyright* by DCW Industries Inc. 515 P

[12] You Qin Wang 2010 Turbulence Modeling Applied to Flow Through a Staggered Tube Bundle *Journal of Thermophysics and Heat Transfer* Vol 24 no 3 pp 534–43

[13] Khaybullina A I, Khayrullin A R, Sinyavin A A, Ilin V K 2014 Modeling the turbulence of a pulsating flow of a heat carrier in a corridor bundle of tubes *Thermophysical foundations of energy technologies* pp 368-72
[14] Khaybullina A I, Khayrullin A R, Ilin V K 2016 Heat transfer in the flow channel in tube bundle corridor type under imposed on the flow liquid of upstream low-frequency asymmetrical pulsations *Energy Problems*,No 11-12 pp 64-75
[15] Ramesh K Dusan P 2003 *Fundamentals of Heat Exchanger Design* Copyright 2003 John Wiley & Sons Inc. 931 P

IOP Conf. Series: Materials Science and Engineering **240** (2017) 012028 doi:10.1088/1757-899X/240/1/012028

# Research processing of tool steels surface in low pressure glow discharge

**D I Israphilov**

Kazan Federal University, 18 Kremlyovskaya street, Kazan, 420008, Russian Federation

diisraphilov@yandex.ru

**Abstract**. Plasma ion nitriding processing in glow discharge occurs on whole surface of product. Structure and phase composition of the diffusion layer defined by the phase diagram of iron - nitrogen. Low-pressure glow discharge in cross-flow working gas treatment takes place mainly by the working gas stream, which will hold required nitriding surfaces. In addition, low-pressure environment in chamber will reduce the effect of residual gas contaminants on processing surface.

## 1. Relevance

Plasma nitriding meets environmental requirements and is operated by a large number of equipment for its realization. A feature of the plasma nitriding is that the process can be controlled much easier than other methods.

Process of ion nitriding in hydrogen-containing environments (ammonia, a mixture of nitrogen and hydrogen) is well studied in literature [1 - 3].

A method of forming a wear-resistant coating on a surface of structural steel product includes plasma ion nitriding reactive gas atmosphere - nitrogen, surface cleaning and nitriding. Purification of surface nitriding is carried out at a pressure of reactive gas 5•10-3-2•10-2 Torr, voltage 300-1000 V and ion current density 2 - 8 mA/cm2 for 30-90 min, purification is carried out in an plasma of inert gas - argon at a pressure of 3•10-4 - 7•10-4 Torr and current density of 3-5 mA / cm2. The method allows intensifying process and increasing operational stability of dies, experiencing high specific load in the process of friction.

Structure and properties of hardened layer of nitriding details depend on the following technical factors: the voltage between the electrodes, the gas composition of the environment, degree of vacuum, operating temperature, process time, the mutual arrangement of parts and electrodes. Nitriding details set in the camera by connecting to negative electrode, chamber is sealed and pumped to a pressure near 1 Torr. Camera view is presented in Figure 1.

IOP Conf. Series: Materials Science and Engineering **240** (2017) 012028 doi:10.1088/1757-899X/240/1/012028

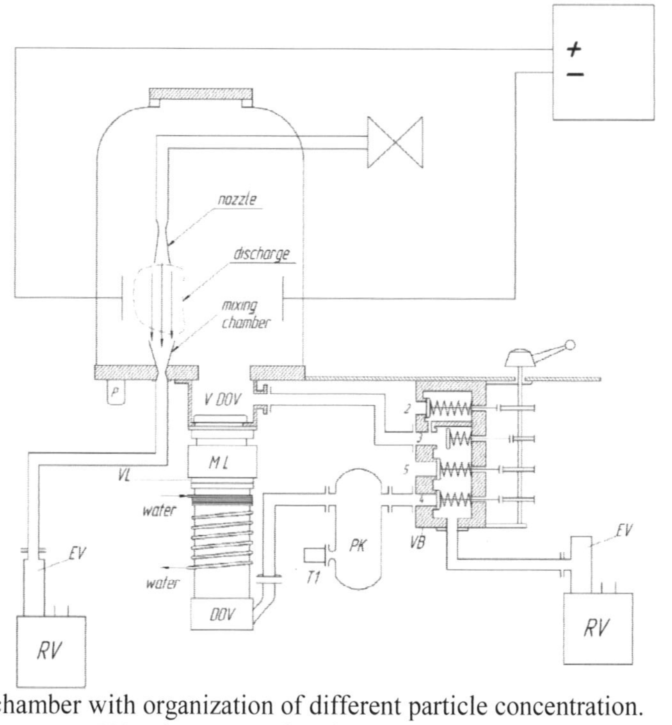


Figure1 Vacuum chamber with organization of different particle concentration.

RV- rotary vacuum pump, EV-valve, VB- valves box, PK- prev vacuum chamber, DOV - diffusion vacuum pump, ML - oil trap, VL - nitrogen trap, R – vacuum chamber, T1, T2 - thermocouple vacuum gauges, P - high vacuum gauge.

## 2. Results

Figure 2. shows a glow discharge at low pressure and the results of experimental studies of gas temperature, depending on the length of the discharge chamber at various currents (Fig. 3).

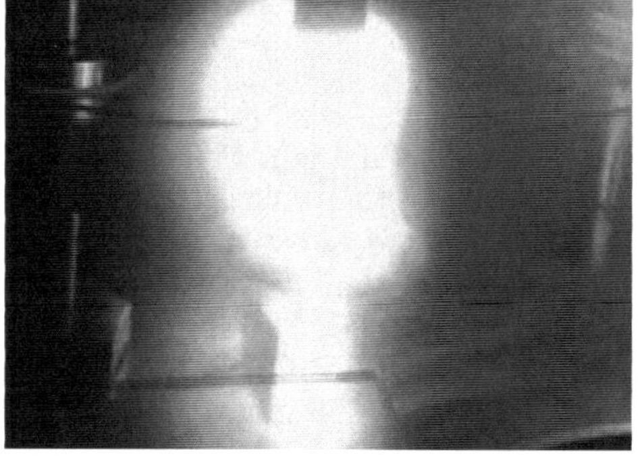

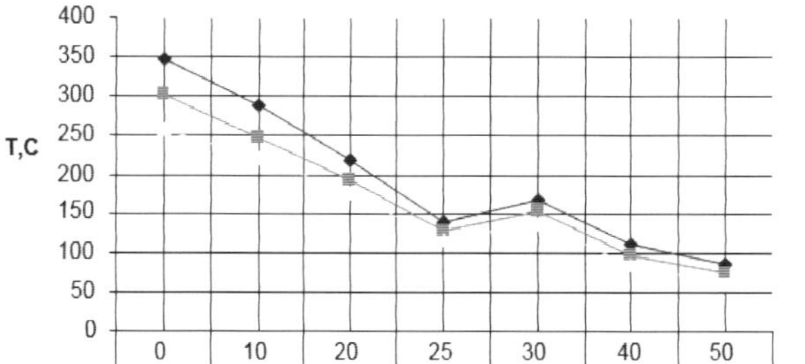


| | 0 | 10 | 20 | 25 | 30 | 40 | 50 |
|---|---|---|---|---|---|---|---|
| 50 mA | 347 | 288 | 218 | 141 | 167 | 112 | 88 |
| 40 mA | 301 | 245 | 192 | 130 | 155 | 99 | 75 |
| 30 mA | 258 | 220 | 176 | 117 | 136 | 87 | 62 |

Z, mm

Figure 2. Glow discharge at a pressure of 3 Pa, working gas - nitrogen.

Figure - 3. Dependence of gas temperature along length of discharge at different currents.

As the chart above shows, gas temperature in the field of supersonic flow drops sharply.

Rational selection of the optimal pressure, which depends on the complexity of configuration and efficient of parts location, since changes in pressure varies with the length of the cathode part of the discharge. With increasing pressure from 1 to 10 Torr cathode area of the discharge is decreased from

10 to 1 mm. This must be considered to ensure the uniformity of the diffusion layer on the workpiece surface.

Nitriding temperature is usually 470-580 ° C, voltage 400-1100 V, vacuum 1-10 Torr. Operating pressure is limited by the properties of glow discharge. At a pressure below 1 Torr ion energy is not enough to heat the workpiece to a working temperature, at a pressure above 10 Torr stability of discharge is disturbed.

Process is carried out for 5-60 minutes at a voltage of 1100-1400 V and pressure of 0.1-0.2 Torr, surface temperature of the parts does not exceed 250 ° C. Operating parameters of the process: U = 400- 1000 V in electrode gap is fed supersonic flow of working gas - nitrogen

After surface treatment for 5-60 minutes in a cathode-sputtering mode, voltage is reduced to 1100-1400 V in working. Cathode-sputtering mode can be performed in a hydrogen atmosphere, the use of which is especially useful when nitriding austenitic steels. Working process temperature (470-580 ° C) achieving in 15-30 minutes. The heating rate is determined by ratio of details surface and weight [7]. After isothermal holding parts are cooled to room temperature under vacuum. In this cooling rate is above than in heating furnace, since only product heated by ionic treatment and heating walls of the working chamber by convection and radiant heat in a vacuum is light.

Steel X40CrMoV5-1 is basis for a variety of stamps for deformation of steel and non-ferrous metals and injection molding forms of aluminum and magnesium alloys with a diameter of 70-80 mm.

Accordingly, resistance of the steel becomes 1.5-2 times higher (Figure 4).

a)

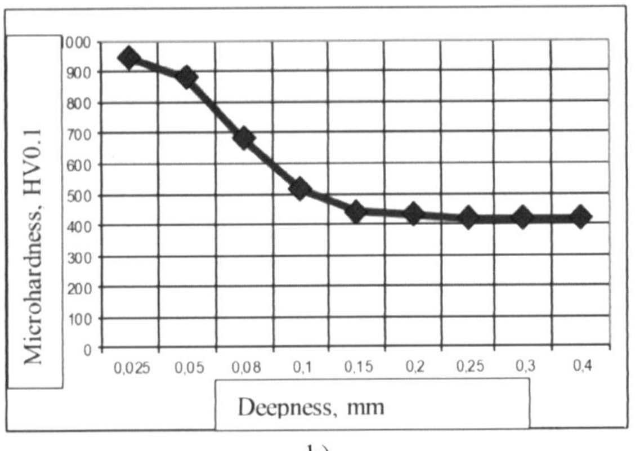


b)

Figure 4. - Microstructure details, ×500; a) - at a distance of 5 mm from the end of rod; b) - plot of microhardness of the nitrided layer of tool steel X40CrMoV5-1.

Thus, nitrided layer consisting of nitride compounds and internal nitriding zone. Figure 4 shows a photography of the microstructure and a plot of microhardness of nitrided layer steel X40CrMoV5-1. Decomposition of austenite in the pearlite and the intermediate region is preceded by the selection of carbides, occurring at the grain boundaries, which reduces viscosity by 30-40%.

## 3. Conclusion

Change in high current density range (0.5-20 mA/cm2) does not affect the nitriding process.

Increasing the technical efficiency of ion-plasma nitriding by optimizing process parameters depending on weight and geometric characteristics of the parts. This is because the performance depends on the data density distribution uniformity of the ionized gas on the workpiece surface. It is also necessary to consider increasing intensity of glow discharge plasma in a nitrogen atmosphere at sharp edges of parts.

## References

[1] Arzamasov B N, Bratukhin A G, Eliseev Y S, Panayiotou T A 1999 *Ionic Chemical Heat Treatment of Alloys in Gas Media* p 400

[2] Israphilov I H, Zvezdin V V, Israphilov D I and Chernova M A Study on parts ion-plasma

nitriding *Journal of Physics: Conference Series* Volume 567 **012030**

[3] Kaplun V G, Shepherd I M 2000 Plasma diffusion processes of surface modification of materials, technology and equipment *Proceedings of the Scientific-practical Symposium "equipment Itehnologii Heat Treatment of Metals and Alloys in Engineering"* pp 145-54

[4] Zvezdin VV etc. 2010 Investigation of parameters influence of pulsed plasma generator on performance quality of the process *Vestnik KGTU of Tupolev* V 2 pp 50-52

[5] Saifutdinov A I, Timerkaev B A and Zalyaliev B R 2014 Control of glow discharge parameters using transverse supersonic gas flow - numerical experiment *Journal of Physics: Conference Series* 567 **0120313**

[6] Timerkaev B A, Zalyaliev B R, Saifutdinov A I 2014 Glow Discharge Characteristics in Transverse Supersonic Air Flow *Journal of Physics: Conference Series* 567 **012032**

IOP Conf. Series: Materials Science and Engineering **240** (2017) 012029 doi:10.1088/1757-899X/240/1/012029

# Ion source surface processing based on glow discharge in supersonic cross gas flow control

**D I Israphilov**

Kazan Federal University, 18 Kremlyovskaya street, Kazan, 420008, Russian Federation

diisraphilov@yandex.ru

**Abstract**. Surface treatment by glow discharge in a transverse gas flow provides a number of advantages over magnetron discharge. Sputtering target at lower pressures, which increases the purity of the deposited material. In addition, more uniform erosion of target and thickness of deposited coating. However is necessary to develop glow discharge control system in cross flow working gas, which allow to obtain products with desired characteristics.

## 1. Relevance

Requirements to quality indicators ion-plasma treatment of surfaces, such as sufficient adhesion, porosity, uniformity, cleanliness, chemical composition and thickness of the resulting layer, no defects on detail surface, mechanical properties, etc. with the development of technology increases.

To achieve desired purity and chemical composition necessary to reduce pressure in the chamber. This reduce concentration of working gas inclusions on the treated surface. For example, during the movement of target material in the vapor phase, impurities existing in the atmosphere of the chamber deposited with target material.

However, reducing chamber pressure reduces efficiency of glow discharge necessary for formation of ions, because increases voltage, with weak growth of current this leads to an unjustified waste of electricity, heat up the electrodes and, most importantly, overheating treated surface, which is unacceptable. And at pressures below 10-2 Torr discharge stops to burn.

There is fundamentally new solution to this problem [1, 2]. It is creation of zones with different concentrations of neutral particles in electrode gap. For example, in cathode area to create a high vacuum with a minimum concentration of neutral particles, which can provide a pumping system, and anode area to provide comparatively high concentration of neutral particles necessary for glow discharge implementation. Electrons in a region with a high concentration of neutral particles may experience tens collisions with neutral particles, while material processing at vacuum provided by pumping system. To satisfy this condition is necessary to create a continuous flow of the working gas particles in part of electrode gap and at the same time maintain vacuum in the discharge chamber lower than 0.5 Pa. To achieve this separation zones in the electrode gap may allow a continuous flow of gas. One type is the continuous stream of supersonic flow. Density of the gas pumped through the work area should provide more than a dozen of electron collisions with neutral particles, which will maintain glow discharge, providing chamber pressure below 0.5 Pa.

## 2. Results

This requires research possibility of technological parameters regulating in vacuum technology complex (VTC) (Table 1) [3]. Developed control system should ensure that the process parameters would automatically make better decisions based on information about the process. Reaching the stable quality of finished product, its cost is reduced and there is a possibility of its mass implementation. Analysis of technological surface treatment processes showed that control of process parameters and analysis of systems working capacity is difficult due to the of ultrafast processes in high-energy technology complex [1, 2]. Maintaining the stability of the process variables is possible due to system response in a short time. To do this, is necessary to identify the process parameters influencing the quality parameters of products, as well as equipment parameters influencing process parameters and bind them into a single technological complex (Figure 1).

Table 1 - Values of factors of units instability VTC, affecting quality of treatment.

| VTC units | Parameters | Factors affecting the stability of parameters | Accuracy parameters |
|---|---|---|---|
| Cathode | Power | Voltage, current density | ±5% |
| | Density of plasma flow | Electrode gap, sprayed material composition | ±5% |
| Plasma distribution environment | Permittivity | Composition of gas environment. | ±2% |
| Detail | Coefficient of thermal energy of the plasma conversion | Thermal material parameters. | ±3% |
| | Dimensions and weight characteristics | Weight, shape, initial temperature of the surface. | ±0,5% |
| Movement gears | Movement speed, electromechanical drive characteristics | Supply voltage, moment of inertia | ±5% |
| Control system | Performance characteristics of units transfer functions | Accuracy approximation method | ±3% |
| | System specifications | Microprocessor bit depth, speed | ±0,1% |

Overall picture of the influence of factors of instability automatic control system (ACS) units of VTC on its parameters, based on the analysis of experimental results and patent information studies, is shown in Table 1 [3]. It should be considered as an aggregate structure of VTC units as a complex system, involved in surface treatment.

The main criterion for evaluating the quality and stability of process parameters serves uniform reflectance coating applied over the entire area of detail that speaks of uniformity (optical characteristics), desired thickness of coating and its chemical composition.

Optimal way of ASC VTC building, leading to stabilization of process quality indicators, is to stabilize the set value sensor readings [4], in this case, the reflection coefficient. Correlation between the parameters of VTK, process parameters and quality indicators is shown in Figure 1.

Therefore, the main task is to develop methods for measuring reflection coefficient and its stabilization by controlling parameters of units VTC.

To realize coating process with given parameters is necessary to make monitoring and control VTC parameters in real-time flow of process.

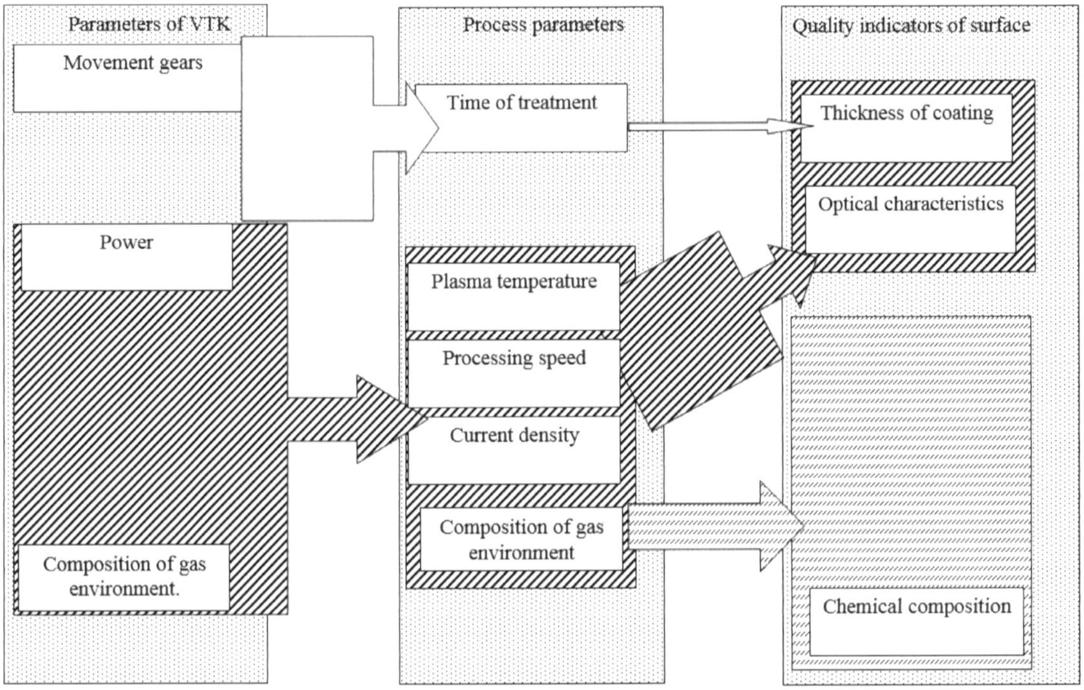


Figure. 1- Correlation between VTC parameters and quality indicators.

In these conditions is actual, on the one hand, calculation of optimal process parameters change, and on the other - development and research of ACS process [4] plasma spraying, based on the implementation of the combined control by introducing feedback informative process parameters [3], which will provide more efficient use of VTC.

## 3. Conclusion

Study of carried out by sputtering metals on material show the instability of the quality indicators of the process. Essential value in these play VTC parameters.

The main criterion for evaluating the quality of process parameter serves reflection coefficient of treated material. However, complexity of temperature measurement is related to screening zone of plasma torch interaction, to measure temperature of the plasma brightness, spectral pyrometer with polarization filtering used.

## References

[1] A I Saifutdinov, B A Timerkaev, and B R Zalyaliev 2014 Control of glow discharge parameters using transverse supersonic gas flow - numerical *experiment Journal of Physics: Conference Series* 567 **0120313**

[2] B A Timerkaev, B R Zalyaliev 2014 Glow Discharge in a Transverse Supersonic *Gas Flow at Low Pressures. High Temperature* 52 No 4 pp 471–74

[3] D I Israphilov 2007 Management of A Plasma Technological Complex for Heat Treatment Ofparts with Specified Strength Characteristics

[4] I Sh Abdullin 2013 Analysis of the automatic control system of vacuum-sputter technology complex for consumer industry *Vestnik KGTU* No 24 pp 160-62.

IOP Conf. Series: Materials Science and Engineering **240** (2017) 012030 doi:10.1088/1757-899X/240/1/012030

# Application of reverse engineering in the medical industry.

**A A Kaleev[1], L N Kashapov[1], N F Kashapov[1], R N Kashapov[1,2,3]**

[1] Kazan Federal University, 18 Kremlyovskaya street, Kazan, 420008, Russian Federation
[2] Laboratory of Radiation Physics, Kazan Physical-Technical Institute of the Kazan Scientific Center of the Russian Academy of Sciences, Kazan, 420029, Russia
[3] Kazan State Medical University, Kazan, 420012, Russia

kashramiln@gmail.com

**Abstract.** The purpose of this research is to develop on the basis of existing analogs new design of ophthalmologic microsurgical tweezers by using reverse engineering techniques. Virtual model was obtained by using a three-dimensional scanning system Solutionix Rexcan 450 MP. Geomagic Studio program was used to remove defects and inaccuracies of the obtained parametric model. A prototype of the finished model was made on the installation of laser stereolithography Projet 6000. Total time of the creation was 16 hours from the reverse engineering procedure to 3D-printing of the prototype.

## Introduction

Reverse engineering is the method obtaining three-dimensional data in computerized form from physical models or products. It has clear benefits of effective use together with additive technologies and vacuum casting in silicone molds. [1]

3D printing, also known as additive manufacturing technology, creates the object layer by layer. 3D printers can produce more complicated shapes than traditional manufacturing techniques. Currently, 3D printing involves areas: manufacture of individual surgical templates, fabrication of individual surgical implants and prostheses, spinal implants manufacturing, manufacturing a variety of orthopedic and prosthetic constructions in dentistry. [2]

Complex geometric form of the medical instrument complicates the process of measuring of the control size and determine of the exact geometry. Tweezers used in microsurgery of the eye are subject to frequent changes in design appearance. On the one hand, the simplicity of the tool, on the other hand the complexity of the geometry of the handle of sticking has necessitated the use of reverse engineering. Therefore, the aim of this work was to explore the possibility of reverse engineering, in this example the speed of obtaining the product prototype.

## Main part

IOP Conf. Series: Materials Science and Engineering **240** (2017) 012030 doi:10.1088/1757-899X/240/1/012030

For the analysis of tweezers scanning features three 3D-scanner were selected: Solutionix Rexcan 450 MP, Smart Optics Activity 850, Artec Spider.

Artec Spider is the professional manual 3D scanner. The device uses an LED backlight in combination with four cameras 1,3Mp resolution. Scan rate up to 7.5 frames (or 1,000,000 measurements) per second. The accuracy of scanning 0,05 mm.

Smart Optics Activity 875 is a desktop optical 3D-scanner with enhanced performance. The device is designed to produce high-precision digital models. The scanner has a high resolution, which reaches 0.01 mm. The scanning process is fully automated and takes about one or three minutes, depending on the size of the object. The obtained data are exported as STL file.

Solutionix Rexcan 450 MP is a professional 3D-scanner that creates highly accurate three-dimensional models of real objects with a resolution of 5 megapixels. The device works on the contactless method of the optical phase triangulation of a structured white light. Variable angle of triangulation and adjustable scanning area provide an opportunity to obtain scans accuracy of 0.01 mm.

Artec Spider was chosen first for scanning. It has the lowest resolution camera consequently the accuracy of the scan he also low. It is portable and does not require calibration. The scanning process showed that for small objects such as tweezers, it is not suitable. It was decided to abandon this type of scanner because the scan has a lot of noise and inaccuracies.

Smart Optics Activity 875 has a high accuracy of scanning, does not need in the calibration, and the scanning process is automated. This scanner is designed to scan objects with a diameter up to 85 mm and tweezers exceeds this size.

Solutionix Rexcan MP 450 has the same accuracy of the scanning that Activity 875 but it needs to be calibrated. Calibration performed to determine the relative position and provide high accuracy of measurements. It is necessary to guideways the intersection point should be at the center of the middle group of the target on the calibration panel (Fig.1). After calibration, the scanning process begins. It became possible to scan deep and narrow zones of complex shapes through the use of triangulation angle equal to 10°. Dead zones, which were difficult to scan with a standard angle, can now be covered by means of the triangulation angle equal to 10 °. The scanner shines with structured light on the product and receives information about the distance to the object surfaces by reflection from the surface. Sectors of surface model are created using this information. After a sufficient number of such sectors, the specialized software ezScan connects them into a single object in the automatic or manual mode. The format of the exported model is SNX and its quality is not suitable for 3D printing. Scan of model need to be translated into STL format and processed in the program Geomagic Studio.

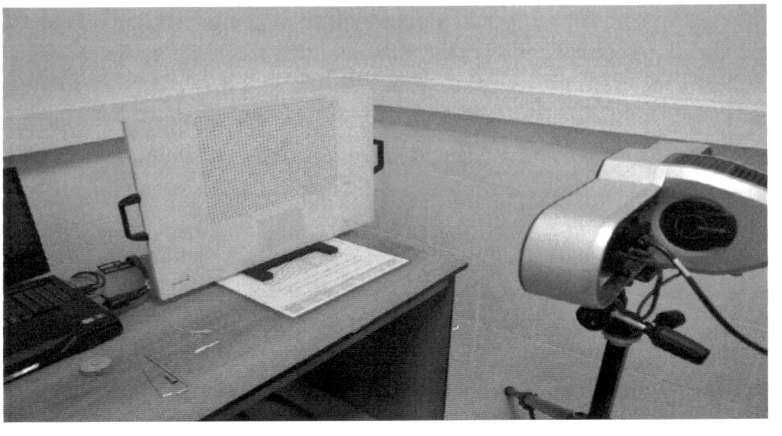

Fig.1. Calibration of scanner.

Even with a perfect surface of scanning (volumetric, white, matt, without tight spaces and sharp edges) 3D scanner captures still various noises. These can be characteristics of the object and the external conditions and the characteristics of the scanner. Eventually unnecessary roughness, tunnels and holes and other artifacts are formed. The resulting surface of the object can be viewed in different ways: as a cloud of points or mesh. In the second case, all the points are connected into triangles, which form a million microsurface. This grid is a complete polygonal model. It can be saved in stl format or other formats [4] (Fig.2).

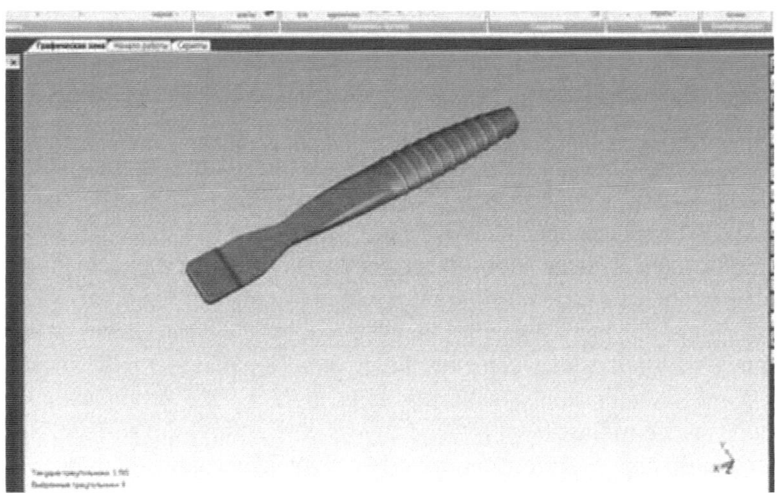

Fig.2. Processing of tweezers in the program Geomagic Studio

At the first stage should be to "clean up" the cloud of points obtained from the scanner, remove excess geometry and reduce the spread of the points from the average value.

Tool "Mesh Doctor" removes all the defects in the automatic mode. After we connect misalignment scans and build surface on the unscanned areas. Also at this stage it is possible to change the number of polygons and their distribution, by optimizing the model what will reduce its size. The total model processing time was 12 hours.

SLA-technology has been selected for printing. This method is based on laser irradiation of the liquid photopolymer resin for creating solid physical models. Construction of the model is produced by layer upon layer. Each layer is sintered by a laser beam according to the program. Laser irradiation leads to polymerization of the material at the points of passage of the laser beam. [3]

After printing is complete the product is washed to remove the liquid photopolymer and is processed in the ultraviolet chamber. Stereolithography requires the use of support structures for the construction of high-angle components of the model. Supports are generated in the "Magics" program, and are made of the same photopolymer material. Support elements are temporary construction elements which are removed manually after the completion of the manufacturing process. Existing technology allows applying fibers with a thickness of 15 microns, which is several times smaller than the thickness of a human hair. The total manufacturing time of tweezers was 2 hours. (Fig. 3)

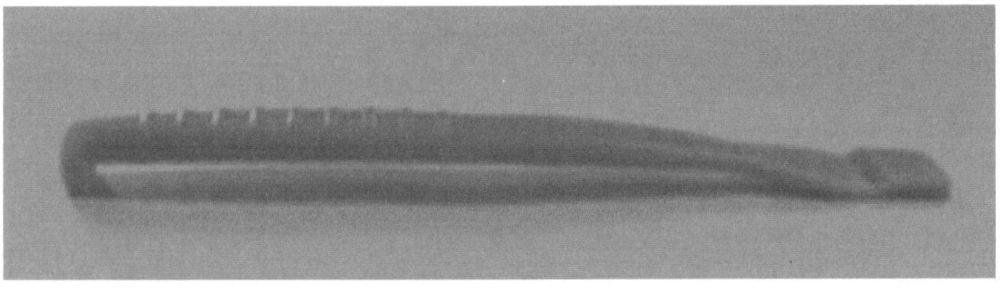

Fig.3. Prototype of tweezers obtained at the SLA installation Projet 6000.

## Conclusion

As a result of the work ophthalmic tweezers for microsurgery was scanned. Virtual model with a new design was created. Prototype of tweezers obtained at the SLA installation Projet 6000. The use of reverse engineering and additive technology has reduced new product development of up to 16 hours, instead of 2 weeks.

## References

[1] Zexiao, Xie, Wang Jianguo, and Zhang Qiumei. 2005 Complete 3D measurement in reverse engineering using a multi-probe system International Journal of Machine Tools and Manufacture V 45 no 12 pp 1474-1486

[2]Kashapov L, Rudyk A, Kashapov R 2014 IOP Conference Series: Materials Science and Engineering Volume 69 Issue 1 012016

[3] Kashapov R, Korobkina A, Platonov E, Saleeva G 2014 IOP Conf. Series: Materials Science and Engineering 69 (2014) 012026

[4] Filiberto Chiabrando, Roberto Chiabrando, Dario Piatti, Fulvio Rinaudo 2009 Sensors for 3D Imaging: Metric Evaluation and Calibration of a CCD/CMOS Time-of-Flight Camera / Filiberto Chiabrando Sensors 9, pp 80-96 doi:10.3390/ s91210080.

IOP Conf. Series: Materials Science and Engineering **240** (2017) 012031 doi:10.1088/1757-899X/240/1/012031

# Simulation of parameters of the vapor-liquid compression cooling system of power machines

**D L Karelin**

Kazan Federal University, Naberezhnye Chelny Institute
Pr. Mira 68/19, Naberezhnye Chelny, Republic of Tatarstan, Russia, 423812

karelindl@mail.ru

**Abstract**. The paper presents a mathematical model of the cooling system with vapor-liquid compression installation modified with an additional allowance for the effect of mass flow rate of ambient air on a heat transfer in the system's heat exchanger-condenser. Also the correction factors considering the difference between the mean log and the mean arithmetic temperature pressure are introduced. The results of numerical simulation showed that the efficiency of the proposed cooling system is 149% and 200% higher than the classical at the ambient temperature plus 10°C and 50°C respectively.

The resolution of the problem of qualitative improvement of cooling systems parameters of responsible for stabilization of the thermal state of power machines, the efficiency of which is largely determined by the external factor of seasonality due to the reduction of heat transfer of heat exchangers when increasing temperature of the environment, is considered by many authors [1, 2, 3, 4].

They offer the way of high-temperature cooling which can not be accepted for cooling systems as a part of the combined internal combustion engine with electric transmission because of the restriction of cooling liquid temperature on the entrance to traction electric motors and power converters [5], or for technological reasons.

In the work [6] it is proposed a method of cooling electrical machines using the liquid-vapor compression system (see figure 1). It gives a description of the design, the mathematical model used for calculations and dependencies of electrical machines exhaust thermal power when the mass flow and temperature of the ambient air circulating through the heat exchanger-condenser changes.

The proposed mathematical model [6] doesn't take into consideration the influence of mass flow air circulating changes of through the heat exchanger-condenser on heat transfer and the model was obtained using the expression for the arithmetic mean temperature pressure, which affects the precision of determined parameters.

Therefore, in this work we propose a modification of the mathematical model of the cooling system [6], taking into consideration the dependence of heat transfer from the dynamics of the air mass flow, and with correction factors $k_{t\_К}$ and $k_{t\_И}$ taking into consideration the difference between the arithmetic mean and mid-logarithmic temperature pressure in heat exchangers the evaporator and condenser respectively.

IOP Conf. Series: Materials Science and Engineering **240** (2017) 012031 doi:10.1088/1757-899X/240/1/012031

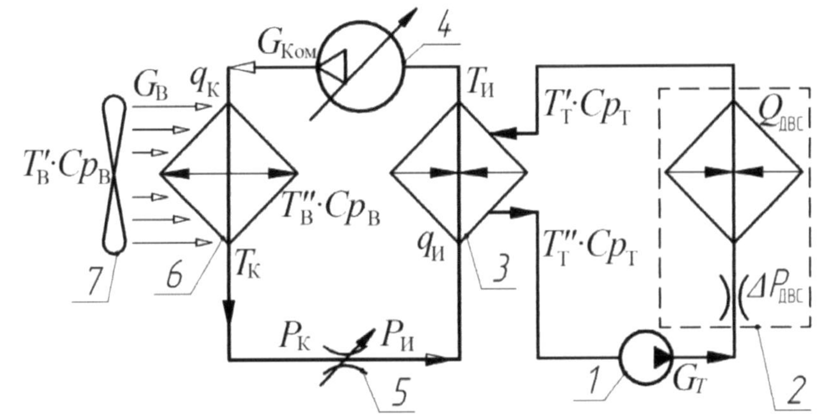


*Figure 1 – Schematic diagram of the liquid-vapor compression cooling system 1 – coolant pump; 2 – cooling jacket of the energy machine (internal combustion engine); 3 – heat exchanger-evaporator; 4 – compressor of the working medium circuit; 5 – expansion device (throttle); 6 – heat exchanger-condenser; 7 – fan.*

Modified mathematical model has the following form:

$$\begin{cases} Q_{\text{И}(i,j,k)} = \dfrac{G_{\text{Т}} \cdot c_{p\text{Т}} \cdot \left(T''_{\text{Т}} - T_{\text{И}}\right) + \left(x_k \cdot \dfrac{q_{\text{И}}}{q_{\text{К}}}\right) \cdot G_{\text{В}_i} \cdot c_{p\text{В}} \cdot \left(T_{\text{К}} - T'_{\text{В}_j}\right)}{\dfrac{G_{\text{Т}} \cdot c_{p\text{Т}}}{k_{t\_\text{И}} \cdot k_{\text{И}} \cdot F_{\text{К}}} + \dfrac{G_{\text{В}_i} \cdot c_{p\text{В}}}{k_{t\_\text{К}} \cdot k_{\text{К}_i} \cdot F_{\text{К}}}} \\ k_{\text{К}_i} = f\left(\alpha_{\text{В}}\left(G_{\text{В}_i}\right)\right) \\ 0 < x_k \leq 1 \end{cases}, \tag{1}$$

where $k_{t\_\text{И}} = \Delta T_{\text{И}\_\log} / \Delta T_{\text{И}\_\text{Л}}$ and $k_{t\_\text{К}} \Delta T_{\text{К}\_\log} / \Delta T_{\text{К}\_\text{Л}}$ – dimensionless correction factors; $x_k$ – the coefficient taking into consideration the energy cost of the compressor 4 (see Fig. 1) depending on the mass flow of the working medium.

Heat emission to ambient air was identified based on design features of the heat exchanger-condenser according to the equation of similarity [7]:

$$\alpha_{\text{В}} = C \cdot \frac{\lambda_{\text{В}} \cdot \text{Re}_{\text{В}}{}^{n}}{L} \cdot \left(\frac{d_{\text{К.н}}}{b}\right)^{-0,54} \cdot \left(\frac{h}{b}\right)^{-0,14} \cdot \psi \cdot \varepsilon_{\text{C}} \cdot \varepsilon_{Z}, \tag{2}$$

Heat emission from the coolant to the working medium in the heat exchanger-evaporator was calculated taking into account design features of the equation of similarity:

$$\alpha_{\text{Т}} = C \cdot \frac{\lambda_{\text{В}} \cdot \text{Re}_{\text{Т}}{}^{n}}{L} \cdot \text{Pr}_{\text{Т}}^{0,36} \cdot \left(\frac{\text{Pr}_{\text{Т}}}{\text{Pr}_{\text{C}}}\right)^{0,25} \cdot \varepsilon_{\varphi}, \tag{3}$$

For calculation characteristics of the cooling system with vapor-liquid compression setting were made as numerical values of the parameters of the cycle and working medium similar in the work [6]: working medium freon-10, the condensing temperature $T_{\text{К}}$=127°C, evaporating temperature $T_{\text{И}}$=67°C, the coolant Tosol-65, which temperature at the outlet of the heat exchanger-evaporator was equal to $T_{\text{Т}}$=70°C and kept stationary, the heat transfer areas of the condenser $F_{\text{К}}$ and evaporator $F_{\text{И}}$ were obtained for the thermal power of the exhaust cooling system 180 kW at ambient temperature equal to 50°C and the air mass flow rate $G_{\text{В}}$=7.4 kg/s.

The result of numerical simulation is presented on the figure 2 a, b.

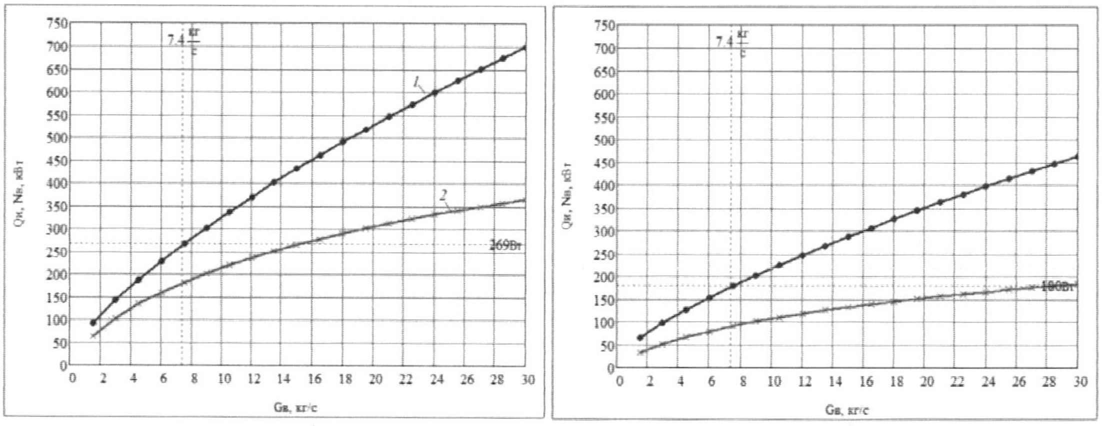


*a)* *b)*

*Figure 2 – Dependencies of the amount of heat dissipated by cooling systems when the temperature and mass flow rate of ambient air:* ***a*** *- at ambient temperature 10°C;* ***b*** *- at ambient temperature 50°C 1 – system with the liquid-vapor compression installation; 2 - a classical cooling system.*

The result of modeling showed that the thermal power which is taken away in the atmosphere in case of use of the cooling system with vapor-liquid compression installation at the same parameters of ambient temperature, mass flow rate the cooling air and identical stuffing of the heat exchanger-condenser (the heat exchanger for the classical cooling system), is much higher than the power taken away by a classical cooling system. So at the air mass flow rate of 7,4 kg/s and temperature of equal to 10°C heat emission of the classical cooling system is about 180 kW. The cooling system with vapor-liquid compression installation in the same conditions can take away 268 kW from the power machine that makes 149%.

With temperature increasing of ambient air to 50°C heat emission the classical cooling system decreases to 90 kW, and in the system offered the vapor-liquid cooling system to 180 kW, at the same time efficiency of cooling increases up to 200%.

**Thus**, the modified mathematical model allows to calculating performance cooling system characteristics with vapor-liquid compression installation taking into account all primary technical and thermal properties of system and environment. Implementation of vapor-liquid compression installation in the cooling system will allow to considerably increasing the average temperature differences due to increase in temperature of the working medium in the heat exchanger-condenser. The consequence of that will be increase of amount of the taken away heat to the atmosphere that in turn will ensure more effective functioning of the power machine at the increased ambient temperature, and will also improve weight and size characteristics of the system that is especially actual important for mobile equipment operated at high loads in harsh conditions.

**References**

[1] Shelest V P 1971 *Locomotives results and prospects* (Moscow: Znanie) p 48
[2] Sklifus Ya K 2015 *The reduction of energy consumption of the cooling system of diesel locomotive of change of the graph of a function and the method of heat transfer:* thesis (Rostov-on-don: RGUPS press) p 159
[3] Liventsev F L 1965 *High-temperature cooling of piston engines of internal combustion* (Moscow: Mashinostroenie) p 204
[4] Mkrtumyan E A 1938 *Cooling diesel engine at elevated temperatures of cooling water* (Proceedings Of The Moscow Red Banner MMI N. E. Bauman) issue 38-39/4, p 72
[5] Altynova N E 1984 *Intensification of heat transfer in cooling systems of powerful elektrovozny*

*conversion devices:* thesis (Novocherkassk: NPI) p 197
[6] Karelin D L, Gureev V M, Mulyukin V L 2015 Simulation of cooling systems with liquid-vapor compression installation *Bulletin of KSTU named after A. N. Tupolev* vol 5 (Kazan: KNRTU-KAI press) p 5-10
[7] Kutatelodze S S Borishansky V M 1958 *Handbook of heat transfer* (Moscow: Gosenergoizdat) p 414

IOP Conf. Series: Materials Science and Engineering **240** (2017) 012032 doi:10.1088/1757-899X/240/1/012032

# Energy-saving technologies of cultivation of sugar sorghum

**N F Kashapov[1], M M Nafikov[1], I R Gilmanshin[1,4], M Kh Gazetdinov[2], M M Nafikova[2] and A R Nigmatzyanov[3]**

[1] Kazan Federal University, 18 Kremlyovskaya street, Kazan, 420008, Russian Federation
[2]Kazansky state agrarian university, Kazan, Russia
[3]Tatarsky institute of retraining specialists agribusiness, Kazan, Russia
[4] Kazan National Research Technical University named after A N Tupolev, Kazan, Russian Federation

arnig76@yandex.ru

**Abstract.** The article describes the main elements of energy-saving technologies of cultivation of sugar sorghum, which, in particular, is the introduction of the combined units. Integrated units allow for one pass across the field to perform up to eight operations, thereby ensuring a significant reduction in energy costs, but there are certain negative points.

The current stage of development of agricultural production is characterized by high rates of growth of farm equipment means of production, in particular energoosnaschennoy technique. At the same time there is a problem of rational use of available technology with sufficient effectiveness. In this connection, in the current economic conditions, the transition to a qualitatively higher level of productivity, profitability, resource and energoberezhlivosti, environmental safety associated with the use of energy-saving technologies.

The introduction of new technologies has become a priority not only because they contain the results of the new achievements of domestic and foreign researches in the field of creation of new equipment and technologies for agriculture, but also because it is necessary to solve at this stage of economic development new challenges: improving the competitiveness of agricultural products on domestic and foreign markets; increasing environmental safety; reduction of soil degradation processes, etc.

Before the formation of market relations in agrarian sector of economy the main direction of increase of productivity was the feeding of the fleet farms in the production of modern mechanization, it was observed a steady trend towards increasing energy intensity of agricultural production. The increase in growth of gross output of agriculture by 1% was achieved by increasing 1,8% and 2,7%

used generating capacity. The analysis shows that since the beginning of 2000-ies of the increased energy consumption of the means of production, as agricultural production along with the traditional energy sources (oil, gas, electricity) began to use solar energy, energy of biomass and waste energy. The energy cost of producing 1 ton of conventional grain units in Russia compared to the United States more than 5 times, this is partly due to the fact that agricultural production in Russia operates in less favorable climatic conditions.

The results of previous years proved that the process of intensification in the agricultural sector is closely linked with the growth of energy consumption. Including energy-intensive methods of cultivation of sugar sorghum significantly increase the value of products received, so priority in the organization of production of sweet sorghum should be allocated to the development of resource - saving technologies.

Currently, the energy intensity of production acts as a factor of competitiveness of made production. In this connection it is necessary to create conditions for the effective formation of a highly mechanized units, require the development of models and methods of substantiation of effective structures of agrarian formations, optimization of production technology and management, rational organization of labor and logistics.

To solve the problem of energy saving in agricultural production it is necessary to solve three tasks: to adopt and gradually implement the organizational-economic and legal activities; to implement energy-saving technologies with wide use of secondary energy resources; to change the machine technology with a dramatic reduction in energy costs.

Technology of cultivation of each culture, like art, is based on a set of techniques aimed at ensuring the most favorable conditions for the growth and development of plants. Increasing the productivity of sugar sorghum is also associated with systemic approach to the cultivation of this crop, which is based on the achievements in the fields of science, engineering, technology and complex application techniques. A set of techniques covers all manufacturing operations performed since the release of the precursor fields to harvest, inclusive, starting with primary and secondary tillage. These techniques need to pay attention as to effects on biological, chemical and physical properties of the soil they are universal. By mechanical action only created the conditions for the development of the root system of cultivated plants, the efficiency of fertilizers etc. the Conducted research allows to claim that up to 25% of the crop is formed by the main processing of the soil. At the same time, the cultivation of sweet sorghum is one of the labor-intensive farming practices on its implementation consumes up to 40% energy and up to 25% of the workforce [1,2,3,4].

Summary tillage involves substantial costs for fuel and lubricants, which constitute up to 30% at various technologies.

Currently in crop production will radically change the approach to the application of technologies of cultivation of crops in order to significantly reduce energy costs. With the aim of reducing energy costs, while the basic processing of soil are used resource-saving soil treatment methods, among which ploskorezy tillage, small share plowing, disking. Studies show that strukturalnych fertile soils conservation treatment compared to the plowing does not reduce the yield of crops, more fuel consumption if you decrease the depth by 1 cm decreases 1 L. In agricultural production in the Republic of Tatarstan introduced the crop combined aggregates, which in one pass across the field is carried out six to eight operations: harrowing, fertilizing, cultivation, leveling the soil, sowing, crop compaction. As a result, combined sowing units in comparison with the separate application of methods of seedbed preparation reduce energy costs at planting. So, fuel consumption on average when sowing with combined sowing machine "Victoria" on stubble background in SPK Alga Alekseevskogo municipal district decreased by 8,1 kg/ha [5,6].

In this regard, the development and introduction of energy-saving systems the main processing of the soil in the cultivation of sweet sorghum focused, in particular, to minimize the cost of fuel and lubricants, providing the economic effect, contributing to the achievement of environmental security is timely. In mixed economy, different security groups agricultural production resources necessary to develop different technology options [7,8,9,10].

Classic technology (moldboard) cultivation of sweet sorghum involves periodic or annual plowing with a turnover of formation, repeated passages of agricultural equipment in the field. The consequence of this technology is at a certain level of soil compaction, destruction of its mechanical structure, and as a result water and air erosion, reduction of fertile soil. Consequently, all this leads to disruption of natural ecological balance, pollution of the environment. [11,12,13].

Currently, in all regions of the country, the intensive search of new technological methods of primary tillage to reduce the impact of these consequences, preserving and improving soil fertility, reducing labor and energy costs. [14].

Developed and widely implemented comprehensive techniques, which are based on substitution of ploughing with other types of cultivation, minimum tillage or partial replacement of moldboard moldboard plowing, as plowing is consumed over 50% of the fuel and lubricants, therefore, this is a significant source of savings.

The cultivation of sweet sorghum were tested the technology of direct seeding, which is based on moldboard plowing chisel, the characteristics given in the table.

Table 1 – Cost analysis for various options of cultivation of sugar sorghum

| | Technological operations | The applied technique | Quantity, pcs | Cost, thous.rub. | Personnel | Cost items, rub. | | | Total, thous.rub. | Cost structure, % |
|---|---|---|---|---|---|---|---|---|---|---|
| | | | | | | Wage | Fuel and lubricants | The costs of service and maintenance | | |
| Classic technology | Disking | Tractor K-744<br>The Disc Header "Дискомастер 9*4 | 1<br>1 | 5500<br>1160 | 1 | 1500 | 10800 | 9720 | 22020 | 6,9 |
| | Moldboard plowing | Tractor K-744<br>Plow depleted PRUNE-8-45 | 3<br>4 | 16500<br>2000 | 3 | 7000 | 57600 | 51840 | 116440 | 36,4 |
| | Harrowing | Tractor MT31221<br>Harrow heavy | 1<br>1 | 1400<br>1000 | 1 | 800 | 6300 | 452 | 7552 | 2,4 |
| | Combined seeding | Tractor K-744<br>Seeding complex "Agrator 12200" | 1 | 2920 | 1 | 4000 | 11700 | 10530 | 26230 | 8 |
| | In total | | 13 | 30480 | 6 | 33250 | 86420 | 77760 | 320000 | 100 |
| The technology of | Disking | Tractor K-744<br>The Disc Header "Дискомастер 9*4" | 1<br>1 | 5500<br>1160 | 1 | 1500 | 6300 | 5670 | 13470 | 6,9 |
| | Moldboard plowing chisel | Tractor K-744<br>Chisel plough "Kizilmescit 6000" | 1<br>2 | 5500<br>1100 | 1 | 800 | 18000 | 16200 | 35000 | 18,0 |

| | | | | | | | | | | |
|---|---|---|---|---|---|---|---|---|---|---|
| | Harrowing | Tractor MT31221 Harrow heavy | 1<br>1 | 1400<br>1000 | 3 | 3920 | 11700 | 10530 | 26150 | 13,5 |
| | Combined seeding | Tractor K-744 Seeding complex "Agrator 12200 | 1 | 2920 | 1 | 4000 | 11700 | 10530 | 26230 | 13,5 |
| | In total | | | 18580 | 5 | 1550 | 56800 | 51120 | 194000 | 100 |

Combining tillage with seedbed treatment allows for a 9-14% reduction in the cost of fuel and lubricants and improve the quality of soil preparation for planting sweet sorghum, this is achieved by aggregation with the plough "Kizilmescit 6000". In General, the combination of technological operations the basic and pre-treatment into a single process allows you to save 25-30% of fuel and lubricants, labor resources compared to separate these operations.

However, when you switch to energy-saving technologies arise and negative aspects, in particular excessive soil compaction, resulting in deterioration of the permeability malagasyrepublic, heavy structureless soils. Therefore, the technology of direct seeding should be used in the crop rotation without row crops with a humus content of not less than 3%, the structural nesoplivye soils. In addition, the problem arises with plant residues. With their large numbers, uneven distribution across the soil surface there is a problem with the placement of seeds at a reasonable depth. In these cases, should be aggregated complexes with disc coulters. Practical experiments show that the disk coulters are less clogged with straw and easier to penetrate the surface. While crop residues and rhizome are located and stored in the upper soil layers and not in the entire arable horizon, thereby enhancing the potential of soil fertility, accumulation of soil moisture.

Overall, the transition to energy-saving technologies in cultivation of sweet sorghum enables to improve its economic indicators:

– to increase productivity, to reduce the need for labor resources;

– reduce cost of fuel and lubricants by 25-30%, and all costs of the whole technological cycle of cultivation of sugar sorghum for 9-14%;

– to reduce the cost of acquisition and operation of tractors and agricultural machines. The cost of service and maintenance with technology of direct seeding in comparison with the classical technology are reduced by 34.3%;

– there is a cost savings to prevent wind and water erosion of soil;

– increases fertility and soil blagodarenie;

– reduced environmental pollution.

Thus, the use of energy saving technologies in agriculture should contribute to the solution of problems of reduction of the direct and total energy costs and increase production.

## References

[1] Nafikov M M, Yakushkin N M, Fomin V N, Talanov I P 2015 Cultivation of single-species and mixed crops of sorghum crops and their performance Moscow 248 p

[2] Kashapov N F, Nafikov M M, Gazetdinov M KH, Nafikova M M, Nigmatzyanov A R 2015 Economic justification of the choice of machines and tools for primary tillage under sweet sorghum (scientific paper) proceedings of the International scientific-technical conference "Innovative mechanical engineering technologies, equipment and materials – 2015" (MNTK "IMTM–2015") Part 1 pp 285-88.

[3] Nafikov M M, Paznikov V V 2013 Forecasting and programming harvest sorghum in the forest-steppe of the Volga region Modern problems of science and education No 1 URL: www.science-education.ru/107-8378 .

[4] Nafikov M M, Valiev I Z and Fomin V N 2013 Evaluation of some elements of farming sorghum in the conditions of the Kama river region of Tatarstan Fodder production No 2

[5] Nafikov M M, Fomin V N and Valiev I Z 2012 Yields of sorghum and depending on the basic soil

cultivation and fertilizers in the conditions of forest-steppe of the Volga region *Advances in science and technology APC* Part 2 pp 25-27
[6] Kashapov N F, Nafikov M M, Gazetdinov M KH, Nafikova M M and Nigmatzyanov A R 2016 IOP Conference Series: Materials Science and Engineering V 134 012013
[7] Malikov M M 2002 System of forage production in the Republic of Tatarstan Kazan 364 p
[8] Nafikov M M 2010 Study of sugar sorghum in comparison with other silage crops Forage production No 4 P 22-24
[9] Nafikov M M Khafizova A R 2010 the Cultivation of forage crops in odnowica-o and mixed crops in Western Zakamye Vestnik of Kazan state agrarian University 2(16) pp138-142
[10] Gabdrashit Z A, Reutov, S P 1986 Climate and the crop (Kazan Tatknigoizdat) 112 p
[11] Nigmatzyanov A R 2016 the Role of plant protection products and fertilizers in yield formation of sweet sorghum Youth science 2016: technology, innovation, all-Russian scientific.-practical Conf. young scientists, postgraduates and students (2016; Perm) pp 73-76
[12] Nigmatzyanov A R, Fomin D V 2016 The application of fungicides and fertilizers and their role in yield formation sweet sorghum will Print Science and youth: new ideas and decisions in agribusiness: proceedings of the All-Russian scientific-methodical conference with international participation-eat Volume 1 pp 68-72
[13] Nafikov M M, Nigmatzyanov A R, Samatov Z A 2016 Justification of cultivation of sugar sorghum depending on the environmental factors in the Republic of Tatarstan Food self-sufficiency of the region in terms of import substitution: issues of theory and practice Issue 10 pp 219-224
[14] Nigmatzyanov A R 2016 Evaluation of the effect of fertilizers and means of plant protection on yield of sweet sorghum Food self-sufficiency of the region in terms of import substitution: issues of theory and practice Issue 10 p 241 - 247

IOP Conf. Series: Materials Science and Engineering **240** (2017) 012033 doi:10.1088/1757-899X/240/1/012033

# Comparative evaluation of different machines for seedbed for sorghum

**N F Kashapov[1] M M Nafikov[1] M Kh Gazetdinov[2] M M Nafikova[2] and A R Nigmatzyanov[3]**

[1] Kazan Federal University, 18 Kremlyovskaya street, Kazan, 420008, Russian Federation

[2]Kazan state agrarian university, Kazan, Russia

[3]Tatar institute of retraining specialists agribusiness, Kazan, Russia

**Abstract**. The purpose of research - selection of the optimal composition of agricultural equipment for seedbed preparation for sweet sorghum. Identified and disclosed to the essential characteristics of technology. Thus, innovation in relation to agribusiness - is a new technology, new equipment, new plant varieties, new fertilizers and plant protection products. A special place is occupied by technical-technological and production innovatsii. In order to optimize the total cost in the cultivation of sorghum it is recommended to review the composition of sowing complexes, in order to optimize the timing of sowing, as well as reducing the complexity of the implementation.

At the present stage of development of agrarian and industrial complex is carried out at an accelerated pace modernization of agriculture. From year to year improved tillage, there are various power-saving technologies of cultivation of crops. At the same time, the RF energy supply (in hp 100 hectares of crops) tends to decrease as well as decreased in comparison with the year 2014 by 1,9%, compared to 2010 in 2015, energy supply - by 13,2%. Similar rates of decline remain and the Republic of Tatarstan: 1,8% in 2014 and by 11,9% compared to 2010. At the same time arable land accounts for 3,3 tractors per 1,000 hectares of grain crops, two combine harvesters and 100 tractors only 5 sowing complexes. At the same time the rate of renewal of agricultural machinery low. In 2015, the rate of renovation of tractors amounted to 3% to the presence at the end of the year, combine harvesters – 5,3%, sowing complexes – 7,2%, while the elimination of factors were as follows: for tractors – 4,6%, grain harvesters – 5,9% and sowing complexes – 2,3%.

In general, in Russia the share of the main types of equipment in operation for more than 10 years is as follows: 60% of tractors, combine harvesters by 45%. Against the backdrop of these trends, the share of imported machinery in Russian agriculture: tractors - more than 65% of combine harvesters - more than 20%, with high productivity.

With the increase in supply of agricultural machinery to grow and the cost of its purchase, use and maintenance. In this regard, an evaluation of the economic efficiency of individual machines or units and tractor fleet as a whole for managing integrated units. Determination of the effectiveness of techniques used in the cultivation of crops in terms of value is currently difficult because of the constant fluctuations in the prices of equipment, fertilizers, fuel and lubricants, and others. Therefore, it is advisable to evaluate the efficiency of machines and tractors park based on innovative technologies in energy costs. The total energy used in the cultivation of the crop, characterizes the effectiveness of a product obtained, and cultivation techniques.

For the production of agricultural products in plant are used as a classical technique and technology with minimum or zero tillage. Technology of cultivation of agricultural crops, including sweet sorghum should be based on a holistic, systemic approach, aimed at obtaining the final results.

Application of these technologies allows for reducing the number of passes of units on the field as a result of combination and with wide machines, which reduces soil compaction, and to reduce labor costs by reducing the volume of process operations or multiple operations in a single pass machine optimization system and the composition of machine, and and staff, reduce the cost of fuel and lubricants and the timing of field work[7,8,9,10,11,12].

Practice shows that the mechanical mixing of different technologies or the use of foreign, with limited use of means of intensification, does not ensure the achievement of goals, on the contrary leads to soil compaction, weed crops and others [13,14,15,16].

Combined machines appeared at the turn of 1970-80-ies. The pioneer in this direction was the company John Deere. The first machines that combine the functions of tillage and sowing, appeared in the 1980-s. They consisted of two or more guns. By 1990 combined machines for soil cultivation and sowing began to rapidly gain popularity. In 1993 another manufacturer Kverneland Group (Norway) suggested that the combined units, representing a harrow in combination with a grain drill, it was one whole machine. In the Republic of Tatarstan modular machines began producing 90-ies of the last century [1].

Improved combined machines in the modern sense began to produce of 2005-2007. In the market of tillage and seeding equipment began to appear combo units providing seed crop at the minimum and zero tillage [2].

Under the sweet sorghum is relatively new culture for the farmers of the Republic of Tatarstan, has not yet been tested and is not recommended resource techniques seedbed. Therefore, on the basis of the foregoing, we have the 2013 field trials and laboratory tests with a set of tillers were established and carried out. The test results are given below [3,4,5].

Maximum yield 44,1 t/ha of sorghum on the average for three years received during seedbed combination unit CBM-10.5 (tab. 1). The increase in control was 11,3 t/ha. Low yields of 28,2 t/ha of sorghum obtained by preplant tillage unit VNIISS-P.

Table 1 - Yields of sorghum depending on seedbed preparation, t/ha

| Unit | Years | | | The average for 2013-2015 | Increase (+, -), t/ha |
|---|---|---|---|---|---|
| | 2013 | 2014 | 2015 | | |
| CPS-4 | 33,6 | 30,3 | 34,4 | 32,8 | – |
| CBM-10.5 | 44,7 | 42,1 | 45,4 | 44,1 | 11,3 |
| VNIISS-P | 30,1 | 26,6 | 28,1 | 28,2 | 4,5 |
| CPG-4 | 33,4 | 31,8 | 32,6 | 32,6 | -0,2 |

Using cultivator CPS-4 for the treatment of the soil allowed to collect 32,8 tons of green matter per 1 hectare. The highest yield of 45,4 t/ha of sorghum produced in 2015 during the preplant soil unit CBM-10.5, the lowest - in 2014. Studied techniques seedbed preparation had an impact on the nutritional value of green mass of sorghum (tab. 2).

Table 2 - Charges feed units, protein and protein supply of forage sorghum units for 2013-2015.

| Unit | Charges feed units, kg/ha | | Collection protein, kg/ha | Availability of 1 kg of feed. u protein, g |
|---|---|---|---|---|
| | years | average over | | |

| | 2013 | 2014 | 2015 | 3 years | | |
|---|---|---|---|---|---|---|
| CPS-4 | 6552 | 5947 | 6708 | 6402 | 486 | 76 |
| CBM-10.5 | 8717 | 8210 | 8853 | 8593 | 670 | 78 |
| VNIISS-P | 5870 | 5187 | 5480 | 5512 | 408 | 74 |
| CPG-4 | 6513 | 6201 | 6357 | 6357 | 477 | 75 |

The largest collection of 8593 kg of fodder units per hectare obtained by the treatment of the soil combined unit CBM-10.5. Slightly inferior to this embodiment tillage cultivators CPS-4 and CPG-4, which is collected from one hectare respectively 6402 and 6357 kg of fodder units. The lowest collecting fodder units 5512 kg/ha was obtained when tillage harrow VNIISS-P.

The basis for calculating the evaluation of economic efficiency of crop cultivation technologies is operating costs. In the developed economic and mathematical model as an objective function is selected criteria specific minimum amount of reduced costs for the j-th unit performs the i-th operation:

$$Z_{ji} =\{[(KM_{ji}\ SM_{ji} + KP_{ji}\ SP_{ji})/\ W_{ji}] + TCM_{ji}\ P_{ji} + [CV_j\ (QTR_j + QKR_j\ ) /GN_j) + \quad + \sum_{l\in N} CV_l\ KOM_{lji}\ (QTR_l + QKR_l\ ) /GN_l)]\}\ T_i\ L_i\ ,$$

where N - set of agricultural machinery belonging to the j-th unit performs the i-th operation;

$KM_{ji}$, $KP_{ji}$ - the number of machine and auxiliary workers the j-th unit performs the i-th job, people;

$SM_{ji}$, $SP_{ji}$ - hourly rate mechanic and auxiliary workers the j-th unit performs the i-th operation, rub/h.;

$W_{ji}$ - hour performance of the j-th unit performs the i-th job, ha/h.;

$TCM_{ji}$ - consumption of fuels and lubricants j-th unit in the performance of the i-th operation, l;

$P_{ji}$ - the price of fuel and lubricants used by j-th unit in the performance of the i-th operation, ths. rub.;

[$CV_j$, $CV_l$ - the carrying amount of the j-th energomashiny and l-th agricultural machines, ths. rub.;

$QTR_j$, $QTR_l$ - cost ratios respectively for the current repairs and maintenance energomashiny j-th and l-th of agricultural machines;

$QKR_j$, $QKR_l$ - respectively the coefficients of the cost of major repairs of the j-th energomashiny and j-th of agricultural machines;

$GN_j$, $GN_l$ - respectively annual load energomashiny j-th and l-th agricultural machines, h.;

$KOM_{lji}$ - the number of agricultural machines l-th type of make up with the j-th energomashinoy when the i-th operation, pcs;

$T_i$ - the number of hours in a change in the performance of the i-th operation, h.;

$L_i$ - shift factor in the performance of the i-th operation.

Data for economic efficiency are shown in table 3.

Table 3 - Economic efficiency of cultivation of sorghum green mass

| Unit | Yield of green mass, t/ha | Gross output value, rub. | The cost of 1 hectare, rub. | The net income from 1 hectare, rub. | The level of profitability,% | The cost price of 1 ton of green mass, rub. |
|---|---|---|---|---|---|---|
| CPS-4 | 32,8 | 4002 | 2886 | 1116 | 38,7 | 88,0 |
| CBM-10.5 | 44,1 | 5380 | 3085 | 2295 | 74,4 | 70,0 |
| VNIISS-P | 28,3 | 3452 | 2862 | 590 | 20,6 | 101,1 |

| | | | | | | |
|---|---|---|---|---|---|---|
| CPG-4 | 32,6 | 3977 | 2884 | 1093 | 37,9 | 88,5 |

The greatest experience in the net income of 2295 rub/ha and the lowest cost per 1 ton of green mass of sweet sorghum to 70 rub/t obtained in the processing of the soil combined unit CBM-10.5.

The lowest in the experience of the net income and the highest production cost was during harrows VNIISS P seedbed preparation and amounted to 590 rub/ha and 101.1 rub/t.

The results of the energy evaluation are shown in table 4. The greatest experience in the energy efficiency ratio of 7.2 and a coefficient of bioenergy potential agro-ecosystems 8.1 sweet sorghum produced in the processing of the soil combined unit CBM-10.5. The lowest ratio of 4.4 energy efficiency obtained in the form seedbed preparation undertaken harrows VNIISS-P. When processing cultivators CPS-4 and CPG-4 energy efficiency ratio was 5.0.

Table 4 - Energy efficiency preplant soil at cultivation of sorghum

| Unit | Yield of green mass, t/ha | Energy savings, GJ/ha | It took power, GJ/ha | Energy efficiency ratio | BEP agro-ecosystems |
|---|---|---|---|---|---|
| CPS-4 | 32,8 | 151,53 | 30,23 | 5,0 | 1,9 |
| CBM-10.5 | 44,1 | 209,75 | 29,18 | 7,2 | 8,1 |
| VNIISS-P | 28,3 | 129,82 | 29,43 | 4,4 | 1,3 |
| CPG-4 | 32,6 | 150,61 | 30,25 | 5,0 | 1,9 |

The energy remaining in the plant residues sweet sorghum was determined on the basis of the linear regression equation derived AM Lykov [6]:

$Y= 0,1\ X+6,27.$

And it was on the options accordingly:

$Y= 0,1 * 32800+6,27= 3286,27$

$Y = 0,1 * 44100+6,27= 4416,27$

$Y = 0,1 * 28300+6,27= 2836,27$

$Y = 0,1 * 32600+6,27= 3266,27$

Analysis of the data of the equation shows that the greatest amount of energy in the plant remains in the version 4416,27 was seedbed preparation, held a combined unit CBM-10.5.

As a result, the comparative evaluation seedbed treatment options, it was found that the low surface evenness and great lumpy 6.0 pcs./$m^2$ were present in the form of tillage cultivators harrows VNIISS-P. Maximum yields of sorghum 44,1 t/ha on average for three years received during the pre-treatment of the soil combined unit CBM-10.5, to control growth was 11,3 t/ha. On average over three years of research the largest collection of fodder units produced during seedbed combination unit CBM-10.5. This option proved to be a cost and energy efficient.

## References

[1] Zelenin A N, Yusupov M L 2014 Automation driving agricultural machinery for tillage, planting, plant care and harvesting (Yekaterinburg: Urga) 152 p

[2] Leachman G I, Belenkov A I 2015 Precision farming (precision agriculture): Questions and Answers Niva Zauralye V 5 pp 56-58

[3] Nafikov M M, Karimov H Z 2008 Influence of ways of crop and seeding rates on the yield of sweet sorghum International academic exchanges as a means of integration of the Russian education in the global cultural space pp 56-57

[4] Nafikov M M 2011 Formation of highly cenoses forage crops in the conditions of forest-steppe of

the Volga region: the dissertation dis. ... Doctors of agricultural sciences: 06.01.01 / Nafikov M M. - Yoshkar – Ola 35 p
[5] Nafikov M M, Fomin V N, Korol'kov V A 2009 Influence of times and norms of crop on the yield and quality of sweet sorghum in the conditions of forest-steppe of the Volga region. Scientific notes of the Kazan State Academy of Veterinary Medicine. Bauman. -Kazan. V 197 pp 292-98
[6] Lykov A M, Tulikov A M 1985 Workshop on agriculture with the basics of soil science (M .: Agropromizdat) 207 p
[7] Alabushev A V 2007 Technological methods of cultivation and use of sorghum Rostov n /D, 224 p
[8] Bolshakov A Z 2002 Sorghum – the culture of the XXI century (memo of sargood). – Rostov-n/Donu: Rostizdat, 2002.
[9] Manhush P A 1993 Developing studies on sorghum Maize and sorghum. No 1 pp 2-3
[10] Nafikov M M 2012 The dependence of the yield of sugar sorghum of seedbed preparation Corn and sorghum. V 3 pp 21-23
[11] Nafikov M M 2012 Effect of methods of primary tillage on productivity of sorghum in the conditions of forest-steppe of the Volga region Corn and sorghum No 4 Pp 8-10
[12] Nafikov M M, Fomin D V, Nigmatzyanov A R 2016 Varieties and processing methods of cultivation of sugar sorghum (Sorghum Bicolor (L.) Moench) Fodder production V 7 pp 29-33
[13] Nigmatzyanov A R, Fomin D V 2016 Impact of fertilizers and pre-sowing seed treatment on the productivity of sugar sorghum Fertility 4(91) pp 10-12
[14] Fomin D V, Nigmatzyanov A R, Chekmarev P A, Nafikov M M 2016 Influence of predecessors and levels of nutrition on impurity, agrophysical properties of soil and productivity of sweet sorghum Agriculture V 5 pp 26-28
[15] Nigmatzyanov A R, Nafikov M M and Korolkov V A 2016 Assessment of the effectiveness of fertilizers and plant protection products in the cultivation of sweet sorghum Ecology, environment and human health XXI century: collected articles on materials of the international.scientific. practice.Conf. pp 133-139
[16] Nafikov M M, Nigmatzyanov A R and Samatov Z A 2016 Bioclimatic resources, and justification for the cultivation of rainfed sorghum in the Republic of Tatarstan Scientifically based system of dry farming in modern conditions: materials of International scientific-practical conference dedicated to the 30th anniversary system the topic of dry farming Volgograd region. pp 213-19

# Obtaining alternative fuel from sweet sorghum in the conditions of the Republic of Tatarstan

**N F Kashapov[1], M M Nafikov[1], I R Gilmanshin[1,3] and A R Nigmatzyanov[2]**
[1] Kazan Federal University, 18 Kremlyovskaya street, Kazan, 420008, Russian Federation
[2]Tatar institute of retraining specialists agribusiness, Kazan, Russia
[3] Kazan National Research Technical University named after A N Tupolev, Kazan, Russian Federation

arnig76@yandex.ru

**Abstract.** In the agro-industrial complex of the Russian Federation the main types of energy resources is the FCM (fuel-lubricating materials), electricity, coal and gas. Priority energy is determined depending on the orientation of the activity of the agricultural enterprise. In the cost of getting products one of the key factors is its energy intensity. Under the energy intensity means the amount of energy expended per unit of finished product. Domestic manufacturers lag behind on this indicator from their foreign colleagues. Greatly influenced by the climatic conditions of production, which affects the amount of energy expended annually becoming more expensive. In the article, the authors address a topical issue of renewable(alternative) fuels from sweet sorghum in the stems of which contains from 14 to 21 % sugar. In the Republic of Tatarstan tested and introduced varieties of sweet sorghum. On the basis of literary data and carried out their own research given a set of equipment and presents non-waste production chain of biodiesel and fuel pellets from stems of sweet sorghum.

The issue of biofuel production from agricultural raw materials is hotly debated agricultural experts. Liquid biofuels considered motor fuels from renewable raw materials derived from rapeseed, maize, sugar, sorghum, soybeans, wheat and waste processing. World the use of biomass during the year, equivalent to consumption of 1 billion tons of oil and is comparable with the level of consumption of natural gas and coal. Since 2003 there has been a steady upward trend in production and consumption of biofuels due to increase in world oil prices, as well as ideas to improve energy security and the desire to reduce greenhouse gas emissions into the atmosphere. Generous state support of alternative energy in the United States and Western Europe has led to intensive growth of production capacities of bioethanol and biodiesel from food and plant materials[1,2,3].

For the fast-growing bioenergy sector economies, an important component of not only environmental, but also energy security (independence). Bioenergetics gives such security, as biofuels are domestically produced, moreover, is a renewable source that is always available. At the same time, a number of EU countries came to the conclusion that you need to go to liquid second generation biofuels produced from straw or waste wood processing [1,4,5,10].

The most economical for the production of biotoplivo are sugar cane, corn and sweet sorghum. By a large margin followed by cereals produced in different regions of the world. A promising source of raw material for biofuel production, would seem to be cellulose, from which it is possible to obtain liquid biofuels of the second generation.

CIS is a pioneer in the development of biofuel production became the Republic of Kazakhstan. There in 2008, there was commissioned a plant for the production of biofuels with capacity of 57 thousand t in a year. Also provides for deep processing of wheat into flour and biofuels with an intermediate stage of extracting dry gluten, as well as the production of fodder yeast and carbon dioxide.

For ethanol production in the United States used in 2008. about 30% of the crop sorghum, and in 2015 on ethanol processed half-grown sorghum.

Despite the financial crisis, President Barack Obama was among the first policy statements announced their commitment to plans to decrease oil consumption by increasing the consumption of bioethanol.

"Obama's Plan" envisages investments in the amount of $ 150 billion. in companies that produce environmental liquid motor fuel. Increasing priority is given to fuels from non-food raw materials to 135 million tons per year, while production of corn ethanol will not exceed 36 million tons per year.

According to the optimistic forecast, the expected consumption volume of bioethanol in the EU in 2020 may reach 10 billion liters, and the compound annual growth rate of the market of bioethanol in the EU in 2008, by 2020. will reach 8.7 percent. According to the pessimistic forecast, the expected consumption volume of bioethanol in the EU in 2020 will be 7 billion years, a compound annual growth rate of volume of the market of bioethanol in the EU from 2007 to 2017 and 5.6%. In both scenarios, the need for the EU bio-ethanol will be partially offset by imports from 10 to 25% of market capacity. China in just three years has managed to make the top four largest producers of fuel ethanol in the world. By 2016, the projected increase in ethanol production in China, mostly from corn, to 3.5 billion L. According to various sources, in Russia there are about 10 projects of construction of plants for the production of bioethanol, but only one project (GK Titan, Omsk oblast) is implemented.

In the Republic of Tatarstan from the 90-ies of the last century, the initiative group engaged in creative search of alternative fuel sources (biofuel) from various plant materials. Stopped for sweet sorghum, a variety of varieties and hybrids which are contained in the stems of 14 to 21 % sugar. In the Republic of Tatarstan you can get from 270 to 600 kg /ha green mass yield of sugar content in the juice of the stems to 21 %[6,7,8,9].

Figure 1. Ecological testing of varieties and hybrids of sugar sorghum in Tatarstan

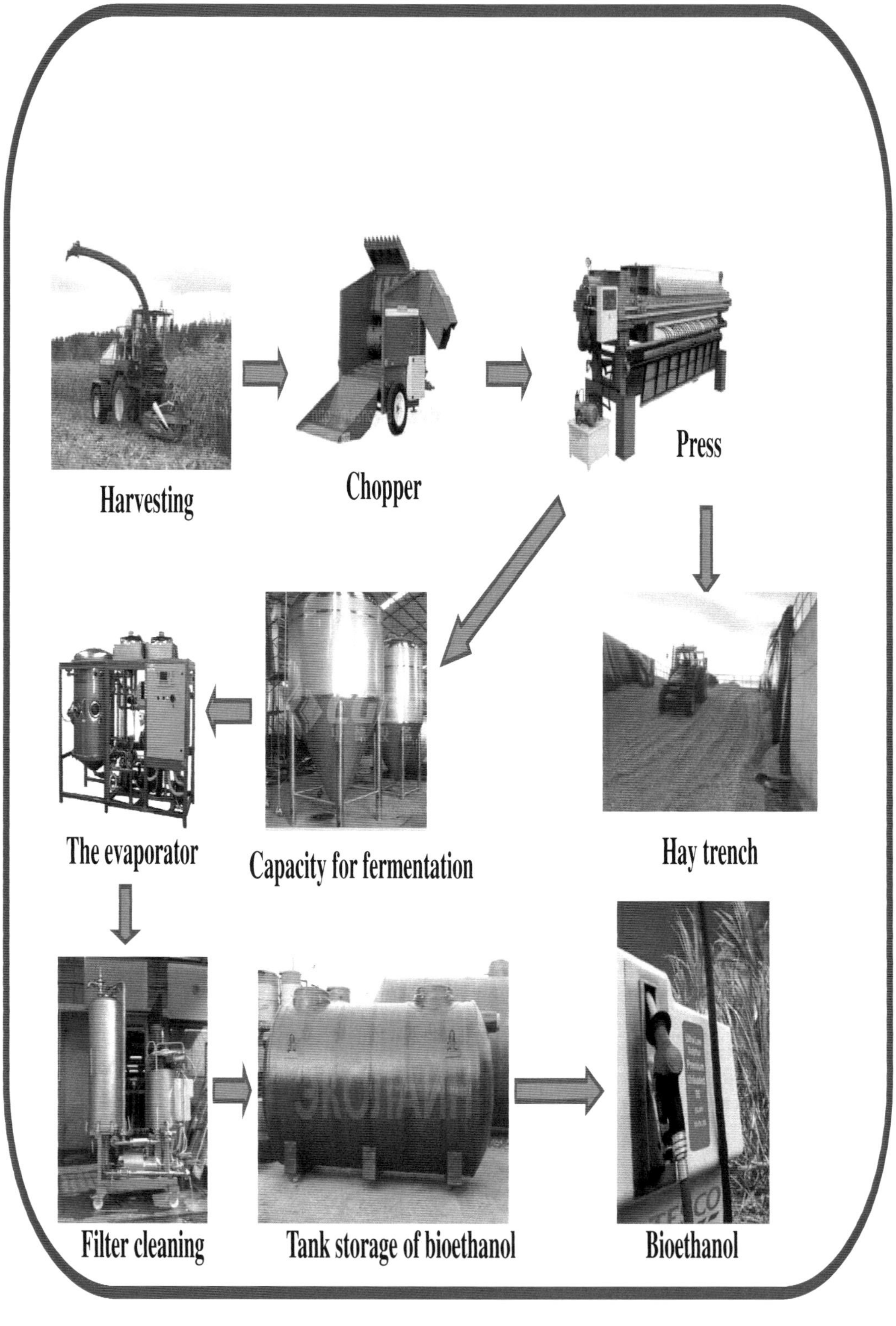
Harvesting
Chopper
Press
Hay trench
Capacity for fermentation
The evaporator
Filter cleaning
Tank storage of bioethanol
Bioethanol

Figure 2. The project of biofuel production from sweet sorghum

The relatively high cost of production and increasing the market value of the grain also limit the spread of bioethanol in Russia. If the market price of grain will continue to increase its use as raw material for bioethanol production is economically disadvantageous. On the other hand, the real alternative may be such the sources of raw materials, such as beet molasses –molasses, etc.

In the short term, Russia is expected to change the legislation in relation to biofuels. Begins to change the legislative framework is a complete list of legislative acts to ethanol became a separate product as fuel component. There was the intention to use biofuels in the domestic market: now in Russia at the state level, standards are being developed gasolines with addition of ethanol.

Experts believe that bioethanol produced in the Russian Federation, has good prospects of promotion in the markets of the European Union, Japan and several other countries in the world.

**References**

*[1]* Pashchenko A I 2010 Economic evaluation of prospects for obtaining of biobutanol from non-grain raw materials *Collection. articles of II international. nauchno-prakt.Conf*

[2] Pashchenko A I 2010 Economic and non-Economic regulation of production of second generation biofuels Chemistry and the environment *The development of science and education*

[3] Tarasov V I 2008 the Technological and economic prospects and regulatory support of production and sales of Russian biofuel *Manufacturer of Russia* No 9

[4] Tarasov V 2010 Prospects for biofuel production from genetically modified raw materials Chemistry and the environment *The development of science and education*

[5] RBC 2010 The Consulting Department *The market of biofuel: biofuel, biodiesel, biogas, pellets* Analytical review Moscow

[6] Nafikov M M, Fomin D V, Ngmatzyanov A R 2016 *Varieties and processing methods of cultivation of sugar sorghum* (Sorghum Bicolor (L.) Moench) Fodder production 7 pp 29-33

[7] Fomin D V, Nigmatzyanov A R, Chekmarev P A and Nafikov M M 2016 Influence of predecessors and levels of nutrition on impurity, agrophysical properties of soil and productivity of sweet sorghum *Agriculture* V 5 pp 26-28

[8] Nafikov M M, Yakushkin N M, Fomin V N, Kashapov N F, Fomin D V and Nigmatzyanov A R 2016 *Sweet sorghum: technological and economic aspects of cultivation in the Republic of Tatarstan* (Tutorial Kazan: Publishing house Snacks) p 40

[9] Kashapov N F, Nafikov M M, Gazetdinov M KH, Nafikova M M and Nigmatzyanov A R 2016 Innovative production technology ethanol from sweet sorghum *IOP Conference Series: Materials Science and Engineering* V 134 **012012**

[10] Pashchenko A 2011 *Market Prospects of liquid biofuels* APK: Economy, management No 4 pp 46-50

IOP Conf. Series: Materials Science and Engineering **240** (2017) 012035 doi:10.1088/1757-899X/240/1/012035

# Automated system of sheet-metal forming preproduction engineering control

**L R Kashapova and D L Pankratov**

68/19, Av. Mira, Apt. 335 (1/18), Naberezhnye Chelny, Tatarstan, Russian Federation, 423812

pishite.lile@gmail.com

**Abstract**. This work presents expert functioning system based on statistic knowledge of engineer and control algorithms. This expert system allows operating cold sheet-metal forming processes. Thus automated control system allows getting high quality products by cold sheet-metal forming.

In the processes of cold sheet-metal stamping we often find defects repeated from time to time (cracks, folding, etc.) as a result of tool wear, insufficient plasticity of the workpiece material, technology violations, etc. Today it is important to have and apply the setting ways of technological processes control at the stage of production technological preparation, which would reduce the defect rate and production costs of the production preparation.

In the field of stamping in automatic modes there is mainly the technological route control of parts manufacturing, but at the present stage of the production, in the conditions of a large amount of the analyzed technical information it's reasonable to develop the automated control system designing technological processes (TP), sheet stamping (SS) on the preproduction engineering (PPE) stage.

The methodology for assessing the reliability of the sheet stamping process described in [1-3], which allows identifying and correcting the critical parameters of the TP at the stage of the TPP can be taken as the basis of this system

The essence of this method is in the automated analysis of reliability factors: the details form; material brand, the mechanical and physical requirements to it; equipment parameters, where stamping takes place; tool characteristics, etc. After that it's possible their adjustment and, consequently, the most rational values of the parameters determination for defect-free sheet-metal stamping in a particular company.

The elements of cold sheet-metal stamping system are presented in the form of so-called system of Material-Process Structure-Tooling-Equipment-Personnel (MSTEP). Parameters systematization was carried out in connection with the fact that the stamping process has a large number of factors, thus to produce one piece, it is usually possible to develop several versions of manufacturing technology, i.e. to create some sets of MSTEP. The result of MSTEP system options formation process is shown in Figure 1.

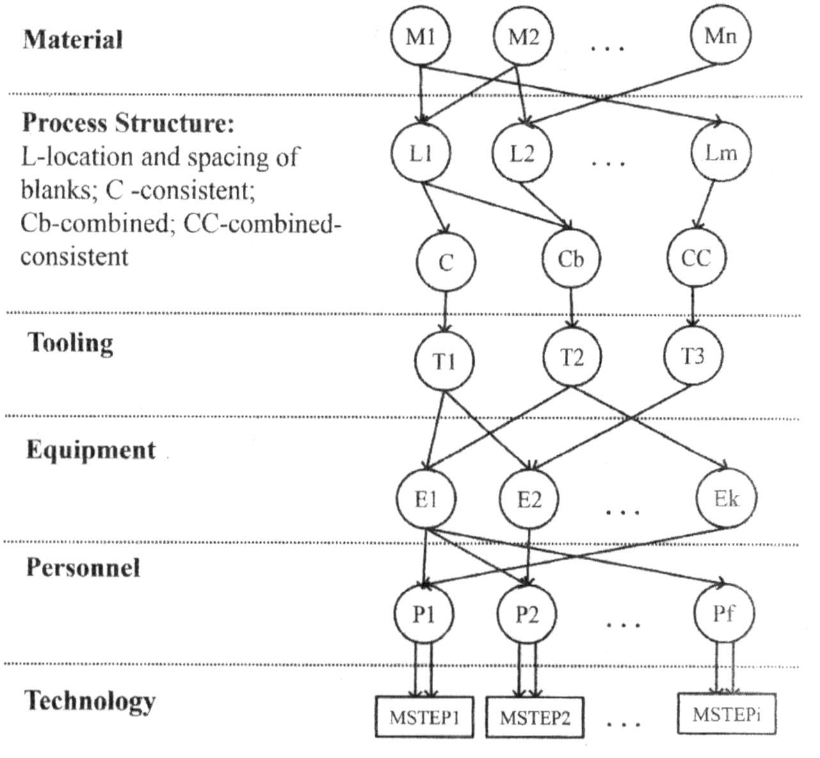


Figure 1 – Problem solving tree of MSTEP system alternatives formation

All MSTEP variants are previously assessed for reliability, ie on the defect-free technology. In case the value of any parameter of the system may lead to a defect, any other possible variants are considered to reduce the negative impact on the cold sheet-metal stamping.

Further, in order to obtain a rational MSTEP variant, the corrected list ranking is carried out by means of its assessment on the key performance indicators (cost of products, saving of resources, processing time, etc.), which are specific for each enterprise.

On the basis of this methodology the expert system (ES) has been developed, which operates on the basis of operator knowledge statistical accumulation (technologist) and control algorithms solutions. ES scheme is presented in Figure 2.

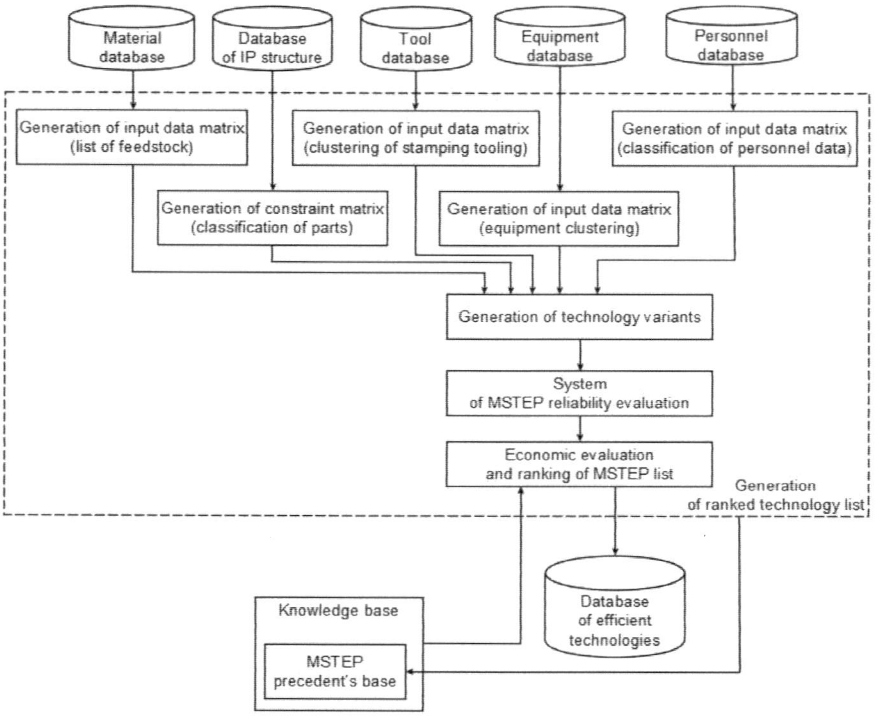


Figure 2. Diagram of interaction between elements of SS control expert system

ES of cold sheet-metal forming process control consists of:

1) Knowledge base including a database (DB) of the material, the structure of the TP, tools, equipment, staff, rational technologies and the base of precedents (BP), technology options;

2)"Technology ranked list forming" module.

The input information for the PPE is the design documentation (DD), being developed at the stage of pre-production design (PPD). According to KD, the product manufacturability, the group of complexity (part type), the thickness and dimensions of a workpiece, the type of steel, the properties of the assortment and the requirements of dimensional accuracy and surface roughness of the parts are being determined. [4-6]

The creation of automated process control system by TP designing at the stage of the PPE will improve the quality of products manufactured by cold sheet-metal forming process, and reduce the cost of the PPE by providing defect-free TP with implementation of the economic efficiency and expediency.

## References

[1] Kashapova L R, Pankratov D L, Shibakov V G and Vinogradov A I 2016 Obespechenie kachestva detalei slozhnoi formy na etape proektirovaniya tekhnologicheskogo protsessa listovoi shtampovki [Quality control for complex parts at the stage of stamping process design] *Journal of Machinery Manufacture and Reliability* no 1 pp 81-86

[2] Kashapova L R, Pankratov D L, Utyaganov R F 2015 Upravlenie nadezhnost'yu tekhnologicheskogo protsessa kholodnoi listovoi shtampovki na baze intellektualnoi sistemy [Control of die stamping process reliability on the basis of intelligent system] *Future of Machine-Building Industry in Russia* pp 182-85

[3] Kashapova L R 2015 Upravlenie nadezhnost'yu tekhnologicheskogo protsessa listovoi shtampovki

IOP Conf. Series: Materials Science and Engineering **240** (2017) 012035 doi:10.1088/1757-899X/240/1/012035

izdelii slozhnoi formy na etape proektirovaniya [Control of stamping process reliability for complex parts at the design stage], *Proc. Russian National Applied Science Conference of Students, Postgraduates and Young Researchers – VII Kama Readings* pp 100-02
[4] Steblyuk V I 1983 *Sheet stamping technology* Kiev: Vishcha school. Head Publishing House 280 p
[5] Meshcherin V T 1976 *Sheet stamping. Atlas of schemes* M .: Engineering 208 p
[6] Romanovsky V P 1979 *Cold stamping Guide* L .: Engineering 520 p

# Mesh three-dimensional arm orthosis with built-in ultrasound physiotherapy system

**R M Kashapova[1], R N Kashapov[1] and R S Kashapova[1]**

[1]Kazan Federal University, 18 Kremlyovskaya street, Kazan, 420008, Russian Federation

reginasharapova13@mail.ru

**Abstract**. The possibility of using the built-in ultrasound physiotherapy system of the hand orthosis is explored in the work. The individual mesh orthosis from nylon 12 was manufactured by the 3D prototyping method on the installation of selective laser sintering SLS SPro 60HD. The applied technology of three-dimensional scanning made it possible to obtain a model of the patient's hand and on the basis of it to build a virtual model of the mesh frame. In the course of the research, the developed system of ultrasound exposure was installed on the orthosis and its tests were carried out. As a result, the acceleration of the healing process and the reduction in the time of wearing orthosis were found.

## Introduction

In recent years, 3D printing technologies are increasingly being used in medicine. Additive technologies are widely used in neurosurgery, dentistry, traumatology, orthopedics and many other areas [1],[2],[3],[4],[5],[6]. One of the widely used orthopedic devices is orthosis.

Orthosis is a special device designed for unloading, fixing, activating and correcting the functions of a damaged joint or limb. It is used for instability of the ligamentous apparatus of joints, trauma in sports, arthrosis and arthritis of joints, in the postoperative period. Until recently, the solution of such conditions was the imposition of a plaster bandage. Orthoses made with the help of 3D-technologies have significant advantages over the old methods, namely before the traditional gypsum bandages. The disadvantages of gypsum dressings are primarily its mass, they are quite heavy. Plaster bandages are hygroscopic, which causes the patient does not come into contact with moisture. Over time, the gypsum grows old and begins to crumble. When wearing a cast plaster, there is a risk of abrasions and diaper rash. With a sharp squeezing of the limb improperly impregnated plaster bandage sometimes there is a breakdown of the nerve function. It is manifested by certain types of paralysis, which are difficult to treat. Finally, a long immobility in the joints causes wrinkling of the bag and ligamentous apparatus and changes in articular ends, which leads to a sharp decrease in mobility and soreness in the joint when moving. Due to inactivity of muscles and insufficient blood supply, their sharp atrophy sets in, and only after a certain time after removal of the bandage return normal working capacity. Orthoses are much easier, less restrict the patient's movements, are not afraid of water and are hygienic. However, orthoses do not affect the speed of healing of tissues and bones and do not shorten the period of treatment. At the moment, the acute issue is to find methods to increase the speed of rehabilitation and return the patient to an efficient state. One of such methods is the use of physiotherapy. The most effective method of physiotherapy is considered to be ultrasound physiotherapy. Physiotherapy by ultrasound is caused by mechanical, physicochemical and thermal effects caused by it. Ultrasound performs microvibration, performs massage at the cellular and subcellular levels. This stimulates the function of cellular elements and the whole cell, leads to an increase in the permeability of cell membranes, rupture of weak bonds, a decrease in the viscosity of the cytosol, a change in microcirculation, loosening of connective tissue, acceleration of diffuse processes. The physicochemical effect of ultrasound is also determined by mechanical resonance, under the influence of which the movement of molecules is accelerated, their decay into ions is increased, new electric fields are formed, free radicals appear and various products of sonolysis of biological solvents. Electronic excited states arise, lipid peroxidation is activated, enzymatic activity and mitochondrial activity are changed, physicochemical and biochemical processes in tissues are stimulated, metabolism is activated, the amount of prostaglandins of the P2a group increases, the pH of the tissues is increased, BAS-histamine, serotonin, heparin are released from mast cells. Ultrasound stimulates tissue respiration and oxidative processes in tissues, has a normalizing effect on carbohydrate, fat and mineral metabolism. These changes largely determine the stimulating effect of ultrasound on the processes of physiological and reparative regeneration. Due to the above properties, ultrasound under certain conditions may have analgesic, antispasmodic, anti-inflammatory and bactericidal effects. The use of ultrasound can be combined with other types of therapy. Ultrasound accelerates the processes of regeneration and repair, restoration of conduction of nerve fibers in peripheral nerve injuries, resorption of infiltrates, traumatic edema, exudates and hemorrhages, has anti-inflammatory (secondary effect), analgesic, ganglion blocking, spasmolytic, metabolic, hypotensive, desensitizing, fibrinolytic, defibrizing and bactericidal action, increases the adsorption properties of the skin and enhances the adaptive-trophic processes in the body and the regional blood flow. At the moment, patients can undergo physical therapy in the hospital and only a specific course.

The aim of the work was the use of additive manufacturing technology in the process of creating an individual model of the orthosis with integrated ultrasound generator. The product was manufactured on an SLS unit (selective laser sintering). As a result, based on the scanning and 3D modeling of the patient's arm, an individual arm orthosis was made. The ultrasonic generator was installed in the orthosis. A study was conducted to describe the results and justify the benefits. The practical application of this model of orthosis by different groups of patients was studied. As a result, it has been established that the orthosis, in which the ultrasonic generator is integrated, significantly alleviates the condition of the patients after injuries and fractures, and allows to shorten the period of the latch wear. The most optimal course of treatment was daily therapy with ultrasound 3 times a day for 15 days.

The process of creating an orthosis model consists of several stages. The first step is to scan the patient's hand. Scanning is performed using the Artec Eva device (Fig. 1). The scanner makes it possible to obtain a mathematical model of a solid object. The technology of three-dimensional scanning, used in this scanner, allows to extract information about the surface of objects (depth) with high accuracy and speed, using the principle of structured illumination. All data is obtained by projecting a special grid at the parallax angle to the scene objects. Distortions of the projection of the lattice, created by the geometry of objects, allow to calculate the exact position of each of its points in three-dimensional space. The resulting points are triangulated to form a polygonal surface that can be represented in any of the common formats. Texture camera, synchronized with the wide-field 3D-sensor, allows you to simultaneously remove the surface texture of the object.

IOP Conf. Series: Materials Science and Engineering **240** (2017) 012036 doi:10.1088/1757-899X/240/1/012036

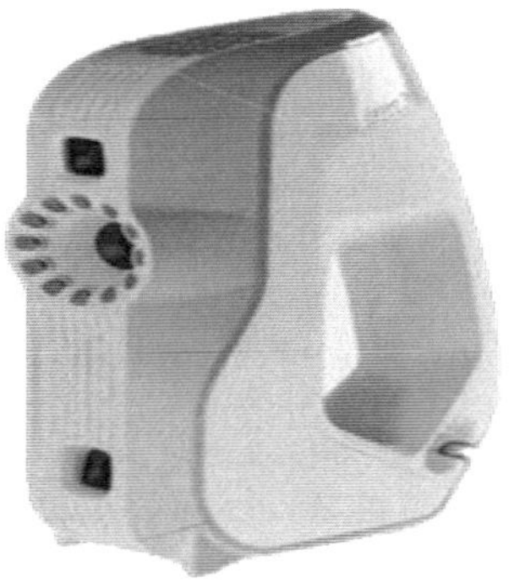

**Fig.1. Artec Eva**

At the second stage, 3D modeling of the orthosis is carried out. For this, 3DS max, Meshmixer, Netfabb were used. We get the 3D model of the patient's hand (Fig. 2). After that, we make the 3D model of the orthosis in the program. (Fig.3)

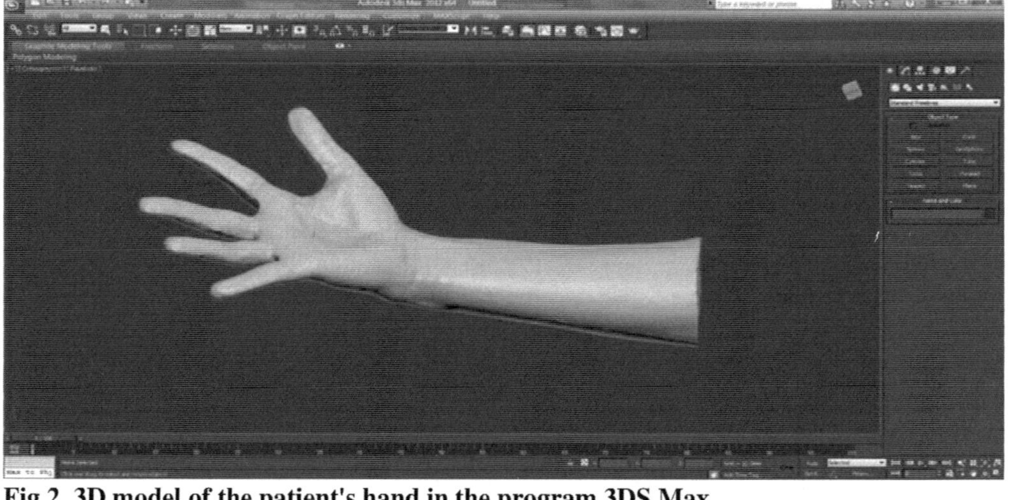

**Fig 2. 3D model of the patient's hand in the program 3DS Max**

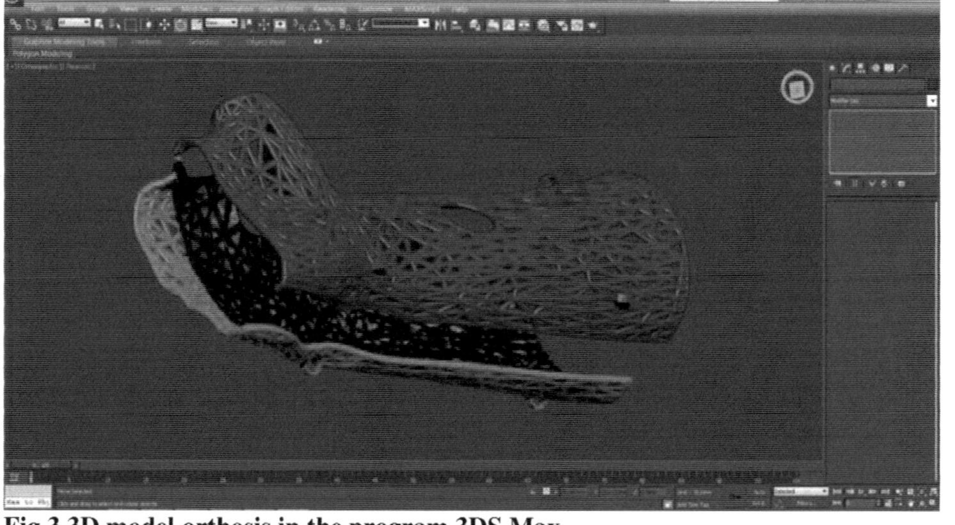

**Fig.3 3D model orthosis in the program 3DS Max.**

The finished 3D model of the orthosis is started to be printed on the SLS SPro 60HD device. The product prints last 7 hours. After this, the finished product should cool down. It takes about three hours. After the product has cooled, it is necessary to get rid of excess powder by sandblasting. We get the finished orthosis model.

The orthosis itself is made of polyamide Nylon 12 (Fig. 4). The dimensions of the product are selected individually for each patient. It consists of two devices: an orthosis for fixing the limb and an ultrasonic generator with a detachable speaker, for directional action of ultrasonic waves. The orthosis consists of two parts connected by two cable ties (Fig.5)

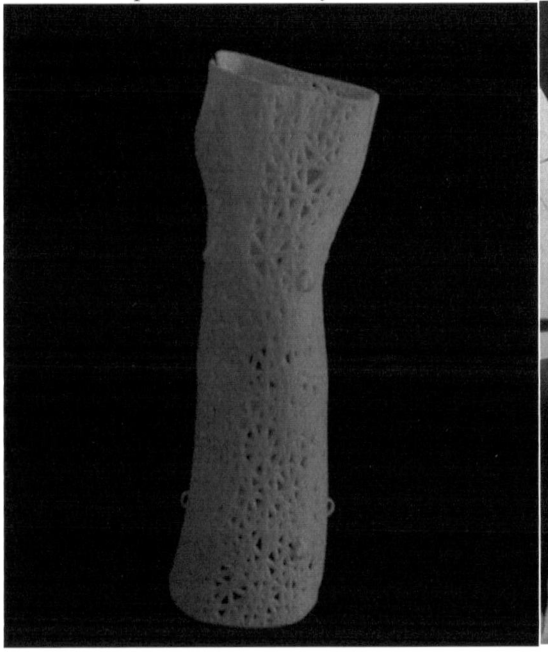

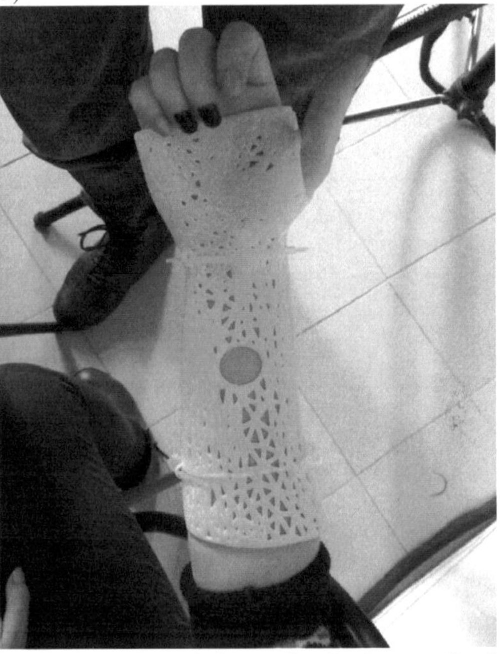

**Fig.4 Orthosis from polyamide Nylon 12 Fig.5 Two parts of the orthosis, fastened with clamps**

To accelerate the healing processes, it was decided to install an ultrasonic generator into the orthesis. The ultrasonic generator is assembled according to the following electrical scheme (Fig. 6).

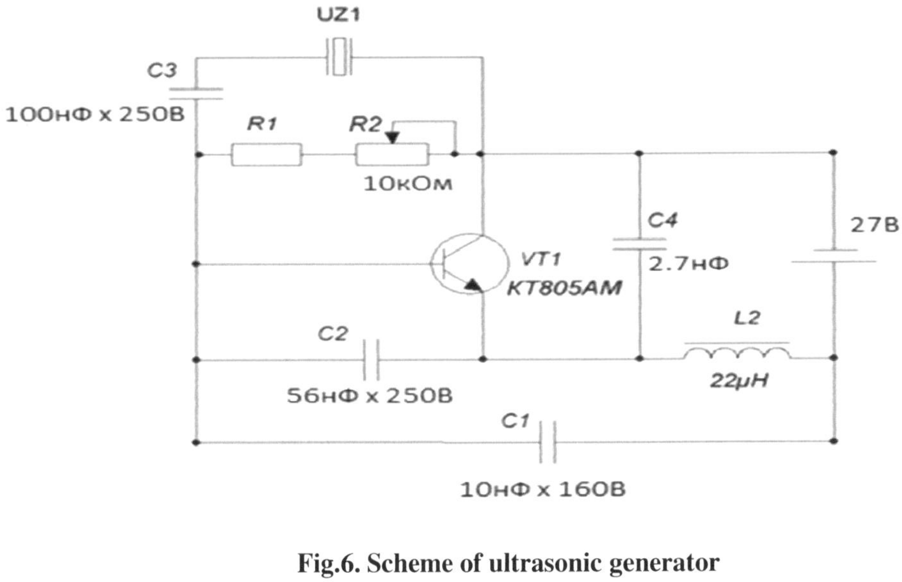


**Fig.6. Scheme of ultrasonic generator**

The radiation power should be 0.2 W / cm2. The pulse frequency is 1.5 MHz. The obtained system is attached to the finished orthosis by means of a special bracelet.

To substantiate the benefits, practical experience in using this orthosis model and describing the results obtained is important. For this, three groups of patients with diaphyseal fractures of the upper extremities were selected. The first group used an orthosis with physiotherapy ultrasound daily in unlimited quantities. The second group of patients used orthosis every day 3 times a day for 15 days. The third group of patients used orthosis every day 1 time per one for 15 days. The fourth group of patients were impregnated with gypsum bandages. Ultrasound physiotherapy in this group was performed in the hospital once a day, for 15 days. The duration of the procedure for the second, third and fourth groups of patients increased gradually every day, ranging from 5 minutes with a gradual increase to 15 minutes. All manipulations took place under the supervision of specialists. For treatment, small doses of ultrasound (up to 1.2 W / cm2) were used.

The first group of patients on the third day of therapy had a feeling of malaise. Disturbed headaches. Patients noted sleep disturbances and lack of appetite. This group discontinued ultrasound treatment immediately. Despite this, patients noted a significant analgesic effect and a decrease in edema at the site of the fracture on the first day of treatment.

The second group on the second day noted a decrease in swelling of the tissue at the site of the fracture, an analgesic effect. Due to the physico-chemical effects of ultrasound, experts noted the intensity of tissue redox processes, the increase in the formation of biologically active substances - heparin, histamine, serotonin. The nutrition of the tissues at the fracture site was improved, which allowed to avoid atrophy of the muscle tissue. Complete fusion of bones in this group of patients occurred in the interval from 20 to 28 days.

The third group achieved similar results. Anesthetizing effect patients of this group felt on the 5th-6th day of therapy. Edema was eliminated on day 5 of therapy. Improvement of blood circulation specialists noted on day 7. The period of complete bone fusion this group of patients achieved more than 45 days.

Patients of the fourth group did not complete the course of treatment with physiotherapy. The treatment was intermittent, due to the complexities of wearing the plaster bandage (crumbling of gypsum, the appearance of diaper rash). Those patients who completed the course fully noted the analgesic effect only after the 7-9 course of physiotherapy with ultrasound, puffiness passed after 10-12 courses. Complete fusion of the bone in the fracture site took more than 60 days.

Data on the results of the study are given in Table 1.

| | **1st group** | **2nd group** | **3rd group** | **4th group** |
|---|---|---|---|---|
| **Number of exposures per day** | Unlimited amount | 3 times a day | 1 per day | 1 per day |
| **Removing the pain syndrome (on what day)** | 1st day | 2nd day | 5-6 day | 7-9 |
| **Reduction of swelling of the tissues (on what day)** | 1st day | 2nd day | 5th day | 10-12 |
| **Increase in the formation of biologically active (on what day)** | No | 4-5th day | 8-10th day | 20th day |

| | | | | |
|---|---|---|---|---|
| **Improvement of blood circulation** | No | 3rd day | 7th day | 10th day |
| **Complete fusion of bone** | No | 20-28 days | >45 days | >60 days |
| **Side effects** | Feeling of malaise, headaches, sleep disturbance and lack of appetite. | No | No | No |

**Table 1. Description of research results.**

Thus, it can be concluded that the orthosis, in which the ultrasonic generator is built in, significantly alleviates the condition of patients after injuries and fractures. It gives an opportunity to conduct a course of treatment without leaving home. The period of wearing the latch is greatly reduced. This model of orthosis allows in the shortest time possible to reduce the swelling of the tissue, reduce the pain syndrome and improve blood circulation. The most optimal course of treatment was daily therapy with ultrasound 3 times a day for 15 days.

**References**

[1] Kashapov R N, Korobkina A I, Platonov E V, Saleeva G T 2014 The method of manufacture of nylon dental partially removable prosthesis using additive technologies *IOP Conference Series: Materials Science and Engineering* Volume 69 (1) **012026**
[2] Kashapov L N, Rudyk A N, Kashapov R N 2014 Applying 3D-printing technology in planning operations of cancer patients *IOP Conference Series: Materials Science and Engineering* Volume 69 (1) **012016**
[3] Kashapov L N, Kashapov N F, Kashapov R N, Pashaev B Y 2016 The application of additive technologies in creation a medical simulator-trainer of the human head operating field *IOP Conference Series: Materials Science and Engineering* Volume 134 **012011**
[4] Akhmetov I D, Kashapov N F, Kashapov R N 2014 Modernization of clamping unit of draft of endosurgical monopolar forceps *IOP Conference Series: Materials Science and Engineering* Volume 69 (1) **012002**
[5] Kashapov R N, Khusaenova E L P, Faizrakhmanov I A, Kashapov L N 2016 Ion-Beam Treatment of Glass Substrates for Creation of Biomatrices *Indian Journal of Science and Technology* 9(18) 5 p
[6] Gavrilova V A, Kashapov N F, Kashapov R N 2011 Plasma application of protective polymer-powder coatings to ultrasonic sensors *Biomedical Engineering* Vol 45 Iss 5 pp 198–200

IOP Conf. Series: Materials Science and Engineering **240** (2017) 012037 doi:10.1088/1757-899X/240/1/012037

# Ways of decrease in the material consumption in case of their separation by the combined methods

**I I Khafizov**

Kazan Federal University, 18 Kremlyovskaya street, Kazan, 420008, Russian Federation

khafizov@kpfu.ru

**Abstract.** When processing hardly processed materials it is necessary to consider the combined methods. Application of an anode component in the combined process considerably reduces cutting forces from mechanical influence at division of materials that will allow to accelerate process without violation of accuracy and quality of a blanket. Electrochemical processing doesn't damage a processed surface, provides a high class of purity.

## Introduction

Electrochemical cell with electronic conductors in contact with ionic conductors has to be considered as uniform system. At the same time we receive the system interesting us — one border has undressed an electrode — electrolyte. In the same way as this the system is a part of big system, itself gra- Nice of the section consists of components, parts or under -systems. As all systems are in the nature in the constant movement, they need to be considered as processes. It means that activity of border of times - has put an electrode — electrolyte develops from active sty components or processes. Further, subsystems also the making processes -interact - in such way of system survive and keep viability. Therefore to understand behavior systems in response to external influence, it is necessary to identify components which make system and to understand interactions between them. Names -but thus function and structure communicate.

Interesting is the method of electro-chemical treatment [2]. What are the components of processes at the border the partition depends on the level of analysis. For example, on a phenomenological level, one can consider the following system components of the boundary:

1) separation of charge, including the relationship between current and the charging of the double layer and changes in the potential and the surface excess;
2) adsorption — desorption, which links the value of the activities and capacities with the use of isotherms;
3) the charge transfer on the electrode, which determines the relation between the Faraday current, potential and activity of the particles at the interface associated with the reaction of charge transfer;
4) mass transfer with (or without) the volume of bytkom (downside), which binds on the top notnow concentration and flow;
5) homogeneous chemical reaction (complexopathy, dissociation, etc.);
6) formation of the space charge;
7) formation of the phase on the electrode (anodefilm, layer O2);
8) diffusion in the electrode;

9) heterogeneous chemical reactions (re the combination of dispro-portioning);
10) the phenomenon of surface migration(surface diffusion etc.).

These process components at the boundary time the case was long known and most of them received mathematical treatment, as specifical and processes. However, since components of the surface section of the complex, these processes function in the system together; one the surface of the section differs from the other number, co-operating processes, i.e. hard-STU interactions[3].

In the late 40's-early 50-ies ripe a detailed understanding of electrode kinetics and electrode processes in General [1]. Was Vyasa- HN mechanisms of solvation of ions, the cathode you division of hydrogen and anodic oxygen evolution; progress was made in the General understanding of the after- coherent reactions; solved the mysteries of the double the layer immediately generalized as multi-layer systems topic, and paving the way for future KINETICAcal research a net surface processes Sov. In particular, since it has been possible figure out the kinetic behavior hemosorbtiontion of intermediate products, soluble Pro-intermediate products in various solutions, or-organic electrochemical reactions. In the lasttion 10 years particular attention has received optical methods: conventional and electronic micro scopy, spectrophotometry, reflection methods, al-elipsometry etc., reflecting microscopic look at macroscopic phenomena. Though that much is already known, just start over to understand the complexity of the electrochemical processes of growth the phases observed during the deposition of metals the use of oxides.

**Options for reducing costs in their dividing by the combined**

1. The application of anodic component in the combined process greatly reduces cutting forces from mechanical action when division of materials, which will accelerate the process without compromising the accuracy and quality the surface of the layer.
2.Control of combined process separation of material is possible by changing the anodic component in a wide range that will give lower cutting forces and eliminate defects (chips, burrs) at the exit from the cutting zone instrument for all materials.
3.Control of combined process can independently change the settings chemical and mechanical component with Ogrenicename their limits. The purpose of the controltion would be to optimize the techno-logical modes, combining all components of the Combirelated process.
4. The process control is possible with using adaptive systems, working on known mechanism of interaction of elements technological systems with constraints on the presensible error surface separation material and acceptable micro-roughness.
5. Adaptive equipment systems-solyut to implement feedback and perform operational adjustment modes is selectedtion effects depending on the properties divided given material and requirements to the product.
6. The precision management division who- through the maintenance of the desired position cutting edge non-rigid external tool effects controlled automatically.

**Insights**

1. Studied a systematic approach to the study combined methods with overlay electriction of the field.

It consists in considering technology the ecological problems of separating materials with a maliniemi losses in the cutting as a single processa, including technological issues, characteristics of equipment technological equipment and adaptive management structuring elements of the structure of the machine - fixture-detail.

2. Examined experimental equipmention, with adaptive control in koordinate that allows you to set a numerical yet the indicators of maximum achievable results for the pointness of separation and the quality of the surface layer.

3. Studied developments of the past years for to solve the problem by creating a resurdobarokamere the process and equipment, ensuring the priority of the country in the creation of competitive products on quality indicators of value and novelty.

The solution to the problem of stabilization parameters the trench separation is possible with adaptive control development of process modes, the position of the cutting of the tool in the groove, which will allow marginal accuracy in the division- Institute of workpieces. The expansion of technology who opportunities process becomes real after calibration of the lateral surfaces of the grooves [8].

**References**

[1] Afonichkin A I 2014 *Management decisions in economic systems* (St. Petersburg: Piter)
[2] Dick V V 2009 IT-strategy as a means of feedback in business development *Application informatics* V 1 pp 22-9
[3] Zelenkov Y A 2013 *Methodology of strategic management development of major industrial enterprises corporate information system in modern conditions* South Ural State University Publ 275 p
[4] Kalyanov G N 2016 *Modeling, analysis, reorganization and automation of business processes* (M.: Finance and Statistics) 239 p
[5] Litvak B G 2016 *Development of the management decisions* Delo Publ. 440 p
[6] Sotnikova A V 2015 *Development of models and distribution of labor resources method in IT-projects portfolio management* (Plekhanov Russian University of Economics Publ) 142 p
[7] Shamsutdinova Z R 2016 Analysis and improvement of environmental management system of oil refineries (using an example of JSC TANECO) *The Republic of Tatarstan economic research journal* V 2 p 85
[8] I I Khafizov 2015 Processing methods with imposing of electric field at low- waste division of materials *IOP Conf. Series: Materials Science and Engineering* 86 012013
[9] I I Khafizov 2016 Economic efficiency and effectiveness of ways of separating materials electro diamond processing *IOP Conf. Series: Materials Science and Engineering* 134 012014
[10] I R Gil'manshin, N F Kashapov 2014 Energy service contracts in regional engineering center for small and medium businesses *IOP Conference Series: Materials Science and Engineering* 69 012010

IOP Conf. Series: Materials Science and Engineering **240** (2017) 012038 doi:10.1088/1757-899X/240/1/012038

# IT-strategy and major aspects of quality management on the market of goods and services

**I I Khafizov and A N Galimov**

Kazan Federal University, 18 Kremlyovskaya street, Kazan, 420008, Russian Federation

khafizov@kpfu.ru, galimov_an@bk.ru

**Abstract**. The article deals with the basic provisions of the formation of IT-strategies and interaction with management quality. Formation of the IT-strategy in a volatile, changing marketing environment is a prerequisite for efficient operation of the company.

## Introduction

It is necessary to understand the meaning of the «strategy» category before studying the interaction between quality management and IT-strategy. Based on the analyze of advantages and disadvantages of different authors definitions [1, 2, 3, 4, 7], it may be concluded that strategy is a specification of a company's long-term goals and tasks based on the study of external and internal environment in order to achieve a competitive advantage.

It is necessary to review the system of different author's perspectives on the investigating category in order to identify IT-strategy against the general strategy definition. Three approaches to IT-strategy were systematized during the study [2]: «IT and business alignment» (implies the use of IT-systems to support business-strategies), «IT governance» (considered as IT-functions implementation plan), «enterprise architecture» based on company's architecture. Consequently, IT-strategy is a specification of long-term goals and tasks based on a company's strategy by developing of the IT-systems and focused on business needs satisfaction. The main goal of IT-strategy is to achieve the competitive advantages by using IT-systems and resources that meet all management levels challenges and respect the innovative management system features, goals of common strategy [5]. Management quality system and IT- strategy interaction is shown at the Figure 1. The formation of IT-strategy within management quality system concentrates on a self-reproduction of new IT-solutions that allow to improve corporate activities in the area of staff motivation into the process of IT-systems creation and functioning, and enhance the interaction with third parties.

Thus, the main task of the unified system is the effectiveness evaluation of IT-projects the quality of which has a direct impact on IT-systems development effectiveness that explains the increasing role of quality management system.

## Standards of the ISO/IEC 27000 series

Standards of the ISO/IEC 27000 series are international standards in the field of IT safety for creation, development and maintenance of system of management of IT safety. The initial purpose of the international standard with is management and servicing of IT service. This standard represents

IOP Conf. Series: Materials Science and Engineering 240 (2017) 012038 doi:10.1088/1757-899X/240/1/012038

requirements to system of management of IT service, including thirteen processes grouped in five groups:

1. Service delivery process:
-budgeting and cost accounting.
-management of level of IT service.
-management of a continuity and availability of IT service.
-IT service safety management.
-IT service capacity management.
-reporting under provision of IT services.
2. Relationship processes:
-management with suppliers.
-management of interaction with business.
3. Resolution processes:
-management of problems.
-incident management.
4. Control processes:
-change management in system.
-management of system configurations.
5. Release process:
-development of management decisions.
-correction of management decisions.

It is important to note that today there are problems of creation of standards in the IT sphere that is caused by a variety of reasons: strengthening of the competition in the IT market, the procedure of adoption of standards contradicts realities of lifecycle of IT technologies, there are difficulties of harmonization of standards, there is no single conceptual framework. In the solution of this problem one of key places occupies implementation of methods of functional standardization which allow to identify groups of basic standards by means of functional standards. The most important factor of development of the IT company are activities for standardization of processes which are aimed at providing the high quality level of the provided services.

Thus, it is possible to conclude that standards play a fundamental role in activities of the IT companies. However, the single system of standardization which will allow to manage the following problems is necessary for the IT companies: coordination is more whole than business and IT, safety and implementation of control measures in compliance with the best practice, to manage the complex environment, to conform to regulatory requirements, and also monitoring of efficiency of activities.

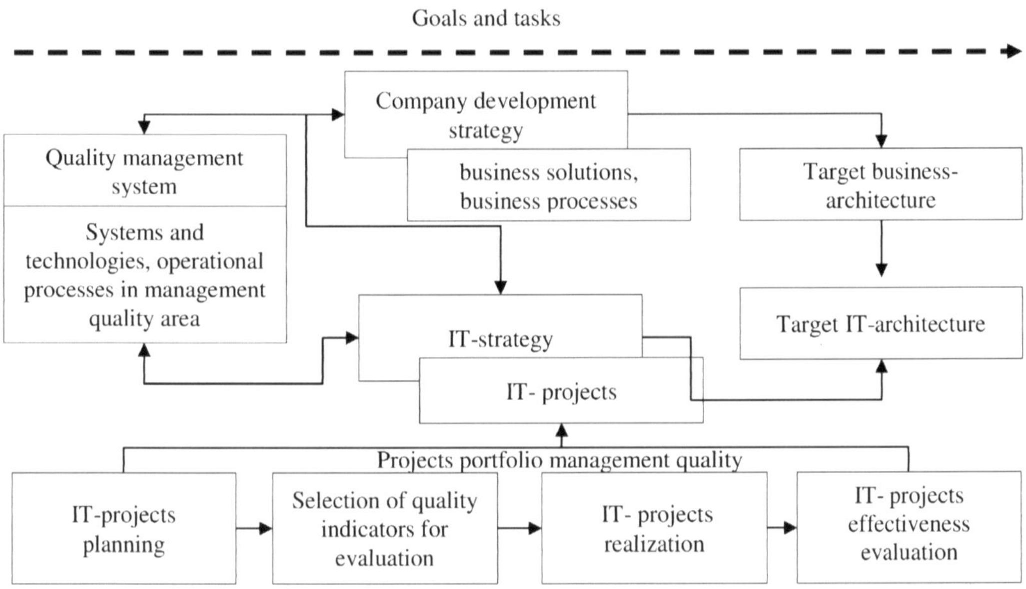


Figure 1. Management quality system and IT- strategy interaction

## References


[1] Afonichkin A I 2014 *Management decisions in economic systems* (St. Petersburg: Piter)

[2] Dick V V 2009 IT-strategy as a means of feedback in business development *Application informatics* V 1 pp 22-9

[3] Zelenkov Y A 2013 *Methodology of strategic management development of major industrial enterprises corporate information system in modern conditions* South Ural State University Publ 275 p

[4] Kalyanov G N 2016 *Modeling, analysis, reorganization and automation of business processes* (M.: Finance and Statistics) 239 p

[5] Litvak B G 2016 *Development of the management decisions* Delo Publ. 440 p

[6] Sotnikova A V 2015 *Development of models and distribution of labor resources method in IT-projects portfolio management* (Plekhanov Russian University of Economics Publ) 142 p

[7] Shamsutdinova Z R 2016 Analysis and improvement of environmental management system of oil refineries (using an example of JSC TANECO) *The Republic of Tatarstan economic research journal* V 2 p 85

[8] I I Khafizov 2015 Processing methods with imposing of electric field at low- waste division of materials *IOP Conf. Series: Materials Science and Engineering* 86 012013

[9] I I Khafizov 2016 Economic efficiency and effectiveness of ways of separating materials electro diamond processing *IOP Conf. Series: Materials Science and Engineering* 134 012014

[10] I R Gil'manshin, N F Kashapov 2014 Energy service contracts in regional engineering center for small and medium businesses *IOP Conference Series: Materials Science and Engineering* 69 012010

IOP Conf. Series: Materials Science and Engineering **240** (2017) 012039 doi:10.1088/1757-899X/240/1/012039

# Product quality as the main factor of increase of competitiveness (on the example of JSC "Kazan helicopter plant")

**I I Khafizov[1], I G Nurullin[1]**

[1] Kazan Federal University, 18 Kremlyovskaya street, Kazan, 420008, Russian Federation

khafizov@kpfu.ru opros16@yandex.ru

**Abstract**. Planning for effective development, and timely research perspectives and opportunities, States, markets, regulations, competition and competitiveness of its production, helps to achieve the stable business success in the market. The relevance of the study of the production of high quality products is that the quality of goods and services should always comply with the requirements established in state regulations, standards, regulations or conditions.

## Introduction

Competitiveness of this multifaceted concept, includes price and qualitative parameters of products, depends on the level of management and control systems of the movements of financial and material flow, as well as innovation and investment activities of the organization.

The issue of quality of manufactured products or services rendered is one of the key factors in the competitiveness of the organization. Quality determines which place thou shalt hold among competitors. In the struggle for influence on the market winner goes, anyone who suggested that the low price and high quality production.

Actuality of the study of the production of high quality products, is that the quality of goods and services must always comply with the requirements set out in government regulations, standards, rules or conditions. But it is equally important to take into account the views of consumers because goods and services are produced for them, so that their needs are met.

## Setting goals and objectives.

The purpose of scientific work was to study the impact of quality on improving the competitiveness of the organization.

To achieve this goal it is necessary to solve the following tasks:

-consider the notion of competitiveness, product quality organization, supplier evaluation and study the relationship between them;

-generate suggestions for improving the process of selection and evaluation of suppliers in the OAO Kazan helicopters.

The object of research is the production of high quality products for improving the competitiveness of.

The subject of research is the process of evaluating suppliers, which will help improve the quality improvement of purchased goods from suppliers.

**The main part.**

In market conditions it is not possible to achieve lasting success in business, if not to engage in effective development planning to examine their own prospects and opportunities, State of markets, the situation on these competitors and competitive products.

To get your piece of the pie in a dynamically developing market, the leadership of the Organization should be able to see the prospects of its development, foresee steps competing with it firms and to take the necessary steps and favorable to sustainable development. It is impossible to market the activities of the Organization, without participation in the competition [1].

If a company wants to take a stable position in the market of goods and services, it must produce products of good quality. Quality determines which place thou shalt hold among its competitors: the first or last in the fight for the consumer. And in the struggle for influence on the market winner goes, anyone who suggested that the low price and high quality industrial products [2].

Consequently, the competitive products, which meets all the requirements of the consumers of this market during the period under review as compared to the competitors ' analogues.

To achieve a high level of competitiveness it is necessary to constantly improve its constituent elements. If they consider them as sophisticated independent control objects contributed to achieving the following objectives: improving the quality, profitability, reduce production costs and increase the speed and efficiency of the after-sale service.

Organization of issuance of required products of sufficient quality in necessary quantity and at the right time is the primary objective of competitiveness management system in the production process.

The issue of quality of manufactured products or services rendered is one of the key factors in the competitiveness of the organization. Because quality is the most important criterion for consumers in determining the long-term orientation of the degree the extent to which the purchased goods satisfies the requested requirements.

The quality of all kinds of manufactured products must comply with clearly defined needs of destination products satisfy the requirements of consumers and comply with accepted standards and specifications [3].

To produce products that comply with international standards, it is necessary to introduce new innovations and use modern equipment. It requires big investments, capable not only of high quality of Russian products, but also to create new jobs. High level of automation and the use of new technologies is a necessary condition in the modern world to receive income from the work of the Organization [4].

However, it is not always high-quality production says that the Organization will always be able to maintain this level. The production of goods of better quality compared to competitors can be CPU-intensive and require large expenses that in the conditions of market relations will lead to inefficiency and a deterioration of the situation of the enterprise.

To improve the competitiveness of products, it is necessary to reduce the cost of production of the goods, as it is cost shows that costs the production company. The management of the Organization are not indifferent to price any of the costs be made products, it is very keen that when lower cost inputs produced more products using raw material of required quality.

Use inexpensive raw materials is not always reducing production costs. For example, use in the production of higher quality and expensive raw materials could significantly reduce waste production and, thus, reduce total costs. In turn, using higher-quality raw materials, the company can significantly improve the quality of their products and, consequently, their competitiveness [5].

The quality and competitiveness of products depend on the stability and reliability of supply, as well as from the quality of supplied materials and components. What would purchase products with high quality characteristics, it is necessary to find reliable and responsible supplier, eat with its reputation and relationships with consumers.

During the manufacture of qualitative characteristics of products is defined by three components, which are closely interact at work:

-quality of the work this product manufacturer;

-quality raw material, materials, components;

-quality equipment.

Analysis of influence of different factors on the quality of the products showed that the level of quality characteristics greatly depends on the quality of raw materials. The strength of this relationship more than reliance on all other factors affecting product quality. Similarly, it may be noted that all these factors operate in conjunction with one another in the manufacturing process.

Organization of input quality control of products supplier is one of the elements of a relationship with a supplier. Input quality control is a verification of purchased materials and components intended for use in the manufacture, repair or operation of the products, as well as the provision of services. The main purpose of which is not allowing the use of materials, raw materials and components, not satisfying the requirements of design and technological documentation, documents for Standardization applicable to products or services, and supply contracts. Not flawless work of this type of control can lead to the fact that the manufacturer of a product and its user to incur substantial losses [6].

At the conclusion of the contract with the supplier without reliable analysis conducted, it may happen that the supplied products are of poor quality, and as a result, we also lose time, money to buy and control input. Choosing must be taken seriously, at first an analysis and assessment of all suppliers, only then purchase products. In this case, the chance of hitting that supplier, seeks to minimize.

As you can see, the choice of supplier is one of the important issues. The importance of it due to the large number of suppliers in the market of analogues of products, as well as the fact that the supplier must be a reliable partner of the Organization in implementing its logistics strategy.

Table 1 provides information on claims (complaints) received from consumer products for 2014-2015 years JSC "Kazan helicopters.

Table 1 - Claims from consumers for 2014-2015 year

| | 2014 year | 2015 year |
|---|---|---|
| 1. Number of notifications from consumers about the call of the representative Company for quality products only: | 332 * | 437 * |
| -of which: for COE | 318 * | 415 * |
| -the fault of the production | 16 * | 22 * |
| 2. Number of actions acts from consumers only: | 195 * | 272 * |
| -of which: for COE | 186 * | 259 * |
| -the fault of the production | 9 * | 13 * |

*-values shown in the tables do not correspond to actual data, and provides an example.

Table 1 shows that 95% of actions acts drafted due to poor quality components, and on wine production at OAO Kazan helicopters is only 5%. That is, the majority of claims arises from the quality of the RFP, but rather the fault of the production only the remaining small part of [7].

Helicopters produced by OAO Kazan helicopters when their real costs and high enough weight parameters, competes and WINS their lighter Western designs.

As you can see, it is necessary to solve the problem of the quality of products purchased from suppliers. Improve the quality component parts received from suppliers, will help reduce the incidence of the appearance of defects and breakdowns. After solving this problem, the situation at OAO Kazan helicopters will strengthen markets, this will keep the current consumers and help get off at newer markets.

The organization has STO XXX.XXX.XXX-2014 "Quality management system vendor evaluation » this standard establishes only general requirements for suppliers and, therefore, the new draft was proposed a new draft STO XXX.XXX.XXX-2015 "QMS. Methodology the preliminary and periodic evaluation of suppliers ", which established the procedure of choice, using questionnaires and methods preliminary and periodic assessment of suppliers. This standard helps rate suppliers until the conclusion of treaties, and periodically in the predefined period of time.

For vendor evaluation method is applied, allowing ratings to calculate rating suppliers. Using this method, for each provider is calculated your rating, compared with the other, and then selects the best supplier, has the highest rate. Based on the assessments and the results of calculations produced a category is assigned to the vendor.

Supplier Evaluation is performed based on criteria at set intervals and at the end of the period of validity of the contract or after renegotiation of the contract for the supply of products.

She designed and installed in to ensure the quality of purchased products during the design, development and manufacture of the NII prototypes and serial helicopters civilian, military and dual appointment, as well as when repairing helicopters.

Thus, improving p Conference vendor selection, using questionnaires and introducing methods of preliminary and periodic assessment will lead to:

-supplement or modify existing on the OAO Kazan helicopters list of acceptable vendors;
-the choice of the most suitable suppliers of purchased products;
-improve the quality of manufactured products OAO Kazan helicopters;
-minimize the cost of purchased goods and supplies in the required quantity and in time;
-reduce the incidence of defects, inconsistencies in the quality of the contract among consumers.

**Results and conclusions.**

The competitiveness of organizations is the ability to effectively work and profitable implementation of manufactured products in a competitive environment. effective implementation and production of competitive products and services is a particularly important indicator of viability of the Organization, the ability to maximize the use of all its industrial, scientific and technical, human and financial capacities.

If the competitiveness of the organization provides the ability to successfully compete for a share of the market, the quality is one of the most important tools of influence on it. But the quality of the products depends on the stability and reliability of supply, as well as from the quality of supplied materials and components. Therefore, for the production of high quality products, it is necessary to cooperate with reliable, responsible and reliable suppliers.

The solution to the problem of improving the quality of purchased goods from suppliers to the OAO Kazan helicopters, reduces cases of occurrence of defects and inconsistencies in the quality of products, terms of contracts, becoming another opportunity for improve the competitiveness of products.

In the short, medium or long term, improving the quality of purchased product providers will give a positive result, and will be one of the most important and effective actions to improve the quality of manufactured products.

**References**

[1] Okovkina O N and Chupajda A M 2010 Ways of improving enterprise's competitiveness [electronic resource] / O.N. Okovkina,. - 2010. - № 19. Mode of access: http://cyberleninka.ru/article/n/puti-povysheniya-konkurentosposobnosti-predpriyatiya. Date: 12.10.2016.

[2] Khafizov I I 2015 Processing methods with imposing of electric field at low- waste division of materials *IOP Conf. Series: Materials Science and Engineering* 86 **012013**

IOP Conf. Series: Materials Science and Engineering **240** (2017) 012039 doi:10.1088/1757-899X/240/1/012039

**[3]** Khafizov I I 2016 Economic efficiency and effectiveness of ways of separating materials electro diamond processing *IOP Conf. Series: Materials Science and Engineering* 134 **012014**

[4] Sadrtdinova R R and Musin U R 2014 *Product competitiveness factors* No 2

[5] Rudzki R E 2008 *Effective supply: simple and reliable ways to reduce costs and increase profits* Minsk: Grevcov Publisher 304 p

[6] Kalyanov G N 2016 *Modeling, analysis, reorganization and automation of business processes* / M.: Finance and Statistics 2016 239 p

[7] Input quality control materials, KI and convenience foods Mode of access: http://www.bassmontage.ru/ tehnology/sberezheniya26.php. Date: 12.10.2016 year.

[8] Sotnikova A V 2015 *Development of models and distribution of labor resources method in IT-projects portfolio management* Plekhanov Russian University of Economics Publ 142 p

[9] Protocol № XX-14 of the XX-XX.XX.XXXX. Report on the analysis of the operation of the QMS of JSC "Kazan helicopters for year 2013 [text].

[10] Gil'manshin I R, Kashapov N F 2014 Energy service contracts in regional engineering center for small and medium businesses IOP Conference Series: Materials Science and Engineering 69 **012010**

IOP Conf. Series: Materials Science and Engineering **240** (2017) 012040 doi:10.1088/1757-899X/240/1/012040

# Estimation of parameters of charge carriers in dielectric materials by CELIV method

**I I Khafizov[1], T N Saveleva[1], V S Lyubtsov[2]**

[1] Kazan Federal University, 18 Kremlyovskaya street, Kazan, 420008, Russian Federation
[2] Ufa State Aviation Technical University, Ufa, Russia

khafizov@kpfu.ru , TNSaveleva@stud.kpfu.ru, vladimir_lyubcov@mail.ru

**Abstract**. Measuring the mobility of charge carriers by the time-of-flight method has been used for several decades to study organic semiconductors and dielectrics. Modern research in the field of polymer semiconductor devices focuses on the properties of single- and multi-layer thin-film structures with thicknesses less than 100 nm. Such structures are of considerable interest for research, since they are the basis for organic light-emitting diodes, organic solar cells and other electronic devices.

## Introduction

Most electronic devices based on organic materials consist of multilayer film structures with thicknesses much less than 100 nm. This significantly limits the time-of-flight method, since it is known from the theory of this method that the thickness of the polymer film must be much larger, and the film should have rather dielectric properties [9]. An important limitation of the time-of-flight method is the small concentrations of charge carriers. Moreover, measurements by the time-of-flight method do not give exact results, since in most cases the dielectric relaxation time $\tau_\sigma = \varepsilon\varepsilon_0 / en\mu$ much less than the time of flight of charges $t_{tr} = d / \mu E$. Thus, measurement of mobility by a time-of-flight method, as a rule, is limited to the concentration of the main carriers $n \ll 5\times10^{22}$ m$^{-3}$, for $E=3\times10^7$ V/m, D=100 nm и $\varepsilon=3$.

In this connection, it becomes relevant to use new experimental methods for estimating the mobility of charge carriers in multilayer film structures.

The CELIV method assumes the presence of a uniform density of charge carrier distribution, with the total charge density equal to zero, i.e. electrons and holes compensate each other [5]. It is important to note that the CELIV method is applicable for analyzing the mobility of charge carriers with a significant difference in the mobility of electrons and holes [8].

## Methods for measuring the mobility of charge carriers

The mobility of charge carriers is an important parameter that determines the kinetic characteristics of charge carriers in the materials under study. In particular, the efficiency of photocells based on conjugated polymers is significantly reduced due to low mobility of charge carriers. Thus, data on the mobility of charge carriers make it possible to determine the efficiency of polymers for photocells and other practical applications.

The time-of-flight method is one of the most widely used methods of investigating the transport of charge carriers, which was used to study a number of conjugated polymers and organic semiconductors [1, 6]. In this method, the charge carrier packet is generated by a light pulse and drifts through the material under investigation under the influence of an external applied voltage. The arrival time of the charge carrier packet is measured when the opposite electrode is reached and is calculated by the formula

$$\mu_{TOF} = \frac{d^2}{U \cdot t_{tr}}$$

where $d$ is the film thickness, $U$ is the applied voltage and $t_{tr}$ is the time of flight.

A time-of-flight method requires a thick film, which limits this method when measuring mobility in real thin organic films used in electronic devices.

**Transport of charge carriers in dielectrics**

It is known that dielectrics (insulators) are substances in which there are practically no free charge carriers. The term "dielectric" comes from the Greek word dia - through, through and the English word electric - electric. This term was first introduced by M. Faraday in 1838 to designate substances in which an electric field penetrates. As you know, there is no sharp boundary between conductors and dielectrics, since all substances are able to conduct electric current to some extent. But if in the substance of free charges is $10^{15}$-$10^{20}$ times less than in metals, then in such cases the weak conductivity of the substance can be neglected and considered an ideal dielectric. Almost all charged particles inside the dielectric are interconnected and are not able to move around the body. They can only slightly shift relative to their equilibrium positions.

According to the band structure of solids, for a dielectric, a completely filled valence band is distinguished, separated from the empty conduction band by an energy gap of several electron volts (the forbidden band). Neither in the filled or in the empty zone there can be no conductivity unless additional carriers are introduced into the dielectric. Carriers can either be generated inside the dielectric (processes limited by volume), or injected into it from the metal electrode (processes limited by injection). Various processes that determine the passage of current through a dielectric placed between two electrodes.

To produce the samples, glass with a transparent layer of ITO (indium and tin oxides) was used as the substrate. The size of the substrates was 1.5x2 $cm^2$.

The cleaning of the substrates was carried out according to standard technology in 3 stages in an ultrasound bath. The first stage - purification in acetone is pure for analysis, the second stage - in ethyl alcohol 95%, the third stage - in distilled water. After that, the substrates were dried in a drying cabinet (SNOL-350/65) at a temperature of 150$^0$C for 45 minutes.

The purified substrate was centrifuged (centrifuge with a variable speed of rotation SM-50) with a 2-5% (weight fraction) polymer solution of polydiphenylenephthalide. The rotational speed of the centrifuge was 1000-2000 rpm, the spinning time of the centrifuge was 20 seconds. After application of the polymer, the samples were dried in air for 45 minutes at room temperature of 24$^0$C (± 1$^0$C) and humidity - 26% (± 1%). Then the samples were placed in a drying cabinet for 45 minutes at a temperature of 150$^0$C (± 5$^0$C).

Metal tracks were produced by the method of thermal evaporation of metals in a high vacuum at the VUP-5M plant. Aluminum (Al) was used as the metal. The time of spraying was 1-2 minutes. The shapes and sizes of the tracks were set using contact masks.

To estimate the mobility in the polydiphenylenephthalide films, experimental samples of different thicknesses were made. The results obtained on samples with a thickness of ~ 0.8 μm and ~ 0.17 μm are presented.

**Conclusion**

IOP Conf. Series: Materials Science and Engineering **240** (2017) 012040 doi:10.1088/1757-899X/240/1/012040

The main task of this work was to assemble a measuring device for carrying out the CELIV study. The method is new to researchers, but at the same time the first mention of it was made in the early 80's of the last century. In the literature, in general, the method is used to estimate the mobility of charge carriers in organic semiconductors. In this connection, the actual task was to measure the mobility in dielectric materials. Studies were carried out on a polymer from the class of polyarylenephthalides, polydiphenylenephthalide [7, 10, 11].

## References

[1] Antoniadis H, Abkowitz M A and Hsieh B R 1994 Carrier deep-trapping mobilitylifetime products in poly(p-phenylene vinylene), *Appl. Phys. Lett.* 65 pp 2030-32

[2] Khafizov I I 2015 Processing methods with imposing of electric field at low- waste division of materials *IOP Conf. Series: Materials Science and Engineering* 86 **012013**

[3] Khafizov I I 2016 Economic efficiency and effectiveness of ways of separating materials electro diamond processing *IOP Conf. Series: Materials Science and Engineering* 134 **012014**

[4] Gil'manshin I R, Kashapov N F 2014 Energy service contracts in regional engineering center for small and medium businesses *IOP Conference Series: Materials Science and Engineering* 69 **012010**

[5] Juska G, Arlauskas K and Viliunas M 2000 Extraction Current Transients: New Method of Study of Charge Transport in Microcrystalline Silicon, *Phys. Rev. Letts.*, 84 **4946**

[6] Kreouzis T, Poplavskyy D, Tuladhar S M, Quiles M C, Nelson J, Campbell A J and Bradley D D C 2006 Temperature and field dependence of hole mobility in poly(9,9-dioctylfluorene), *Phys. Rev.* B 73, 235201(1)-(15)

[7] Rafikov S R, Tolstikov G A, Salazkin S N, Zolotukhin M G 1981 Polyheteroarylenes for the manufacture of heat-resistant materials and the method for their production. 734989 USSR. / *Bul. Invention.* - No 20 p 259

[8] Sebastian Bange 2009 Transient Optical and Electrical Effects in Polymeric Semiconductors *Dis…Potsdam*, pp 107-120

[9] Tsung K K and So S K 2008 Carrier trapping and scattering in amorphous organic hole transporter, *Appl. Phys. Lett. 92* **103315**

[10] Zolotukhin M G, Skirda V D, Sedova E A, Sundukov V I, Salazkin S N 1993 Gelation in the homopolycondensation of 3-arel-3-clorphtalides *Macromol. Chem* V 194. No 2 pp 543-49

[11] Zolotukhin M G, Kovardakov V A, Salazkin S N, Rafikov S R 1984 Some regularities in the synthesis of polyarylene phthalides by homopolycondensation of n- (3-chloro-3-phthalidyl) - biphenyl [Text] *High-molecular. Soed.* T 26a No 6. pp 1212-17

# Plasma spray coating with ferromagnetic powder by thermo-electric plasma equipment

**A A Khafizov[1], R I Valiev[1], Yu I Shakirov[1]**

[1]Branch of Kazan Federal University in Naberezhnye Chelny, 68/19 Mira Street, Naberezhnye Chelny, 423810, Russia

almazok75@yandex.ru, rivaliev87@mail.ru, inekashakirov@mail.ru

**Abstract.** Plasma spray coating is the most preferred and readily available method of carrying out repair and restoration works in comparison to other methods at present. Though the method of plasma spraying has been known for a long time, there are still a number of unsolved issues related to the choice of optimal deposition regimes. The thermo-electric plasma system for plasma spray coating with a liquid electrode is discussed in this article. The process and optimal parameters of plasma spray coating regimes are described.

## 1. Introduction

Development of repair technological processes is actively increasing, and control methods of new and restoration of worn parts appear, which can effectively solve a number of problems – tear and wear of friction parts, friction reducing, etc. Leading manufacturers use the cover of various functionalities in there products for a long time [1].

On the margin of the fatigue strength many details meet the requirements of reliability and, with the restoration of the initial dimensions, are quite efficient. The main cause of failure of these products is usually a wear of the contact surfaces, and they can be easily restored by different ways of surfacing. If we proceed from the cost of new components and the cost of restoring worn ones, we can make a choice in favor of repair. Plasma spraying represents the most preferred and available way of carrying out repair and rehabilitation works in comparison to other methods. The advantages of this method are: short time of repairs; preservation of geometric shapes of the parts; possibility of spraying a wide range of materials, etc. [2-5].

If you consider that most of the restored parts are made of structural steel or cast iron, it is advisable to use as the deposited materials, powder [6].

The main process parameters in the coating process, most efficiently influencing the value and quality of coverage, are the energy of the electric discharge, which, in turn, depends on the magnitude of discharge current, the magnitude of the working gap, the particle size of ferromagnetic powder and its feed into the gap [7]. However, these parameters should be interlinked also with the speed of the relative jet and detail movement.

## 2. Research methodology

Fig.1 shows the structural scheme of the thermo-electric plasma system intended for electric discharge [8].

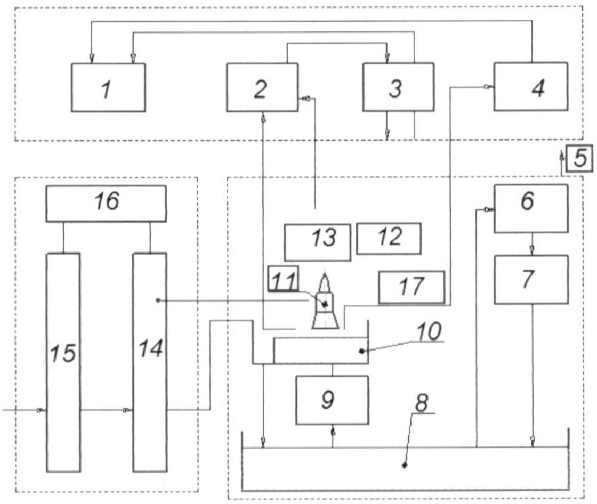


**Figure 1.** Structural diagram of thermo-electric plasma system:
1-voltage regulator; 2-interelectrode gap (IEG) control system; 3-electrolyte parameters control system; 4-control system of electric discharge parameters; 5-exhaust ventilation; 6-pump pumping the electrolyte; 7-filter for electrolyte purification; 8-container with an electrolyte; 9-feed pump of the electrolyte in the electrolytic bath; 10-electrolytic bath; 11 - head of plasma torch (shown in Fig. 2); 12-coordinating device; 13-workpiece; 14-rectifier unit; 15-transformer; 16-voltage regulator; 17-mounting of plasma torch.

Electrical discharge occurs between liquid and solid (metal) electrodes [9] in the range of interelectrode distance $l$ = 1÷100 mm, current intensity I = 10÷100 A, voltage $U_d$ = 40÷300 V. Technical water was used as electrolytic electrode.

Workflow when used as metal electrodes of the discharge chamber in the form of a truncated hollow cylinder is shown at Fig. 2.

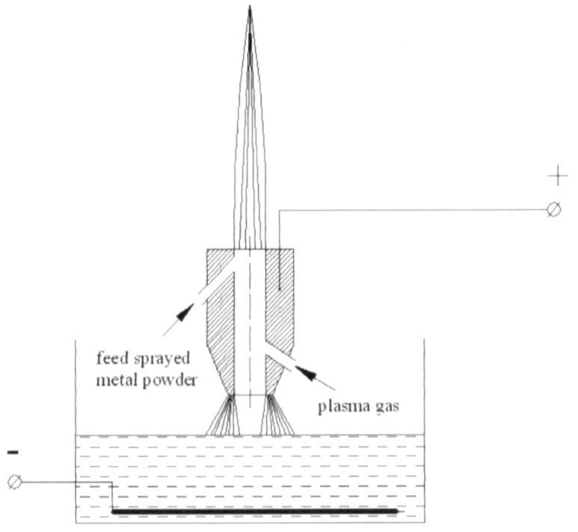


**Figure 2.** A device for producing plasma jet

The studies were carried out with the ferromagnetic powder coating [10]. During working the speed of movement of the selected samples was 5 cm/sec. These mechanical parameters are defined as a result of numerous experiments and provide maximum productivity in obtaining a quality coating on the samples and the details in a simple form. Ferro powder feeding was at the range of 6-7 g/min. A lower limit would not provide a sufficient amount of coverage continuity, and higher – would contributed to the "clogging" of the powder in the working gap, which reduced the stability of the process and dramatically increased wasteful consumption of powder.

The working gap is selected 4-5 mm with the ferromagnetic powder granulation varying 20-80 micrometer. An electric discharge is active under these limits, the process of saturation is stable.

The discharge current renders the greatest influence on the magnitude and quality of the coating while maintaining the optimal regime of hardening. The size ranged between 10 and 40 A. At lower currents the saturation was not active; the coating was formed with insufficient quantity and quality. At higher currents, there is a strong overheating of the surface even to the melting plots, the process resembled arc welding. Granulation of ferromagnetic powder in the selected limits for the total size and composition of the coating is not affected. However, the study of the continuity and roughness of the coating (Fig.1.) showed a direct dependence of these characteristics of granulation of the powder used.

Table 1. Indicators of continuity and roughness of coatings

| Granulation of ferro powder, micrometer | Percentage of the continuity of the machined surface, % | Indicators of roughness | |
|---|---|---|---|
| | | projections height limits of the cone contour, micrometer | the average value of the layer in the depressions, micrometer |
| 10-30 | 50-60 | 35-40 | 10 |
| 30-50 | 70-90 | 40-45 | 11 |
| 50-70 | 70-95 | 30-45 | 12 |
| 70-90 | 60-70 | 30-45 | 6 |

## 3. Discussion of results

According to the table, we can conclude that the optimal granularity of ferromagnetic powder is in the range of 30 to 70 micrometer. With less granulation emission of the powder from the working area takes place at the time of discharge, frequent "sticking" of the gap appears, the layer is characterized by great irregularity in the contour of a hardened surface, imperfections are observed during the process. At a higher granulation an incomplete melting of powder particles takes place, the surface roughness of the hardened surface increases, and imperfections are observed.

Studies have been conducted of the influence of discharge current on the qualitative characteristics of the coating. So, the increase in the discharge current leads to the increase of the applied layer value, but to values not exceeding 120 micrometer. Small currents do not provide the sufficient depth of the layer, although the continuity and roughness may be satisfactory. Too high currents increase the discharge power, it is possible to obtain a sufficiently large layer, but it can happen in reflow of surface area of the products. The surface is very rough, can contain the lapping, severe overheating, etc.

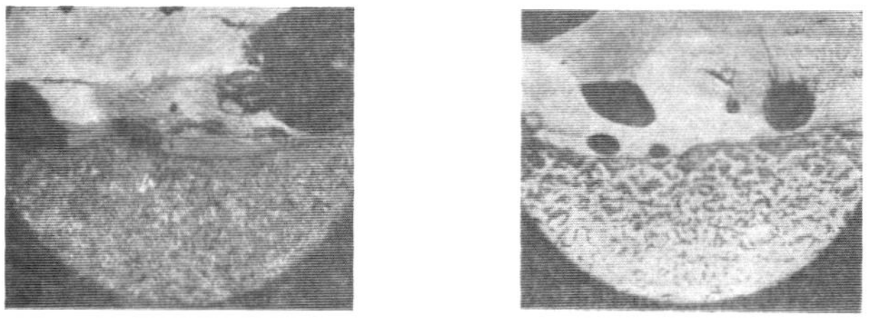

a) b)

**Figure 3.** Microstructure of layers obtained in the jet of the electric discharge for 0,45 carbon steel (a) and 0,3 carbon steel (b) I=50-70 A (x100)

Fig. 3 shows the microstructure of the layers obtained in the jet of the electric discharge between solid and liquid electrodes. The most even layer with the greatest continuity, the optimal roughness of sufficient magnitude was obtained at discharge currents of 50-70 A. The smaller currents do not provide the necessary stable results on these characteristics, and currents more than 70 A are applicable for items to which virtually no requirements of roughness and denseness are made, and cover is valid that is similar to the electric arc surfacing. In subsequent studies the discharge current is 50-60 A, as ensuring optimum quality of coating.

The effect of duration of coating process on the qualitative characteristics of the layer was examined at one, two, and three consecutive passes of the jet on the treated surface. Coating was carried out in the following optimal mode: the speed of movement of the treated samples was 5

IOP Conf. Series: Materials Science and Engineering **240** (2017) 012041 doi:10.1088/1757-899X/240/1/012041

cm/sec; feed rate of ferromagnetic powder - 6-7 g/min; granulation of ferromagnetic powder - 30-50 micrometer; working gap was 4-5 mm; discharge current - 50 to 60 A.

## 4. Conclusions

Studies have established that the complete formation of the layer is already done at the first pass of the sample. The secondary passage provides only a slight decrease of the defect, but at the same time slightly increases the surface roughness due to the "gripped" not melted particles of ferromagnetic powder.

## References

[1] Matrenin S V and Ovechkin B B 2010 *Nanostructured materials in engineering* (Tomsk Polytechnic University) p 186
[2] Kudinov V V 1977 *Plasma coatings* (Moscow: Nauka) p 184
[3] Dautov G Y, Nail K F, Viktor L, Renat G Z, Raphael G T, Fayrushin I and Kashapov R N 2013 Low Temperature plasma in the processes of functional coating preparation *J. of Phys.: Conf. Ser.* **479 (1)**
[4] Kashapov N F and Sharifullin S N 2015 Hardening of the surface plasma jet high-frequency induction discharge of low pressure *IOP Conference Series: Materials Science and Engineering* **012021**
[5] Khafizov A A, Valiev R I, Shakirov Yu I and Valiev R A 2014 Steel surface modification with plasma spraying electrothermal installation using a liquid electrode *J. of Phys.: Conf. Ser.* 567 **012026**
[6] Panteleenko F I, Lyalyakin V P, Ivanov V P and Konstantinov V M 2003 *Restoration of machine parts* (Moscow: Mashinostroenie) p 672
[7] Heifetz M L, Gretskiy N L and Kozhuro L M 2008 Electrical pulse of overlaying coatings of ferromagnetic powders *Strengthening technologies and coatings Scientific and technical journal "engineering"* 3 pp 51-57
[8] Shakirov Ju I, Valiev R I, Hafizov A A and Shakirova G Ju 2011 Mnogokanal'naja plazmennaja ustanovka s jelektroliticheskim katodom *Avtomobilnaja promyshlennost* 2 pp 36-38
[9] Gajsin F M, Son Je E 1990 *Himija plazmy* (Moscow: Atomizdat) 16 p 120
[10] Valiev R I, Gajsin Al F, Gajsin F M, Gumerov A Z, Nasibullin R T, Sadriev R Sh, Sarimov L R and Khafizov A A 2014 Some features of fine powders production of iron oxide in low-temperature plasma electric discharge with liquid cathode. *News of higher educational institutions. Physics* (Tomsk: Izd-vo Tomsk State University) 57 3/3 pp 66-69

IOP Conf. Series: Materials Science and Engineering 240 (2017) 012042 doi:10.1088/1757-899X/240/1/012042

# Practical-oriented teaching of gifted youth in the field of natural sciences

**F D Khalikova and S I Gilmanshina**
[1]Kazan Federal University, 18 Kremlyovskaya street, Kazan, 420008, Russian Federation

fidaliya.halikova@mail.ru, gilmanshina@yandex.ru

**Abstract**. In the article it is presenteds the author's concept of practice-oriented teaching of gifted adolescents to natural-science subjects on the example of chemistry. The main provisions of the concept are substantiated, on the basis of which individual educational trajectories have been developed. The essence of practice-oriented learning is revealed. Particular emphasis is placed on the formation of practical experience in applying theoretical knowledge to solve specific problems.

## 1. Introduction

Today, in the pedagogical science and practice of education, the problem of identifying, developing and supporting gifted youth is increasingly recognized. Educational institutions for gifted adolescents are being established [5], developing programs and technologies for teaching in schools [4] and universities [1, 3, 6] are being introduced, the results of scientific research of young scientists are published [2].

However, there is clearly not enough publications on the scientific generalization of the experience of practice-oriented instruction in the natural sciences of gifted young people. Working with gifted young people requires a clear scientific and pedagogical concept, the development of pedagogical conditions for the effectiveness of an individual practice-oriented educational trajectory. This article can be considered as one of the attempts to find ways to solve the above problem.

## 2. The basic part

To identify and further successful training of gifted youth requires a comprehensive study of their characteristics and abilities. This is necessary for the development of individual training programs, both for gifted adolescents, and for improving the skills of teachers working with them.

The methodological basis of our studies is the competence-activity approach, which assumes general cultural and personal development of the individual, the construction of the educational process, taking into account the individual age and psychological and physiological characteristics of gifted youth. This approach has allowed to develop the concept of practice-oriented training of gifted adolescents, based on the theses: "acceleration, deepening, enrichment, problematization". The concept was implemented in the IT Lyceum of the Kazan Federal University in the development of individual educational programs for teaching gifted young people to natural-science subjects.

The essence of the concept will be considered on the example of a curriculum in chemistry.

The first thesis of the concept - "acceleration" allows to take into account the needs and opportunities of the lyceum students, distinguished by the accelerated rate of development. In the Lyceum with gifted children, the teacher-trainer is engaged in the subject. He accompanies the training of a

IOP Conf. Series: Materials Science and Engineering **240** (2017) 012042 doi:10.1088/1757-899X/240/1/012042

teenager, helps to highlight his characteristics and develop an individual trajectory of learning. In consequence, he tracks the results of the activities of a teenager, creates a data bank. In addition, the teacher-trainer participates in the development of an innovative educational program. The program includes the following components: goals, thematic planning, informative (Olympiad) component of the program, use of innovative pedagogical technologies in the learning process. The developed subject (profile) individual trajectory of the development of a gifted adolescent involves monitoring the development of a teenager, taking into account his personality.

Realization of the second thesis "deepening" implies a more in-depth study of topics or different fields of knowledge, taking into account the interests (profile) of a gifted teenager. The deepening of the material is achieved by working in small groups, where dialogue is the basis of activity. Dialogue serves as a means of identifying the problem and the ways to solve it. At this stage, a full cycle of research activities is carried out: from obtaining in-depth subject knowledge and highlighting the educational and research problem to its applied solution.

The third thesis of the concept - "enrichment" - requires the use of innovative pedagogical technologies. The study of many topics takes place in accordance with the principle of integrating the profile subject with other subjects of the natural and humanitarian cycle. For example, integrated lessons of chemistry and English are conducted.

The implementation of the fourth thesis of "problematization" actualizes the development of the personality of a gifted teenager through the search for new meanings, alternative interpretations, the development of reflection. The lyceum should be aware of the importance of the problem of professional and social self-determination of young people in the context of continuity of secondary and higher education. He must be informed that for the implementation of the innovative individual educational trajectory for him all conditions are created.

An important factor in the implementation of the developed concept is the availability of a practice-oriented educational environment. Practically oriented environment, in our opinion, influences the formation, realization, disclosure, self-improvement of gifted youth. In general, the practice-oriented education is aimed at acquiring practical experience in order to achieve professionally and socially significant competencies. This ensures the motivation and involvement of students in the study of theoretical material for the successful solution of a practical problem.

In general education, the experience of activity implies a greater degree of experience in teaching and cognitive activity in lessons. The acquisition of experience is carried out within the framework of the traditional didactic triad of "knowledge-skills-skills" by developing practical skills in students. In practice-oriented learning, this triad is supplemented by a new didactic structure: "knowledge-skills-skills-experience".

Undoubtedly, the knowledge acquired by gifted young people in chemistry is characterized by the so-called "situational dependence". Otherwise, the students' knowledge does not belong to them, but to the situation in which they are usually required. In order for knowledge to become own, "subjectivation" is necessary. This means applying knowledge in non-standard situations. In general, the application of knowledge involves teaching students how to solve problems, both in typical and in new non-standard situations. In the conditions of practice-oriented learning, the skills of using new knowledge on the subject are formed to solve problems of various difficulties in non-standard situations.

The essence of practice-oriented learning is to build an educational process on the basis of the unity of the emotional-figurative and logical components of content in order to acquire new knowledge. An important point is the formation of practical experience in using this knowledge to solve specific vital problems. In this case, the inclusion of subject knowledge in the system of value knowledge, freely functioning in human life.

Thus, the developed concept is based on the following principles: training in the context of vital problems, professional and applied orientation, intersubject communications. The concept of "acceleration, deepening, enrichment, problematization" considers the interaction of the teacher and the student as two interlocutors, partners who have the right to make their own decisions. As a result, motivational, synthetic (constructive), innovative stages are always highlighted in the lessons. The presence of these interconnected stages allows satisfying the interests of students with left-brain thinking (the material presented in the form of schemes, tables, algorithms is preferable), and with right-brain thinking (the form of presentation of the material in the form of images, associations, analogies).

Thus, in the developed concept of training, logical and figurative thinking is used, which helps to increase the personal status of the adolescent. Training becomes motivated, individually oriented. The social function of education is realized, students are provided with applied knowledge and skills. The formation of a practice-oriented educational environment is provided by the developed practical-oriented situational tasks. These tasks begin with the problematic "why". The process of solving such problems significantly increases the interest of students. They are easily involved in the discussion, the dialogue.

The third year in the IT-lyceum for gifted adolescents of the Kazan Federal University in the lessons of chemistry is practiced practice-oriented training. The work program was compiled by the teachers of the Chemical Education Department named after A.M. Butlerov Kazan State University. Students of the 10th, 11th grades of the lyceum of the chemical and biological profile visit the university's laboratory once a month for practical classes. Students of 10 classes are engaged in the analysis and synthesis of organic compounds, solve experimental problems, identify organic compounds. Lyceum 11 classes receive gases and study their properties, get acquainted with the methods of purification of substances.

Thus, there is a systematic formation of practical experience in the use of theoretical chemical knowledge in solving specific vital problems. At the same time, the subject knowledge is included in the value system of the individual educational trajectory of the lyceum student.

***The conclusion.*** On the basis of the competency-activity approach, the author's concept of practice-oriented teaching of gifted adolescents to natural-science subjects on the example of chemistry was developed and implemented. The main provisions of the concept are substantiated, which can be presented in the abstract as "acceleration, deepening, enrichment, problematization". The essence of practice-oriented learning, consisting in the construction of the learning process on the basis of the unity of the emotional-figurative and logical components of the content with the purpose of acquiring new knowledge, is disclosed. Particular emphasis is placed on the formation of practical experience in solving specific vital problems.

## References

[1] Gilmanshin, I.R., Gilmanshina, S.I. 2016 The formation of students' engineering thinking as a way to create new techniques, technologies, materials *IOP Source: International Scientific-Technical Conference on Innovative Engineering Technologies, Equipment and Materials* Vol 134 **012006**

[2] Gilmanshin I R, Kashapov N F, Gilmanshina S I, Galeeva A I 2016 Landfill energy complex based on the renewable energy installations *IOP Conference Series: Materials Science and Engineering* Volume 134 Number 110 **012007**

[3] Gilmanshina S I, Gilmanshin I R 2015 Building axiological competence of graduate students by means of project-based learning *IOP Conference Series: Materials Science and Engineering* Vol 86 (1) **012029**

[4] Gilmanshina S I, Gilmanshin I R, Sagitova R N, Galeeva A I 2016 The Feature of Scientific Explanation in the Teaching of Chemistry in the Environment of New Information of School Students' Developmental Education *International Journal of Environmental and Science Education* Volume 11 Issue 4 pp 349-58

[5] Gilmanshina S I, Khalikova F D 2016 Teaching Gifted Adolescents in Terms of the Transforming Natural Sciences Education *Ifte 2016 - 2nd International Forum on Teacher Education* Vol 12 pp 50-54

IOP Conf. Series: Materials Science and Engineering **240** (2017) 012042 doi:10.1088/1757-899X/240/1/012042

[6] Gilmanshina S I, Sagitova R N, Gilmanshin I R 2016 Innovative Components for Preparing Graduates and Master's Degree Holders of Natural Sciences, for Education at the University Level *Ifte 2016 - 2nd International Forum on Teacher Education* Vol 12 pp 55-59

IOP Conf. Series: Materials Science and Engineering **240** (2017) 012043 doi:10.1088/1757-899X/240/1/012043

# Engineering Background: Modern Formats and Challenges of Conceptual Engineering

**A F Khamidullina, M A Kuzmina, E M Khusnutdinova and I A Konakhina**
Kazan Federal University, 18 Kremlyovskaya street, Kazan, 420008, Russian Federation

rr-088@mail.ru

**Abstract**. This paper describes the analysis of problems and development perspectives of engineering education in our and other countries. Special attention is given to modern formats of education that motivate creative efforts of engineers-to-be as well as issues of conceptual engineering taking the challenges of modernity into account.

As the global economy has been declining over last 10-20 years, which overcoming requires complete economic conversion worldwide and escalates the competition between the key players, this dramatically changes the policy of a country related to engineering shape and skilled personnel training. Recent years have focused the national education policy of the Russian Federation on the mainstream development of engineering education. The reference figures stating the number of enrolments to study engineering sciences in higher educational institutions grew steadily and exceeded 50% of the total reference amount by 2015. The RF Ministry of Education and Science records demonstrate that the issue related to the deficiency in the number of engineers has been solved by that time, while the issue of their skills is acute until now.

In general, the issue of education quality is typical of both the human and engineering fields while training new generation of specialists for national industry and applied science. The engineering field is going to be exactly that breakthrough that will be able to push the country out of the slump, bring our industry to a new technological level and reliably protect from any aggressive attempts from outside through introduction of new types of equipment.

It should be stressed that lack of highly skilled engineers is true not only for Russia. In late 1990$^{th}$, 'brain drain' has started meaning emigration of gifted specialists in economically developed countries of Europe and America. Inadequate attention to such a complicated process as training of engineering personnel and efforts to allocate the productions in the areas with low labour cost, required to move first the training centres behind the technologies while this, in turn, resulted in such a situation that the young people has less and less interest to do these jobs. The resulting shortage has been covering by attraction of foreign specialists.

Within a long period of time, we, in Russia, faced a phenomenon called bleeding of our science as the young professionals did not want to commit the works they have been trained to while the teachers at universities and institutes became aged. Our young specialists went abroad or did other, more challenging businesses – trade, commerce etc. This was encouraged by a drastic shortfall of backing higher institutions followed by drop in creditability of academicians. Growth in backing observed for the last 10-15 years could not make it possible to completely recover the losses for the time being; however, at least it is noticeable that the situation started to improve.

According to increased number of publications by immigrants of Indian, Chinese and other Asian and Middle East descents, building up new global engineering and scientific centres is another worldwide trend. This may be considered because of the above described processes brought to a fundamentally new level from the quality point of view, which cannot be ignored any more.

Therefore, Russia is currently facing two challenges: short-term (extremely aggressive) – the western countries losing their benefit and searching for a possibility to find a way out of the protracted downfall for the account of others, and long-term, which is associated with the growth of new centres of economical and technological influence.

Not to fall behind the others, our country should have an integral philosophy of engineering education as a system of training qualified human resources in demand who can contribute to a new round of development.

This appears to happen only based on a method of conceptual engineering and strategic planning [1]. Particularly, [2] demonstrates that "...conceptual engineering of the systems is specifically actual in terms of their innovation-driven growth". In other words, conceptual engineering makes conditions for creation of qualitative transitions between the systems (including the system of engineering education) resulting in modification of structural organization of these systems.

The purpose of conceptual engineering is extremely complicated and requires a comprehensive approach [3], study of philosophy components and their degree of contribution to achieve the goal. As is written in the papers of those who studied this problem, one of the basic components of the engineering education philosophy is the development of an enduring interest to be shown by young specialists in choosing this domain of interest, that is of engineering. This is important to be done in time as the creativity, artistic skills, inventive thinking are mostly formed by 15 years of age [3]. And then we have to do with what we have and make efforts not to lose, but to improve and realize.

Popularization of engineering achievements and applied science plays a valuable role in this matter. Not for nothing is this sphere observed to actively look for non-conventional forms and non-trivial solutions. This includes up-to-date technology parks, exhibitions where the people can be directly engaged in the creative process, contests, and studio performances. The movements of stand-up interactions in groups of likeminded persons becoming more popular where the speakers have live communication with the audience at regular club meetings, on which only 5-10 minutes are given for discussion of a single topic to activate thinking process, concentrate the people on it and prevent from feeling bored.

One of the active players is the Ingenium, a studio of technically creative young people at the Institute of Engineering of the Kazan Federal University, which successfully apply new formats of popularization of creative engineering and grow high potential staff for national industry, science and pedagogics.

The studio has been established by a group of enthusiasts namely students, postgraduates and lecturers of the Institute who are inspired to engineering innovations and popularization of applied researches.

However, the matter of subject the Studio is specializing on has demanded not only the active intellectual and publication activity, but also involvement of significant means to collect material resources to implement and test research and engineering ideas. These ideas may include effective and ineffective ones, which is normal practice when searching for new solutions. Even in case of failure, realization of an engineering idea cannot be considered useless as it gives invaluable practical experience and feeling of possibility to improve and optimize someone's ideas and plans.

In 2015, a project developed by Ingenium Studio was granted by PAO Lukoil and OAO RITEK. There were three groups of targets to be addressed:

1 . Material support
2 . Organisational issues
3 . Creative tasks

The first group of targets covered the issues related to the procurement of modern equipment and consumables for individual projects of the Ingenium Studio members.

The second group dealt with how the Ingenium Studio is organized as the number of members has noticeably increased over the last period; the range of its interests and capabilities has extended, however, the matter of how to employ emerging technology and get skilled in using the sophisticated technological means was not gone.

The third group relates to the creative endeavor aimed to popularize technical achievements of the Ingenium Studio and involve new participants and partners as the studio's members focus on how to arrange for networking to distribute and popularize the best successful experience and provide guidelines of how to arrange alike studios in the Republic of Tatarstan.

To support Ingenium Studio in finding solutions to the organizational problems, skilled lecturers and employees of the Institute of Engineering at Kazan Federal University trained the studio's participants, advised them in terms of how to draft a design of training and laboratory trials and explained the sequence, targets, purpose and methods of engineering and design works. From creative process point of view, the implementation of the Ingenium Studio's design encouraged the aspiring designers to have a technologically sophisticated way of thinking, commitment to results, decision-making discretion as well as innovations, inventiveness, skills required not only for an individual, but also for teamwork.

**References**

[1] Aleksandrov E A 1975 *Foundations of heuristic solution theory* (M.: Sovetsloye radio) p 254

[2] Butenko D V 2003 *Relation between strategic planning and conceptual engineering* (XXX Anniversary conference and discussion club IT+SE`2003 New information technologies in science, education, telecommunication and business p 107

[3] Kondakov I M 2016 *Problems of engineering education and university-level science: engineer is a human being who invents* (Russian engineering society [Electronic resource]. http: //rnto.club/biblioteka/kondrakov-i.m./Stati/problemi-inzhenernogo-obrazovaniya.html (Accessed date: 22.10.2016)

IOP Conf. Series: Materials Science and Engineering **240** (2017) 012044 doi:10.1088/1757-899X/240/1/012044

# Mathematical modeling of a running cycle of the diesel engine with the additive of water at the intake

**V B Khlyupin**

The Naberezhnye Chelny Institute (branch) of Kazan (Volga) Federal University, 423800, Russian Federation, Republic of Tatarstan, Naberezhnye Chelny, Mira Av., 16b

VBHljupin@kpfu.ru

**Abstract.** Mathematical modeling of a running cycle of a diesel engine with the additive of water at the intake.

Modern internal combustion engines (ICE) are high-powered engines with an account on average effective pressure. This leads to an increased thermal stress of components. One of the ways to reduce the thermal stress on engine components is the water injection into the intake pipe. Injected water due to vaporization, cools the surrounding surface. A modern vehicle that uses a water injection system in the intake pipe, is the BMW M4 GTS [1]. The system is applied in six-cylinder internal combustion engine with forced ignition. Its main task is to improve fuel efficiency, reduce emissions of carbon dioxide $CO_2$ and thermal stress of engine components. During testing it was found that water injection in the intake pipe helps to reduce $CO_2$ emissions by 4% and fuel consumption by 13%. The consumption of water is several hundred milliliters per hundred kilometers. The car has a storage tank of 5 liters of water, additional pouring of the water is required every 3 thousand kilometers. When the engine stops working, the water flows into the tank. At negative ambient temperatures, water may freeze without damaging the tank. After the engine resumes operation, the water gradually thaws and flows back into the intake pipe. If the water supply cannot be filled-up, the engine will operate with reduced cyclic fuel delivery and reduced torque [2].

The thermal stress on the parts of diesel engines is even more than in internal combustion engines with forced ignition. Modern values of maximum combustion pressure $p_z$of the diesel engines reach 20-25 MPa. So it is necessary to increase the metal consumption of the designin orderto increase the strength of components. Therefore, from this point of view, the study of the working cycle of the diesel engines with the addition of water at the intake is quite an urgent task.

When adding the water at the intake it is necessary to make the following changes in the calculation of the working cycle of a diesel engine.

The mass of fresh charge is increased by the mass of injected water $G_{water}$,kg

$$G_1 = \alpha \cdot l_0 \cdot G_f + G_{water}$$

where α – coefficient of excess air; $l_0$– is the mass of air required for complete combustion of 1 kg of fuel, kg of air/kg of fuel;$G_f$ – the weight of fuel, kg; $G_{water}$–is the mass of the added water intake, kg.

The heat transfer coefficient is determined by G.Voshni'sformula [3], W/(m$^2$·K)

$$\alpha_T = 819{,}5 \cdot D^{-0{,}2} \cdot p^{0{,}8} \cdot T^{-0{,}53} \cdot W_s^{0{,}8}$$

where D – is the cylinder diameter, m; p – pressure in cylinder, MPa; T – temperature in the cylinder, K; $W_s$ – the speed of the working body, m/s.

The amount of water in the exhaust gas is changed to the value of water supply $G_{water}$, kmol/kg

$$M_{H_2O} = \frac{H}{2} + \frac{G_{water}}{\mu_{water}}$$

where H – is the amount of hydrogen contained in the fuel, kmol/kg; $\mu_{water}$– is the molar mass of water, kg/kmol.

Heating of the fresh charge at the intake $\Delta T = T - T_{wall}$ is determined by the formula of Newton-Richman

$$dQ = F \cdot (T - T_{wall}) \cdot \alpha_T \cdot dt$$

where $dQ$– is the heat supplied to the fresh charge, J; $F$ – the area of the walls of the cylinder, m$^2$; $T$, $T_{wall}$– is the temperature of the fresh charge and the cylinder walls, accordingly, $K$; $dt$– is the time during which a fresh charge is in contact with the walls of the cylinder, s.

The exponent ofcompressionpolytrope$n_1$

$$n_1 = 1 + \frac{8{,}314}{mc_v + d \cdot mc_{vH_2O}}$$

where$mc_v$, $mc_{vH2O}$– heat capacity of air and water vapor, kJ/(kmol·K); $d$ – the water content in the fresh charge, kg of water/kg of air.

When injecting the water, part of the heat resulted due to combustion of fuel, will be wasted on heating the water, turning it into steam and steam overheating. Thus, calorific value of fuel $q_z$, kJ/kg, isexpressed as

$$q_z = \frac{Hu \cdot G_f}{(1+\gamma_g) \cdot \alpha \cdot l_0 + 1} - \frac{c_{water} \cdot t_{water} \cdot G_{water}}{(1+\gamma_g) \cdot G_1} - \frac{q_v \cdot G_{water}}{(1+\gamma_g) \cdot G_1}$$

where$Hu$– is the low calorific value of fuel, kJ/kg; $\gamma_g$– is the coefficient of residual gases; $c_{water}$– specific heat of water, kJ/(kg·K); $t_{water}$– water temperature, K; $q_v$– is the specific heat of vaporization of water, kJ/kg.

The fuel burn $x$ is determined using the updated formula of I.I. Vibe for a two-phase combustion process [3]

$$x = 1 - \exp\left[-6{,}908 \cdot \left(\frac{\varphi}{\varphi_C}\right)^{m_1 - m_2} \cdot \left(\frac{\varphi}{\varphi_Z}\right)^{m_2+1}\right]$$

where$\varphi$ – the current angle from beginning of ignition of the fuel, the degree of crankshaft angle($^0$d.c.a.); $\varphi_c$ – the angle corresponding to the end of the kinetic phase of combustion, $^0$d.c.a.; $\varphi_z$– duration of combustion.$^0$d.c.a.;$m_1$, $m_2$ – indicators of the kinetic and diffusional nature of fuel combustion.

The period of the ignition delay is calculated by the formula of A. Tolstov [4]

$$\tau = \sqrt{\left(\frac{V_\Theta - V_c}{V_{zw}} + \frac{1}{\varepsilon}\right) \cdot \frac{T_0}{p_0}} \cdot \frac{B}{100} \cdot \exp\left(\frac{V_\Theta - V_c}{V_{zw}} + \frac{1}{\varepsilon}\right)^{n_1 - 1} \cdot \frac{E}{R \cdot T_0}$$

where$V_\theta$, $V_c$, $V_{zw}$– specific volumes of the working fluid in the cylinder at the beginning of the process of fuel injection, at the end of compression process and at the closure of the intake valve, accordingly, m$^3$/kg; $n_1$ – the exponent of compressionpolytrope; $\varepsilon$ – compression ratio; $p_0$, $T_0$–the atmospheric pressure and air temperature, MPa and K; $R$– is the universal gas constant of air, kJ/(kmol·K); $E$ – the conditional activation energy, MJ/kmol; $B$– is the coefficient associated with the frequency of rotation of the crankshaft.

The change in molecular weight $\mu$ of a water-air mixture in the combustion process

IOP Conf. Series: Materials Science and Engineering **240** (2017) 012044 doi:10.1088/1757-899X/240/1/012044

$$\mu = 1 + \frac{\left(\frac{H}{4} + \frac{O}{32}\right) \cdot \mu_{wa}}{\alpha \cdot L_0 \cdot (1 + \gamma_r)}$$

whereO – is the amount of oxygen contained in the fuel, kmol/kg; $\mu_{wa}$– is the molar mass of water-air mixture, kg/kmol; $L_0$– the amount of air required for complete combustion of 1 kg of fuel kmol of air/kg of fuel.

The specific heat ratio of the working fluid for any intermediate point of the combustion process will be determined by the formula

$$k = k_{wa} - (k_{wa} - k_g) \cdot x$$

where$k_{wa}$– is the specific heat ratio of water-air mixture; $k_g$– specific heat ratio of the productscombustion; $x$– is the proportion of the burnt fuel to the moment time.

The analysis of the computational studies revealed that significant impact on the performance of the working cycle of a diesel engine is due to a change of the heating fresh charge at the intake of ΔT. With water injection into the intake pipeof the diesel evaporation (vaporization) of water leads to a decrease in intake air temperature. The evaporation of 10 g of water per 1 kg of air reduces its temperature by 25°C. As a result, cylinder filling with fresh charge improves and the value of ΔT reduces. If you do not take this into account, the power and fuel-economic indicators of diesel with the addition of water intake will worsen.

**References**


[1] «BMW» [website]. URL. http://www.bmw.ru/ru/all-models/m-series/m4-coupe/2014/m4-gts.html (accessed 20.09.2016)

[2] Group of companies «Bosch» [website]. URL. http://www.bosch-mobility-solutions.com/en/powertrain-electrified-mobility/water-injection (accessed 20.09.2016)

[3] Kavtaradze R Z 2008 *Theory of piston engines* Special Chapters: the textbook for high schools. M.: MSTU named after N E Bauman Publishing p 720

[4] Orlin A S 1983 *Internal combustion engines. Theory of piston and combined engines* Moscow: Mashinostroenie p 372

IOP Conf. Series: Materials Science and Engineering **240** (2017) 012045 doi:10.1088/1757-899X/240/1/012045

# Modeling of forming technological errors in processing by gear shaping machine

**R M Khusainov[1] and R R Khaziev[1]**

[1]Department of Design and Technology Software Engineering Industries. Naberezhnye Chelny Institute of Kazan Federal University, Naberezhnye Chelny, 423812 Mira Avenue, 68/19, the Russian Federation

rmh@inbox.ru

**Abstract.** In the paper, the mathematical model for evaluation the accuracy of the cut gear depending on the initial error of the technological system, is presented. The model is based on the coordinate transformation matrices, variations matrices and considers the deviations of the base and the real surfaces. The technique allows evaluating the possibilities of the technological system to achieve the specified accuracy.

Keywords: parametric failure, technological reliability, gear, mathematical model, error.

**1. Introduction**. The main purpose of mathematic modeling of technological systems is to evaluate their performance, as well as the possibility of providing the specified parameters [1,2]. When considering the processing of gear wheels, especially by the generating method, in the first place stands the problem of evaluating the output accuracy of the technological system [3]. This problem is exacerbated by the fact that it is difficult to reveal any clear correlation between geometric and kinematic errors of the technical system, and indicators of accuracy of the machined gear wheel [4,5,6]. To solve this problem it is necessary to apply a direct mathematical modeling of the formation of errors.

**2.Modeling of generation of geometry of evolvement surface.** Generation of geometry in this case can be represented as the transformation of movements of the tool through the corresponding coordinate transformation matrices of the technological system's elements [7]. The matrix-column that describes a perfect treated surface can be written in the form:

$$r_0 = A^D \cdot A^X \cdot A^Z \cdot A^C \cdot r_{cs} \quad (1),$$

where$r_0$is the coordinate of the perfect (nominal) surface of the work piece; $A^D$ – transformation matrix from the tool rotation around its axis, $A^X$ - transformation matrix on the X-coordinate (distance between the tool axis and the workpiece), $A^Z$ - transformation matrix on the Z- coordinate (vertical movement of ram),$A^C$ - transformation matrix from the rotation of the workpiece around its axis; $r_{cs}$- vector of coordinates of the surface of the cutting tool [8].

Calculating the product of the above matrices, we get a vector of the perfect surface:

$$\text{where } r_0 = \begin{bmatrix} X\cdot\cos(C) - r_b\cdot\sin(C) - C\cdot i\cdot r_b\cdot\cos(C) \\ X\cdot\sin(C) + r_b\cdot\cos(C) - C\cdot i\cdot r_b\cdot\sin(C) \\ z \\ 1 \end{bmatrix} \qquad (2).$$

**3. Modeling of errors of the technological system.** During operation, the machine-tool is influenced by lots of geometric, force, temperature factors. This leads to disruption of the relative position and movement of machine units, including working bodies. Consequently, the actual coordinates of the point of the treated surface are different from the position established by equation (2). Simulation of the resultant error on the output of the machine-tool is done by adding to the equation a perfect surface of the matrices describing the errors of each element of the machine-tool:

$$\Delta r = (\delta A^D)A^D\cdot A^X\cdot A^Z\cdot A^C\cdot r_{cs} + A^D\cdot(\delta A^X)A^X\cdot A^Z\cdot A^C\cdot r_{cs} + A^D\cdot A^X\cdot(\delta A^Z)A^Z\cdot A^C\cdot r_{cs} + A^D\cdot A^X\cdot A^Z\cdot(\delta A^C)A^C\cdot r_{cs} \quad (3),$$

where$\delta A^D$ – the matrix describing the effect of the error of the workpiece's rotation; $\delta A^X$-the matrix describing the effect of the error on the relative positions of the centerlines of spindle of the tool and spindle of the workpiece; $\delta A^Z$- ;the matrix describing the effect of the movement of the ram along the Z axis; $\delta A^C$-the matrix describing the effect of the error of the drive of the shaping cutter rotation.

The output of the model is obtained the matrix-column that describes the deviation of parameters of the treated surface.

**4. Bringing the output error of the machine tool to the geometry of the formed surface.** The errors obtained by forming the gear can be regarded as the error of shape and position of the involute surface. For their definition it is necessary to define the base surface $q_0$and form a vector of dimensional parameters of this surface. Properties of involute are adequately defined by a single parameter – a diameter of the base circle, for convenience, we reduce it to the parameter of the radius of the base circle $r_b$.

The offset of the radius of the base circle of the workpiece $\Delta r_b$ is defined as the sum of the error vectors of position and size [7]:

$$\Delta r_b = \varepsilon_b\cdot r_0 + dr_0 \qquad (4),$$

where$\varepsilon_b$ is the matrix of error of the position of the coordinate system associated with the base surface relatively the system in which the equation was set.

$$\varepsilon_b = \begin{pmatrix} 0 & -\delta B_b & 0 & 0 \\ \delta C_b & 0 & -\delta A_b & 0 \\ 0 & \delta A_b & 0 & 0 \\ 0 & 0 & 0 & 0 \end{pmatrix} \qquad (5).$$

The components of this matrix are small displacements of the coordinate system of the base surface relative to the coordinate system of the nominal surface for the corresponding axes.

$dr_o$ - the total differential of the radius vector$r_0$, taken for the components of the vector $q_0$ of dimensional parameters of the surface:

$$dr_0 = \frac{\partial r_0}{\partial r_b} \qquad (6),$$

where- $\frac{\partial r_0}{\partial r_b}$ - the vector of the partial derivative of the matrix nominal surface (2) with variable $r_b$.

Due to the small values of errors of moving machine units' position in a shape-generating system of the machine, $\Delta r_b$ there is a total differential $r_b$ for the components of the vector $q$, and therefore the formula (9) can be represented as:

$$\Delta r_b = G\cdot\Delta q \qquad (7),$$

Where $G$is the matrix of the 4-th order composed of the matrix-columns of partial derivatives $\frac{\partial r_b}{\partial q_i}$

$$G=\begin{bmatrix} C\cdot i\cdot r_b\cdot\sin(C)-r_b\cdot\cos(C)-X\cdot\sin(C) & 0 & -\sin(C)-C\cdot i\cdot\cos(C) \\ X\cdot\cos(C)-r_b\cdot\sin(C)-C\cdot i\cdot r_b\cdot\cos(C) & -z & \cos(C)-C\cdot i\cdot\sin(C) \\ 0 & r_b\cdot\cos(C)+X\cdot\sin(C)-C\cdot i\cdot r_b\cdot\sin(C) & 0 \\ 0 & 0 & 0 \end{bmatrix} \quad (8).$$

The next phase of construction of the base surface is the calculation of the components of the vector $\Delta q$ of errors in position and size. For the average quadratic base surface, the parameters are found by minimizing the sum of squares of deviations of the actual surface $r$ from the base surface $r_b$.

The components of the vector $\Delta q$ can be found from the matrix equation [7]:

$$\Delta q=\frac{d}{H} \quad (9),$$

where $H$ is the matrix of 4-th order with the elements $h_{ji}$:

$$h_{ji}=\int_0^C\int_0^z f_j f_i\cdot n\,\mathrm{dzdC} \quad (10),$$

$fk$, $fi$ – $k$ and $I$ are the coordinates of the vector $f$ of normal transfer coefficients.

The normal vector to the nominal surface is determined by the equation:

$$n=\frac{\partial r_0}{\partial C}\times\frac{\partial r_0}{\partial z}=\begin{bmatrix} \dfrac{r_b\cdot\sin(\mathrm{C})-\mathrm{X}\cdot\cos(C)+i\cdot r_b\cdot\sin(C)+C\cdot i\cdot r_b\cdot\cos(C)}{\sqrt{r_b^{\,2}\cdot(i+1)^2+(X-C\cdot i\cdot r_b)^2}} \\ \dfrac{r_b\cdot\cos(\mathrm{C})+\mathrm{X}\cdot\sin(C)+i\cdot r_b\cdot\cos(C)-C\cdot i\cdot r_b\cdot\sin(C)}{\sqrt{r_b^{\,2}\cdot(i+1)^2+(X-C\cdot i\cdot r_b)^2}} \\ 0 \\ 0 \end{bmatrix} \quad (11).$$

Determine the vector of normal transfer coefficients:

$$f(z,C)=G^T\cdot n \quad (12).$$

Using the components of the vector $f(z,C)$ we calculate the components of the matrix $H$.

Define $\Delta r_n$- the projection of the vector $\Delta r$ on the normal to the involute surface:

$$\Delta rn(z,C)=\Delta r(z,C)^T\cdot n \quad (13).$$

Determine the matrix-column $d$ according to the formula:

$$d(z,C)=\begin{bmatrix} \int_0^C\int_0^z f(z,C)_0\cdot\Delta rn(z,C)dzdC \\ \int_0^C\int_0^z f(z,C)_1\cdot\Delta rn(z,C)dzdC \\ \int_0^C\int_0^z f(z,C)_2\cdot\Delta rn(z,C)dzdC \\ \int_0^C\int_0^z f(z,C)_3\cdot\Delta rn(z,C)dzdC \end{bmatrix} \quad (14).$$

On the basis of calculated values, determine the vector $\Delta q$ of deviations of the base surface relative to the nominal [7]:

IOP Conf. Series: Materials Science and Engineering **240** (2017) 012045 doi:10.1088/1757-899X/240/1/012045

$$\Delta q(z,C) = H^{-1} \cdot d(z,C) \tag{15}$$

As estimates of the errors in the size and position of the treated surface we use elements of the vector $\Delta q$ determined from the formula (15). These components of the vector can be counted into standard indicators of accuracy of the cut gear wheel.

**5. Conclusion.** Thus, by using this mathematical model one can calculate the deviation of the real surface of the cut gear from nominal considering the initial errors of the technical system.

## References

[1]Govorkov A S and Zhilyaev A S 2016 The Estimation Technique Of The Airframe Design For Manufacturability *IOP Conf. Ser.: Mater. Sci.Eng.*vol 124 1 **012014**

[2] Khisamutdinov R M and Khisamutdinov M R 2014 Automation System Goals For The Creation And Operation Of The Tool *IOP Conf. Ser.: Mater. Sci.Eng.* vol 69 **012021**

[3] Chemborisov N A and Kondrashov A G 2008 End Machining of Teeth by a Single-Turn Worm Mill *Russ. Engin. Res.* vol 28 p 809-11

[4] Golovko A N and Golovko I V 2013 Errors In The Gear-Tooth Profile In Shaving *Russ. Engin. Res.*vol 33 p 309-11

[5] Golovko A N and Golovko I V 2013 Optimal Design Parameters Of A Shaving Hob For Compensation Of The Systematic Error In The Gear-Tooth Profile*Russ. Engin. Res.* vol 33 p 427-8

[6] Petukhov Yu E, Kolesov N V and Yurasov S Yu 2014 Geometric Shaping In Cutting *Russ. Engin. Res.* Vol 34 p 374

[7] Reshetov D N and Portman V T 1986 *Accuracy Of Machine Tools* (Moscow: Engineering)

[8] Fetvaci C2010 Generation Simulation of Involute Spur Gears Machined by Pinion-Type Shaper Cutters *Strojniškivestnik - Journal of Mechanical Engineering* vol 56 p 644-52

IOP Conf. Series: Materials Science and Engineering **240** (2017) 012046 doi:10.1088/1757-899X/240/1/012046

# Comparative Analysis of Thermohydraulic Performance of Enhanced Viscous Oil and Oil Product Heaters

**E M Khusnutdinova, I A Konakhina and A F Khamidullina**

Kazan Federal University, 18 Kremlyovskaya street, Kazan, 420008, Russian Federation

rr-088@mail.ru

**Abstract**. This paper covers the analysis of positive and negative effects produced when enhanced helical hollow beams are installed into oil and oil product heaters if compared to the baseline designs. Special attention is given to the consideration of effect on the complex rheology of high-paraffin oils.

Performance indicators that show actual expenditures spent for fabrication and operation of new heat exchanging equipment are generic indicators of how such equipment, specifically oil heaters, are efficient in use. However, difficulties are usually faced in practice when it is required to estimate benefits, either positive or negative, of several competitive enhanced designs. For getting inputs, substantial material and power consumption, its in-place tests, comprehensive performance tests are required to produce a heat exchanger. Therefore, the decision-making stage demands for methods to adequately represent actual figures of thermo-hydrodynamic processes in heat exchangers and select most suitable options.

The heat-exchange efficiency evaluation methods used depend on a thermal resistance ratio of the fluids and can be uni- and bidirectional subject to relations of these measures. Unidirectional means when the rate of heat exchange in the surface heat exchangers is much less on one side of the medium than another one. The side with lower heat transfer coefficient should be taken into consideration. Such cases may include almost all heat exchangers-preheaters of viscous fluid being limiting factor, it means oil or oil products.

For evaluation of how the geometry of an enhancer influences the performance and efficiency of a heat exchanger, a method proposed by M.V. Kirpichyov can be applied [1]. It is based on the determination of the energy coefficient $E_0$ that shows relation between two types of energy – specific heat production $Q = \alpha\Delta t$, W/(m$^2$), and specific energy consumptions to overcome hydraulic friction $N_0$ that are produced for an unit of heat-exchange surface in upgraded thermal elements when the basic design is compared with those of the below ratio, W/m$^2$:

$$N_0 = \frac{G\Delta p}{(\rho F)} \tag{1}$$

where $G$ is the fluid mass flow, kg/s; $\rho$ is the fluid density, kg/m$^3$; $\Delta p$is the hydraulic losses produced by the friction between the fluid flow and the surface, Pa/m; $F$ is the heat exchange area for 1 m of thermal element length, $m^2$.

Thus, a dimensionless energy coefficient can be calculated by the formula:

$$E_0 = \frac{\alpha \Delta t}{N_0} \tag{2}$$

where α is the coefficient of heat transfer from the surface towards the fluid flow, W/($m^2$ K); $\Delta t$is variation of the inlet and outlet flow temperature averaged by the cross-section, K.

This method allows evaluation of how the enhanced surfaces of heat-exchangers-oil and oil product heaters are effective both at the design stage and during the commercial tests.

However, the Guckman criteria calculation should be performed to estimate effect of particular type of operating expenses for the entire performance of the heat exchanger. These criteria are used to measure the following:

1) Saving of heat energy (hot water or vapour) consumed for heating purpose;
2) Reduction in power consumption for carrying fluids through the heat elements;
3) Lowering specific amount of metal per unit of equipment due to decrease of heat exchange surface required.

Conditioning the competitive conditions of the options being discussed is an important factor to achieve an objective correlation. [1, 2, 3]. For this, three characterizations of heat exchange surface are used: heat flow $Q$, kW; wattage required to flush the coolant $N$, kW; heat exchange area $F$, $m^2$. Three dimensionless criteria of efficiency are applied to achieve comparative conditions of thermo-hydraulic processes going on the enhanced heat exchange surfaces of different configuration. These are:

1) Owing to heat flow $K_q$;
2) Owing to energy used for flushing the coolant $K_n$;
3) Owing to heat exchange area $K_f$.

Generally, the above criteria are described by the following ratios:

$$\mathrm{K}_q = \frac{\alpha_e F_{e\Delta} t_e}{\alpha_0 F_{0\Delta} t_0}; \tag{3}$$

$$\mathrm{K}_n = \frac{G_e \Delta p_e \rho_0}{G_0 \Delta p_0 \rho_e}; \tag{4}$$

$$\mathrm{K}_f = \frac{F_e}{F_0}, \tag{5}$$

where indices "0" and "e" stand for the type of heat exchange surface of the respective parameter, — baseline or enhanced; $\alpha$is the heat transfer coefficient, kW/($m^2$ K); $\Delta t$is the difference between the inlet and outlet flow temperatures averaged by the cross-section of a 1-m long heat element run, K; — related hydraulic losses, Pa; $F$ is the heat exchange area, $m^2$, for 1 m of the element length; $G$ is the mass flow of the coolant, kg/s; $\rho$is the fluid density, kg/$m^3$ .

Since the benefits of the enhancer may be obvious not within the entire range of the oil or oil product heater performance, the entire range is subject to calculation and the output will be visualized in charts and reduced to the analytical or criterial ratios: .

$$\mathrm{K}_q = f(Re); \tag{6}$$

$$\mathrm{K}_n = f(Re); \tag{7}$$

IOP Conf. Series: Materials Science and Engineering **240** (2017) 012046 doi:10.1088/1757-899X/240/1/012046

$$K_f = f(Re). \tag{8}$$

If screw-type enhancers are in use, the rate of heat transfer and hydraulic losses in the enhanced laminar flow passage matches the rate of heat exchange and hydraulic losses occurred in the turbulent flow of bare tube passage. Therefore, the analysis should consider all typical points of flow regime variations and select appropriate analytical or criterial equation. If the analysis is based on the charts, such kind of issues is usually not true, but the charts cannot be used for computations. Anyway, they are subject to conversion to either criterial or analytical equations.

Medium paraffin oil produced for the Yuzhno-Khylchuyuckoye oil field with 4.61% mass per cent of paraffin wax is considered as a simulation fluid to attack the above targets. The fluid rheology is described in [4].

The intensifying vortex generators consist of a twisted tape insert and helical round-shaped wall ledges. The vortex generators have the following data: $S/D$ = 2-10; $d/D$ = 0.6-0.95 ($D$ = pipe ID, $d$ = reduced diameter of pipe flow area with vortex generator, $S$ = twisting step) [4]. The baseline model that is used for correlation of intensifying regimes is bare tube type.

The main benefit that contributes to heat and hydraulic characterization of enhancement flow passages is the reduction on hydraulic friction (internal flow) and wattage consumed to carry the fluid due to its pseudoplastic rheology. This benefit is measured at a level of 10-15% thanks to localized shear effect.

## References

[1] Nazmeyev Yu G and Konakhina I A 2001 *Organization of power technology complexes in petrochemical industry* (M.: MEI publishers) p 364

**[2]** Konakhina I A, Khusnutdinova E M , Khamidullina G R and Khamidullina A F 2016 *IOP Conference series: Materials Science and Engineering* Vol 134 Is 1 **012019**

[3] Nazmeyev Yu G and Konakhina I A 1993 *Enhancement of heat transfer in the flow of viscous fluid in helical pipes* (M.: Teploenergetika) pp 59-62

[4] Nazmeyev Yu G 1996 *Hydrodynamics and heat exchange between vortex flows of rheologically complex fluids* (M.: Energoatomizdat) p 304

# Investigationsof Nanocrystalline Alloy Electrospark Coating Madeof Nanocrystalline Alloy Based on 5БДСР Ferrum

**A V Kolomeichenko[1], I S Kuznetsov[1], A Yu Izmaylov[2], R Yu Solovyev[2], S N Sharifullin[2,3]**

[1]Federal State Budgetary Educational Establishment of Higher Education „Orel State Agrarian University named after N.V. Parakhin", Russia
[2]Federal State Budgetary Scientific Institution "Federal Scientific Agroengineering Center of All-Russian Institute of Mechanization", Russia
[3] Kazan Federal University, 18 Kremlyovskaya street, Kazan, 420008, Russian Federation

Ivan-654@yandex.ru

**Abstract.** The article describes the properties of wear resistant electrospark coating made of nanocrystalline alloy of type 5БДСР (Finemet). It is proved that electrospark coating has nanocrystalline structure which is like amorphous matrix with nanocrystals α – Fe. Coating thickness is 33 μm, micro-hardness is 8461 - 11357 MPa, wear resistance is $0{,}55\times10^4$s/g. Coating ofnanocrystalline alloy of type 5БДСР can be used to increase wear resistance of machinery working surfaces.

## 1. Introduction

One of the promising directions of improving the wear resistance of working surfaces of details is their hardening due to the creation of surface layers with high physical-mechanical properties. Analysis of scientific and technical information showed that among the commonly used methods of hardening of working surfaces can distinguish heat treatment, chemical heat treatment, laser and plasma hardening and others. A promising method of hardening the working surfaces with complex geometric form, is an electro spark processing (ESP) [1]. There are many different ways to method development ESP. One of the ways is the use of new materials with nanocrystalline structure. The use of such materials will allow to obtain multifunctional electro spark coating, can improve wear resistance of working surfaces of parts [2-12].

## 2. Obiectiv.

Explore the properties of electric-reinforcing coating made of nanocrystalline alloy (NCA).

## 3. Methods.

As an electrode material 5BDSR nanocrystalline alloy was selected mark (Finemet) (Fe-70% B - 9.2%, Si- 6.3% Nb- 2.21%, 0.8% Cu-, Mo- 0,2%). The structure of the resulting coatings was investigated with a scanning electron microscope Hitachi TM - 1000. The reinforcing coating is applied to the samples of steel 65G install EDM brand BIG - 4 (mode: №2, K = 0.8). The thickness of the experimental electric-spark coatings was measured on transverse sections. The measurements were performed on a microscope MIM-8. Base surface served as the boundary between the coating and the substrate. Microhardness was measured at a load of 50 g indentation of diamond tips on a computerized Hardness PMT-3M-01.Protsess studied mass transfer was investigated by measuring geometric parameters of single erosion marks left electrode. Hardness measurements were performed PMT-3M. Microhardness was equipped with a screw ocular micrometer MOV-1-16H, and MA-25 lens (epiobektiv plan-achromatic F = 25,0 mm, A = 0,17) .Tribotehnicheskie tests performed on the machine friction MTU-01, at the outer load of 2.5 N, with relative sliding speed - 1.0 m / s. Material for manufacturing served kontrobraztsa steel grade 65G, hardened to HRC 58 ... 60. Depreciation is determined by gravimetric method using weights Sartorius Competence CP64 with an accuracy of 0.0001 g.

## 4. The results of the research.

Scanning electron microscopy made it possible to state that-tension coating obtained by treating the steel substrate electrode of the NCA brand 5BDSR, it has a homogeneous structure to the level of 1 micron. However, in the transition zone between the coating and the substrate are microcrystalline incorporating substrate material (Fig. 1). X-ray studies have shown that the coating alloy 5BDSR has a nanocrystalline structure, is an amorphous matrix with nanocrystals $\alpha$ - Fe. It can be seen from Figure 2, in which there is reflection from the crystal phases $\alpha$-Fe [10].

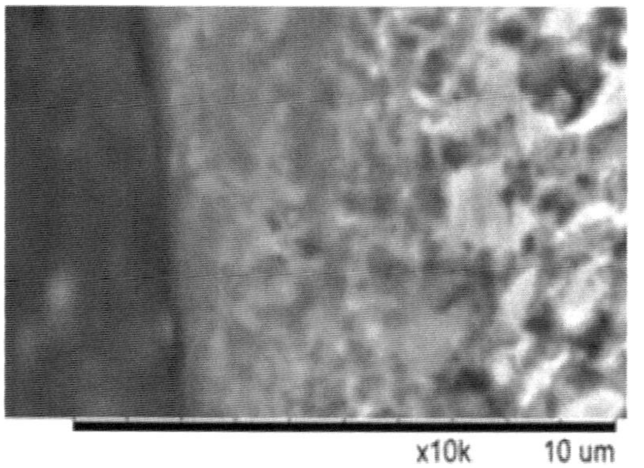


**Figure1.** Electrospark coating structure

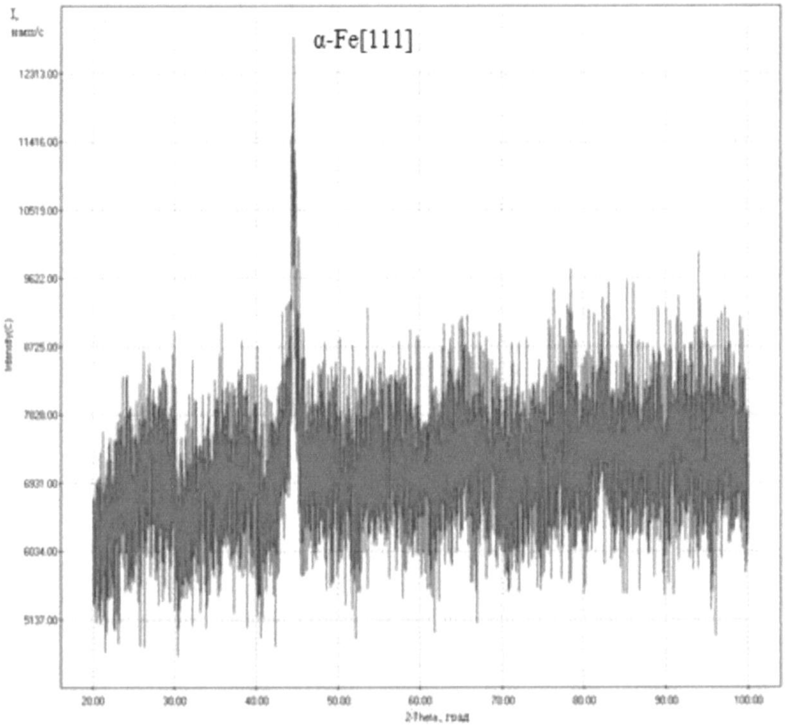


**Figure2.** Diffraction pattern of electric spark coating

The thickness of the electric-cover changes nonlinearly. The experimental dependence of the thickness of the specific processing time provided in Figure 3 As it set the threshold of brittle fracture is 2.1 min / sm2i critical threshold of brittle fracture is equal to 5.2 min / cm$^2$. In ESP at № mode 2, k = 0.8 of the maximum thickness of coating grade 5BDSR NCA - h = 33 m, which is approximately 1.5 times greater than the thickness of the alloy coating-grade VK6 OM obtained in a similar mode.

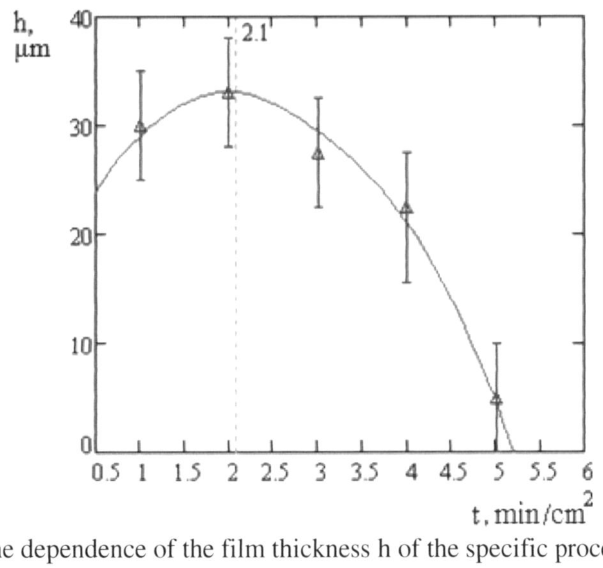


**Figure3.** The dependence of the film thickness h of the specific processing time t

Microhardness strengthening coating is largest dispersion. The maximum value of the microhardness of coatings NCA brand 5BDSR - $N_{\mu max}$ = 11357 MPa minimum - $N_{\mu min}$ = 8461 MPa. A characteristic feature indicative of plastic deformation of the coating is the presence of "crown" on the printout of the indenter [1, 8].

During the mass transfer studies have demonstrated the dependence of the mass of the electrode material transported from the anode to the cathode, the initial stress - one of the main technological parameters of ESP. Figure 4 shows the regression power law, which can be used to develop the process of hardening of details. Using the gravimetric method was established and mass transfer direction generally ESP efficiency. This method allowed us to determine a number of mass transfer parameters necessary for the process to ensure ESP process (Table. 1).

**Table 1.** Mass-transfer parameters

| Mass-transfer average ratio | Threshold fragile destruction coatings, $t_x$, min/cm $^2$ | The critical threshold brittle fracture surfaces, $t_{кр}$, min/cm $^2$ |
|---|---|---|
| 0,54 | 4,0 | 6,5 |

Endurance tests yielded dependence on wear test duration, represented as a function of logarithmic regression. After differentiation its wear rate dependence on the time (Fig. 5) was obtained.

Tribological tests found that the hardening electric spark coating NCA brand 5BDSR has wear rate Wsr = 18 × $10^{-5}$ g / s and wear resistance U = 0,55 × $10^4$ s / g. Low wear rate and high wear resistance indicate that the considered electric spark coating should be used in hardening of working surfaces of parts.

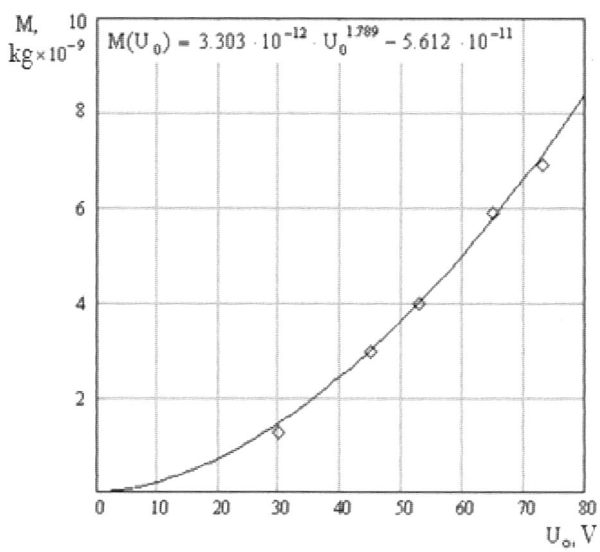


**Figure 4.** The dependence of the mass of the electrode material is transferred from the anode to the cathode, the initial voltage for ESP

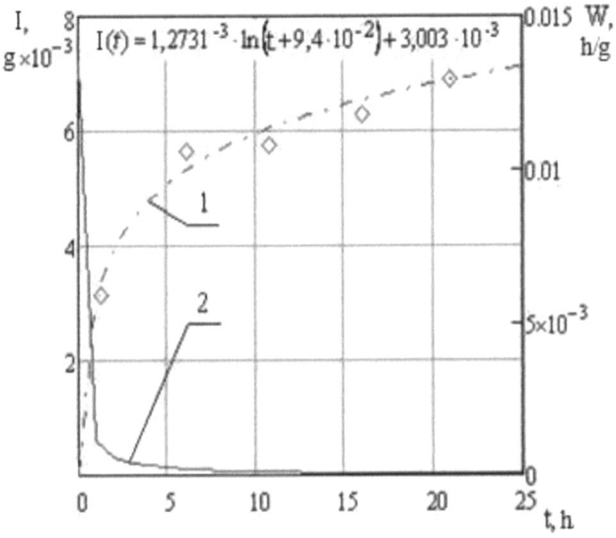


**Figure5.** Dependencies wear (1) and wear rate (2) electric spark coating on the duration of the test

## 5. Conclusion.

Dura-tension coating of nanocrystalline alloy brand 5BDSR a nanocrystalline structure, the thickness of 33 microns, the maximum microhardness of 11357 MPa, the wear resistance of $0.55 \times 10^4$ s / g. Coverage of this alloy can be used to improve the wear resistance of working surfaces of machine parts.

## References.

[1] Kolomeychenko A V, Kuznetsov I S 2015 Theory and practice of electric-hardening cutting machine parts amorphous and nanocrystalline alloys Monograph *Eagle. Publishing house Orel State Agrarian University* pp 174

[2] Ageev E, Mizobata K, Nakajima T, Zen H, Kii T and Ohgaki H 2015 Time-resolved detection of structural change in polyethylene films using mid-infrared laser pulses *Applied Physics Letters* V 107 No 4 **041904**

[3] Ageeva E V, Khor'yakova N M, Ageev E V 2015 Morphology and composition of copper electrospark powder suitable for sintering *Russian Engineering Research* V 35 No 1 pp 33-35

[4] Cadney S, Brochu M 2008 Formation of amorphous Zr41.2Ti13.8Ni10Cu12.5Be22.5 coatings via the electrospark deposition process *Intermetallics* Vol 16 No 4 pp 518-23

[5] Brochu M, Heard D W, Milligan J, Cadney S 2010 Bulk nanostructure and amorphous metallic components using the electrospark welding process *Assembly automation* Vol 30 No 3 pp 248-56

[6] Milligan J, Heard D W, Brochu M 2010 Formation of nanostructured weldments in the Al-Si system using electrospark welding *Applied surface science* Vol 256 No 12 pp 4009-16

[7] Cadney S, Goodall G, Kim G, Moran A, Brochu M 2009 The transformation of an al-based crystalline electrode material to an amorphous deposit via the electrospark welding process *Journal of alloys and compounds* Vol 476 No 1-2 pp 147-51

[8] Kolomeichenko A V, Kuznetsov I S, Kravchenko I N 2015 Investigation of the thickness and microhardness of electrospark coatings of amorphous and nanocrystalline alloys *Welding International* Vol 29 No 10 pp 823-25

[9] Kolomeichenko A V, Kuznetsov I S 2014 Tribotechnical properties the electrospark coating of amorphous and nanocrystalline alloys based on iron *Friction and wear* Vol 35 No 6 pp 501-504

[10] Kolomeychenko A V, Kuznetsov I S, Rodichev A Y, Penyashki T G 2015 On the question of getting in electric-spark coatings of amorphous and nanocrystalline structures *Repair. Recovery. Modernization* No 5 pp 33-36

IOP Conf. Series: Materials Science and Engineering **240** (2017) 012047 doi:10.1088/1757-899X/240/1/012047

[11] Adigamov N R, Sharifullin S N, Podgaraya K S 2013 Efficient technology for the modernization of the high-pressure fuel pumps, automotive diesel engines *Proceedings GOSNITI* V 113 pp 206-14

[12] Adigamov R N, Lyalyakin V P, Soloviev R Y, Sharifullin S N 2016 Plasma technology in improving the efficiency of the diesel engine high pressure fuel pumps *Welding Engineering* No 2 pp 49-51

IOP Conf. Series: Materials Science and Engineering **240** (2017) 012048 doi:10.1088/1757-899X/240/1/012048

# Technical and Economic Analysis of Injecting Enhancers into High-Paraffin Oil Heaters

**I A Konakhina, G R Khamidullina and E M Khusnutdinova**

Kazan Federal University, 18 Kremlyovskaya street, Kazan, 420008, Russian Federation

rr-088@mail.ru

**Abstract.**This paper covers an algorithm of how to optimize the selection of preferred solutions to upgrade heat-exchange equipment. The algorithm is discussed in terms of high-paraffin oil heaters as compared with other heat-exchange surface features.

Usually, a technical and economic analysis is performed whenever it is required to differentiate between several solutions available and pick an optimum one to upgrade the heat-transfer equipment. The analysis uses the calculation of a single or multiple performance indicators (payback period, reduced cost, cost-effectiveness etc.), which show a degree of preferability of one or another solution [1].

Particularly, with consideration of a profit tax $N_{pr}$, the net annual income of an undertaking gained from equipment upgrading is determined by the formula, kRUB per annum:

$$\sum E'_{res} = \sum E_{res} \cdot (1 - N_{pr}) \ , \qquad (1)$$

where:$E_{res}$is the resulting economic benefit from implementing technical solutions, which can be defined by the formula:

$$\sum E_{res} = E - S \, , \qquad (2)$$

E is the effective (revenue) side of technical solution, kRUB per annum;

*S* is the additional OPEX of technical solution, kRUB per annum.

However, the main procedural troubles are related to correct determination of positive and negative effects coming from changes in operating expenses to be spent for running the equipment throughout the year (or service life between overhauls).

As is shown in [2, 3], the comparison of how the heat exchangers with clean surfaces are efficient is not exactly correct for the cases when the surfaces become less contaminated while in service. Seasonal periodicity, weather conditions or other changes in initial conditions of heating up or cooling down the high-viscosity fluids in terms of their effective viscosity versus temperature are also important.

All these aspects can be fully related to the paraffin oil heaters as well, where both the surface contamination and the initial temperature fluctuation of the environment cannot be ignored.

In addition to operating expenses, the innovation introduction stage will also require significant capital expenses for the heat exchanger itself as well as for its foundation, room, if required, etc. Mainly, all these factors depend on particular operating environment in a particular undertaking. There is, however, a general indicator, which shows material content of new (or upgraded) equipment. The metal content is such an indicator for heat exchangers [4]. Therefore, increase or decrease in allowance for depreciation of equipment should be related to either positive or negative benefits.

Considering all the above said, the algorithm of technical and economic analysis of injecting enhancers into high-paraffin oil heaters looks as follows:

**Phase 1.** Analysis of thermohydraulic performance of exchangers-preheaters of oils with neat surfaces.

Heat power stability is typical for heat exchangers running at rated values. Owing to the Kirpichyov correlation analysis, it means that the thermal effectiveness coefficient is constant, kW/(m$^2$ ·K) [2]:

$$K_Q = \frac{Q}{F} = K \cdot \Delta t = const\,. \tag{3}$$

The factor of power required for circulating the coolant is defined by the formula, kW/m$^2$:

$$K_N = \frac{\sum N}{F}\,. \tag{4}$$

For heat exchanger, the energy efficiency ratio is:

$$\overline{E}_0 = \frac{K_Q}{K_N}\,. \tag{5}$$

Considering the ratio (3), the equation (5) demonstrates that the correlation analysis of thermohydraulic efficiency of enhanced preheaters versus the baseline model is limited to the comparison of pressure loss ratio (the power required for carrying the coolants) in such kind of machinery, i.e.

$$\frac{\overline{E}_0}{\overline{E}_{0i}} = \frac{K_{N_i}}{K_N}\,. \tag{6}$$

Where the index $i$ shows the type of an enhanced heat exchanger.

If the heat exchanger is running (or planned to be run) under alternating load conditions, the measures of formulas (3)-(6) should be averaged taking typical modes of operation and equipment service life into consideration.

**Phase 2.** Analysis of thermohydraulic regimes as a function of initial temperature of oils (oil products).

**Phase 3.** Calculation of hot coolant rate as a function of initial fluid temperature.

**Phase 4.** Determination of heat transfer ratio of a preheater depending on fluid and hot coolant rates $K_{\text{Т.А.}} = f\left(\text{Re}_{\text{н}}, \text{Re}_{\text{т.н.}}\right)$.

**Phase 5.** Calculation of heat transfer area and verification of its compliance with the actual heat exchanger area.

**Phase 6.** Determination of how the heat transfer in an oil heater depends on transfer surface contamination. These dependencies are also reduced as the re-configuration of tube bundle surface reduces flow throat area. Changes in surface on the hot coolant side reduce build-up rate on the outer walls, while changes in surface on the oil (oil products) side reduce the build-up rate on the inner walls. Particularly, it is shown in [2] that spiral or annular grooves drop build-up rate up to 3 times in rated duty. A temperature drop variation coefficient is also determined considering contamination and variation coefficient of hot coolant flow rate to prevent from this negative effect. The measures of item 4 are re-calculated to ensure that the fluid is heated up as required.

**Phase 7.** Calculation of integral technical and economic magnitudes as per formulas (1)-(2) followed by conclusions.

Estimating calculations performed for two types of surfaces – bare tube and enhanced surface with spiral grooves – have shown that the enhanced surface approach is compensated less than within 1.5 years.

## References

[1] Nazmeev Yu G and Konakhina I A 2002 *Heat and energy systems and power balances of production plants* (M.: Publishers of Moscow power engineering institute)

[2] Konakhina I A and Voropayev A N 2010 The question of how to select high-performance heat exchangers-oil coolers considering actual conditions of their operation *Thermal physics and heat power engineering: collection of scientific articles – Magnitogorsk* pp 146-41

[3] Voropayev A N and Ukhanov K V 2009 Evaluation of effect of seasonal temperature variations of cooling water on the efficiency of heat exchangers-viscous fluid coolers *Izvestiya vuzov. power engineering issues* V 5-6 pp 137-32

[4] Simonov V F 1985 *Increasing power consumption efficiency in petrochemical industries* M.: Chemistry p 238

# Aerosol fluxing in electro arc metallization

**V N Logachev[1], A V Kolomeichenko[1], Yu A Kuznetsov[1], R Yu Solovev[2], V I Denisov[2], S N Sharifullin[3]**

[1]Federal State Budgetary Educational Establishment of Higher Education „Orel State Agrarian University named after N.V. Parakhin", Russia
[2]Federal State Budgetary Scientific Institution "Federal Scientific Agroengineering Center of All-Russian Institute of Mechanization", Russia
[3] Kazan Federal University, 18 Kremlyovskaya street, Kazan, 420008, Russian Federation

kolom_sasha@inbox.ru

**Abstract.** The main structural and technological factors of electric arc metallizers, which significantly affect the quality of their operation, are considered. The results of experimental studies on the adhesion strength, porosity, microhardness and comparative wear resistance of coatings obtained by electroarc metallization are presented. Optimal modes of coating with electroarc metallization are determined. It is established that the use of aerosol fluxing during electric arc metallization provides an increase in adhesion-cohesive strength by 25 ... 30%, an increase in microhardness to 40%, a decrease in porosity by 20 ... 25%.

## 1. Introduction

In the world practice of restoring the details of machinery and corrosion protection of aluminum and zinc pipes, equipment and metal structures, electric arc metallization (EM) has become very popular, as one of the most technologically productive, high-quality and universal methods of metal coating. However, along with clearly positive sides, EM has certain drawbacks. This perennial problem is the speed and oxidation of dispersed metal in a heterophase stream. Without overcoming these EM problems, it is impossible to further improve the physical and mechanical properties of metal coatings. Despite the extensive volume of studies carried out to modernize EM over the historical period of its use (over 100 years), these problems remain unresolved.

## 2. Materials and methods.

To solve technical problems, the method of aerosol fluxing and the laws of gas dynamics were used. The design developments were based on TAFA equipment. The following filler materials (wire) were used in the experiments:

1. Sv-08, flux № 1 - Na2CO3, Na3AlF6, Na2B4O7
2. Sv-08, flux № 2 - Na2CO3, Na3AlF6, H3BO3
3. Sv-08, flux № 4 - Na2CO3, Na3AlF6, Na2B4O7
4. Sv-08, flux № 5 - Na2CO3, Na3AlF6, Na2B4O7, NH2CON2
5. Sv-08, flux № 6 - Na2CO3, Na3AlF6, Na2B4O7, H3BO3
6. Sv-08 (without flux)

7. Sv-08 + X20N80 (without flux)
8. Steel 45 (without flux)

## 3. Results of the research.

We have developed two models of supersonic metal sprayers – EAM-11SHD-machine-tool (electro arc metallization) and EAM-12SHD-manual (electro arc metallization) are developed. They are considered a new generation of electro arc metallization (EM) equipment.

Designers and creators of electro arc metallization (EM) equipment of TAFA firm followed up one and the same objective, i.e. supersonic outflow of metal and air flow and contraction of spray-cone angle of spray pattern, obtaining increase of physical and mechanical properties of coatings without application of arc activation complicated devices, i.e. burning of hydrocarbon fuel or flammable gases.

The work was begun from gas dynamic calculation of air jet outflow via shaped nozzle (Fig. 1).

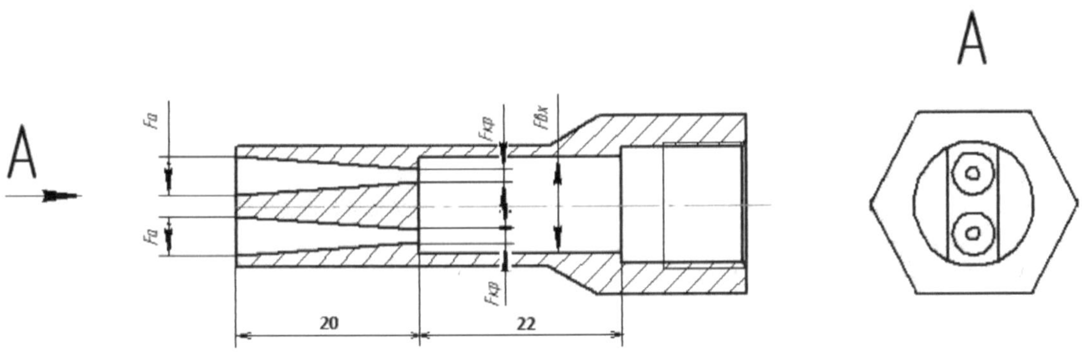


**Figure 1.** Supersonic nozzle of metal sprayer EAM-11SHD (electro arc metallization)

1. Basic data for calculation:
$P_O$ – total pressure in main (kg/sm$^2$);
$F_{mp}$ – minimum passage section at gas outflow (sm$^2$);
μ – jet flow coefficient (μ≈0,85…0,9);
$T_O$ – air temperature (To ≈ 290 °K);
R – gas constant (for air R=29,3);
K – adiabatic k-value (K=1,4);
A – flow coefficient (A ≈ 0,023 1/s);
g = 9,8 m/s$^2$.
2. Calculation of sonic velocity or gas critical velocity at supersonic outflow from nozzle:

$$C = \sqrt{K \cdot g \cdot R \cdot T_o} = \sqrt{1{,}4 \cdot 9{,}8 \cdot 29{,}3 \cdot 290} = 340 m/s.$$

This condition is realized at correlation of receiver pressures (main) and ambient medium $P_H$: $\frac{P_O}{P_H} \geq 1{,}89$. If $\frac{P_O}{P_H} < 1{,}89$ – outflow is subsonic and calculation is done according to other dependences.

In commonly used schemes (Fig.1) correlation $\frac{F_{en}}{F_{mp}} > 1{,}5 \ldots 2$ must be provided. When this condition is not observed the occurring losses result in decrease of total pressure $P_O$ by 20…30% because of energy losses at transfer of supersonic flow to subsonic flow.

Let's examine section parameters $F_{en}$.

In outlet it corresponds to: $F_{en}=0{,}785 \cdot 7{,}5^2=44$ mm$^2$.

IOP Conf. Series: Materials Science and Engineering **240** (2017) 012049 doi:10.1088/1757-899X/240/1/012049

Passage section in nozzle F with account of shadowing of electrode wire ∅2mm:

$$F_{mp}=0{,}785\cdot 7^2-2\cdot 7=24{,}5\ mm^2.$$

There is correlation: $\frac{F_{en}}{F_{mp}}=\frac{44}{24{,}5}=1{,}8$, that satisfies normal conditions of jet system operation.

Air consumption:

$$G=\mu\cdot A\cdot P_O\cdot F_{KP}=0{,}85\cdot 0{,}023\cdot 7\cdot 0{,}245=0{,}0335\frac{kg}{s}\approx 33\frac{g}{s}.$$

Taking into consideration the design features of nozzle EAM-11SHD (electro arc metallization) (electrode wire location in it) gas flow in the nozzle changes from sonic (at $F_{KP}$=24,5mm$^2$) to supersonic, because at outlet section we have $F_a=0{,}785\cdot 7^2=38{,}5mm^2$

With application of table of gas dynamic functions (A.A. Dimentyeva and others) [1] we obtain: q(λ)= $q(\lambda)=\frac{F_{mp}}{F_a}=\frac{24{,}5}{38{,}5}=0{,}63$, from which λ=1,6.

$V_a=340\cdot 1{,}6=540$ m/s (without taking into consideration heat supply for the sake of arc burning)

This calculation of supersonic velocity of air jet from nozzle is done without consideration of heat supply from electric arc. Supersonic velocity is $V_a$=540 m/sec – of cold air.

It is known that arc temperature is about 6000°C [2]. On the ground of gas dynamic calculation it is possible to suggest that $V_a$ increases no less than 3 times.

In connection with development of supersonic metal sprayers the task of metallization distance optimization ($H_D$) was assigned. For this task the experimental tests of function of dependence of adhesive-cohesive resistance of EM (electro arc metallization)-coatings from metallization distance at supersonic and subsonic mode were carried out.

As optimal value of supersonic metallization distance is accepted maximum result of adhesive-cohesive resistance of EM (electro arc metallization)-coatings.

At subsonic metallization $H_D$ was accepted 120 mm, at supersonic metallization $H_c$ it was improved to 160 mm.

Increase of maximum velocity of air flow being measured device PDGP-1 [3] provides the metallization quality improvement at reconditioning of engines crankshafts and other elements which metallization is possible only at the distance more than 120 mm.

Comparative laboratory tests of supersonic metal sprayer EAM-10SHD (electro arc metallization) showed at metallization distance 160 mm. increase of adhesive-cohesive resistance by 25%, coating density by 23%, operation ratio of filler material by 15% in comparison with subsonic level of air jet outflow.

To solve the problem of decrease of oxidation potential at EAM (electro arc metallization) and metal alloying being melted by arc, the method of aerosol fluxing [4, 6, 7], was applied, the essence of which is in the following.

The essence of aerosol fluxing (AF) consists in that in the dispersed metal jet at EM (electro arc metallization), aerosol in the form of different substance water solution is introduced.

At substance solution in water their predissociation takes place, for example:

$Na_2CO_3 + H_2O \Leftrightarrow NaHCO_3 + NaOH$

Getting into the high temperature zone of electric arc and dispersed metal jet the predissociated sodium carbonate is subjected to thermal dissociation.

$NaHCO_3 \rightarrow Na + OH + CO_2$

$2CO_2 \rightarrow 2CO + O_2$

$FeO + CO = Fe + CO_2$

$2CO \rightarrow 2C + O_2$

Further, hard phases are formed at the contact with melted ferrum:

$Fe + C \rightarrow FeC$ (cementite)

$2Fe + C \rightarrow Fe_2C$ (carbide)

Thus, in aerosol fluxing (AF) process the carbothermic process and reconditioning process take place. At aerosol fluxing (AF) some other substances compositions to obtain analogical effect are used, for example, at lithium aluminum hydride solution its dissociation takes place:

$Na_3AlF_6 + H_2O \Leftrightarrow NaOH + AlF_6$

Further thermodissociation:

$AlF_6 \rightarrow Al+6F$

Metallothermy runs with heat emission because of energy-releasing reaction of deoxidization of ferrous oxide by aluminium with emission of about 200 kkal. of heat.

Aluminothermic process running is seen from the flowing reaction:

$3FeO + 2Al = Al_2O_3 + 3Fe$

At application of aluminothermic effect, i.e. FeO deoxidization reaction by aluminium, increase of temperature of metal particles in heterophase flow takes place because of the same reason of energy-releasing reaction of deoxidization of ferrous oxide by aluminium with heat emission.

It is known [3], that the temperature in contact of melted metal particle at its collision with substrate characterizes energy condition of atoms (T):

$$T=f(T_{П}, T_{K}, E_{K}, E_{Г}),$$

where $T_{П}$ – template temperature;

$T_K$ – metal drop temperature;

$E_K$ – kinetic energy of metal drop;

$E_{Г}$ – kinetic energy of heterophase flow.

That is why to obtain the hardening effect of coating adhesion with substrate, as well as particles in coating is necessary to increase T and E. This is proved by the works of V.V Kudinov on research of interaction separate particle with substrate, from them it follows that adhesion resistance mainly can be increased by increasing the temperature of metals particles in contact with substrate and velocity V of particles [5].

Increase of adhesive-cohesive resistance of electro arc metallization EM – coating at the account of carbothermic (C) metallothermic (Al) ferrum deoxidization is the first solvable problem by means of aerosol fluxing.

The second application of method of aerosol fluxing (AF) is alloying of metalize layer to increase its resistance, hardness and wear resistance. It is obtained by application of some additional substances to prepare aerosol, containing boron, nitrogen, chrome.

The process runs identically to the described above reactions. For it the substances soluble in water, for example borax ($Na_2B_4O_7$), boric acid ($H_3BO_3$), carbamide ($NH_2CON_2$), chromic nitrate $Cr(NO_3)_3{\cdot}9H_2O$, carbonate ($N_2CO_3{\cdot}10H_2O$) and others are used.

$Na_2B_4O_7 \rightarrow 2N + B_4O_7$ (dissociation on dissolving in water)

$2B_4O_7 = 8B + 7O_2$ (thermodissociation)

$2B + Fe = FeB_2$ (ferrum boride is an extra strong phase)

$H_3BO_3 \rightarrow H_3 + BO_3$ (hydrodissociation)

$2BO_3 \rightarrow 2B + 3O_2$ (thermodissociation)

$2B + Fe = FeB_2$ (ferrum boride)

$NH_2CON_2 \rightarrow NH_2 + CON_2$ (hydrodissociation)

$NH_2 \rightarrow N + 2H$ (thermodissociation)

$CON_2 \rightarrow 2N + CO$ (thermodissociation)

$N + 4Fe = Fe_4N$ (ferrum nitride is a hard phase)

For experimental tests according to aerosol fluxing (AF) the developed hydro-disperser presented in Figure 2 was used.

It is a steel cylinder where by means of air fed into metal sprayer it is dispersed to midst state of flux water solution. In this form as it shown in the figure heterophase flow moves into electrical circuit, arc, metal sprayer, where thermodissociation of the dissolved substances and other phase transformations [8-12].

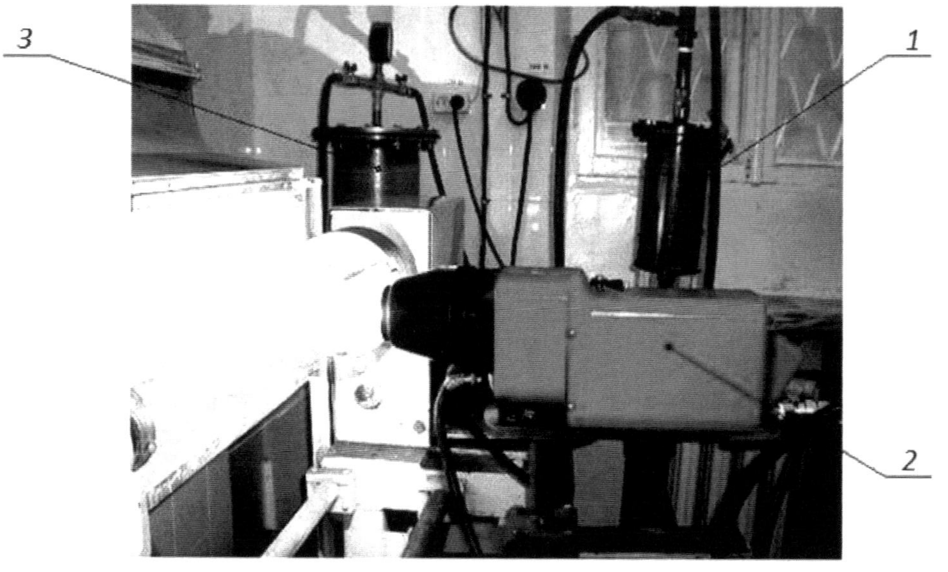


**Figure 2**. Electro arc metal sprayer working together with hydro-disperser for aerosol fluxing (AF): 1 – hydro-disperser; 2 – metal sprayer; 3 – filter for compressed air cleaning.

Results of the experimental tests

The generalized results of experimental tests are presented in the table.

Table – Experimental tests results

| № | № flux | Concentration, g/l | Solution consumption, mm/min | Adhesion, cohesion, kgs/mm$^2$ | Porosity, % | Microhardness, HV | Comparative wear resistance, %* |
|---|---|---|---|---|---|---|---|
| 1 | 1 | 60 | 10 | 5,4/5,2 | 3,8 | 560 | 100 |
| 2 | 2 | 50 | 5 | 5,4/5,3 | 3,6 | 540 | 50 |
| 3 | 4 | 60 | 10 | 5,5/5,3 | 4,1 | 560 | 100 |
| 4 | 5 | 60 | 10 | 4,4/2,5 | 4,5 | 560 | 110 |
| 5 | 6 | 60 | 5 | 5,0/4,8 | 5,2 | 580 | 90 |
| 6 | Sv -08 | | | 2,3/2,1 | 6,0 | 212 | 35 |
| 7 | Sv -08+ X20H80 | | | 3,4/4,0 | 6,1 | 280 | 45 |
| 8 | Steel 45 | | | 3,0/2,8 | 6,2 | 510 | 100 |

*Wear resistance of coating from Steel 45 without aerosol is 100% conventionally

From the table is seen that the maximum results of adhesive-cohesive resistance, microhardness and wear resistance are obtained at application of fluxes № 1, № 4, № 5, № 6 at concentration 60 g/l. Not only flux content but also substance concentration in aerosol and flux consumption influence considerably on the coating properties.

Application of aerosol fluxing also positively influences the reaction of metal coating in friction couple. Figures 3, 4 present diagrams of counterface (block) wear and friction coefficient for different metal coatings at their testing for wear resistance on friction machine.

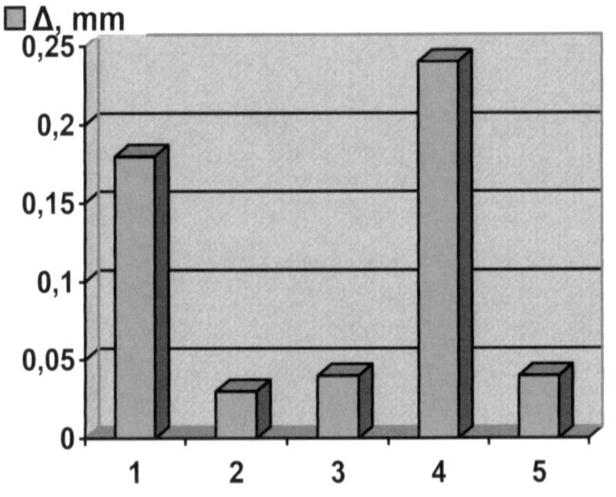


**Figure 3**. Block wear for different metalized coatings: 1 – Sv-08 (without aerosol); 2 – Sv-08 + aerosol № 5; 3 – Sv-08 + aerosol № 4; 4 – Sv-08 + X20H80 (without aerosol); 5 – Steel 45 (without aerosol)

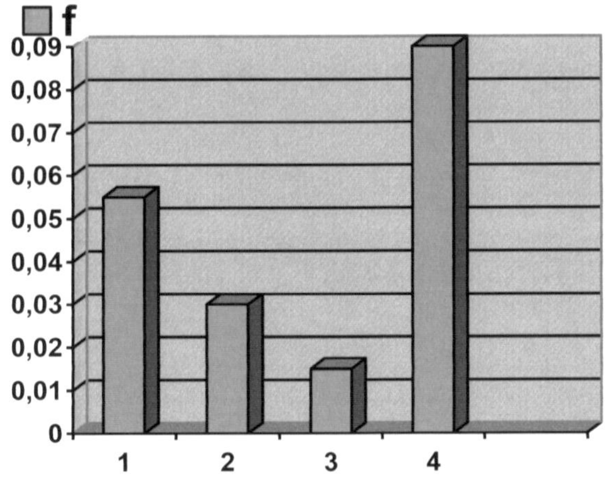


**Figure 4**. Friction coefficient for different metalized coatings: 1 – Sv-08 (without aerosol); 2 – Sv-08 + aerosol № 4; 3 –Sv-08 + aerosol № 6; 4 – Sv-08 + X20H80 (without aerosol)

**4. Conclusions.**

1. According to the gas dynamic calculation of the developed nozzle of electro arc metal sprayer supersonic velocity of air flow is 540 m/s without taking into account arc burning temperature.
2. Experimentally by means of the developed in State Scientific Institution of All Russian Research 3. Technological University device PDGP-1 maximum velocity of air flow at the distance of 160 mm from the point of crossing electrode wires is defined. On this basis the accepted earlier metallization distance from 120 to 160 mm was adjusted. It extends the applicability of electro arc metallization (EM).
3. Improvement of physical and mechanical properties of electro arc metallization (EM) coatings is achieved at the account of deoxidization of ferrum by aluminium and carbon (metallo-carbothermy) and its alloying with boron, nitrogen, carbon at aerosol fluxing.
4. Application of aerosol fluxing at electro arc metallization (EM) provides increase of adhesive-cohesive resistance by 25…30%, microhardness increase to 40%, porosity decrease by 20…25%.

IOP Conf. Series: Materials Science and Engineering **240** (2017) 012049 doi:10.1088/1757-899X/240/1/012049

**References**

[1] Dimenteva A A, Rekstin FS, Ryabov V A 1966 *Tables of gas dynamic functions* Leningrad, Mechanical Engineering
[2] Kretschmar E 1966 *Sputtering of metals, ceramics and plastics* Mechanical Engineering
[3] Soloviev R Yu, Litovchenko N N, Petryakov B I 2010 *Device for measuring the pressure of heterophase flow* Patent No 2385452
[4] Litovchenko N N, Shevchenko V P 1985 Author's certificate No **1183562**
[5] Kuzenkov V V 1979 Influence of superheating of sputtered particles and oxide films on the surface of metal substrates on the thermal processes between the particle and the substrate during sputtering *Physics and Chemistry of Material Processing* No 6
[6] Litovchenko N N, Logachev V N 2013 Ways of improving equipment and technology of electric arc metallization *Tractors and agricultural machinery* No 11 pp 52-54
[7] Vorobiev P A, Yusim M Yu, Litovchenko N N, Denisov V I 2008 Method of aerosol fluxing in electric arc metallization *Proceedings of GOSNITI* V 101 pp 201-04
[8] Litovchenko N N, Petryakov B I, Tolkachev A A, Blokhin S A 2013 Influence of the flow rate of heterophase flux on the physico-mechanical properties of the electrometallic flow *Welding production* No 6 pp 43-47
[9] Solovev R Y, Vorobev P A, Litovchenko N N 2013 Metal-carbothermal methods of reducing the degree of oxidation of dispersed metal in electric arc mettalising *Welding international* V 27 No 5 pp 423-27
[10] Denisov V I, Litovchenko N N, Logachev V N, Tolkachev A A 2015 Activation of the process of electric arc metallization with liquid hydrocarbon fuel *Proceedings of GOSNITI* V 120 pp 160-65
[11] Denisov V I, Litovchenko N N, Logachev V N, Tolkachyov A A 2016 Method for measuring the pressure of a heterophase flux in supersonic arc-metallization *Proceedings of GOSNITI* T 122 pp 163-66
[12] Kolomeichenko A V, Kravchenko I N, Logachev V N, Litovchenko N N, Puzryakov A F 2015 Improvement of physical and mechanical properties of coatings obtained by electroarc metallization. // *Construction and road machines* No 7 pp 25-29

# Multicultural University Education and Museum Pedagogy

**G F Melnikova[1] and S I Gilmanshina[1]**
[1]Kazan Federal University, 18 Kremlyovskaya street, Kazan, 420008, Russian Federation

ms.valitova1989@mail.ru, gilmanshina@yandex.ru

**Abstract.** The specifics of the educational process in the museum are revealed. The experience of using the multicultural educational space of the museum for developing non-humanitarian directions of the university of general cultural competencies is expounded. The emphasis is on the formation of the ability to tolerate social, ethnic, confessional and cultural differences.

## Introduction

Today it is obvious that humanity is developing along the path of expanding the interconnection and cooperation of various countries, peoples and cultures. The existing expansion of international cooperation causes the interest of young people in multicultural university education. A special role in this process belongs to museums that take the individual beyond the borders of society, civilization into the world of culture. However, the problem of using the pedagogical potential of the university's museums in the multicultural education of youth has not been finally resolved.

Moreover, multicultural education is inseparable from the present historical period of the development of mankind. Today, according to the Federal State Educational Standard of Higher Education of Russia, as a result of mastering the bachelor's program, students, along with professional competences, should have general cultural competences. For example, the ability to analyze the main stages and patterns of historical development for the formation of patriotism and civic position, the ability to work in a team, tolerantly perceiving social, ethnic, confessional and cultural differences, etc. In our opinion, museum pedagogy could contribute to the formation of these competencies among university students .

The term "multicultural education" is defined as the formation of a person capable of active and effective life in a multinational and multicultural environment with a developed sense of understanding and respect for other cultures, the ability to live in peace and harmony with people of different nationalities, races, beliefs.

Multicultural education involves the adaptation of one ethnic group to another, the awakening of interest in the culture of neighbors. The reference to the values accumulated and sacredly preserved by mankind in world culture presupposes the inclusion of an individual in the cultural and historical space, which creates the ground for awareness of himself as a subject of culture. In modern socio-economic conditions, we view multicultural education as one of the main tasks of universities, in the solution of which a significant role is assigned to the museums of the university. And we view museum pedagogy as an effective approach to the education of young people in conditions of multinationality.

The purpose of the research: to reveal the pedagogical potential of the university's museums in the multicultural university education of youth.

**The basic part**

A distinctive feature of all forms of work of the museum institution at the present stage of the development of society is that the museum is no longer just a scientific institution, as it was in the XIX century, or a scientific and educational institution, which it was in the 60-70s of the XX century [5]. Today, the museum realizes its functions as a specific mechanism of socio-cultural communication, contributing to the reproduction of culture and the interaction of various cultural communities. The museum has a variety of ways to influence the audience. The process of acquiring knowledge is stimulated by sensory perception and emotional experiences, thanks to familiarization with the rarities of the national and world culture. Museums play a key role in the life of modern society, remaining a center for the preservation of cultural traditions, the formation of moral values and an indicator of the development of civilization in general.

A special category of museums are museums of universities. Let's turn to the history of museum business in Russia, where you can see the role and importance of university museums in the educational process.

For this purpose, we will dwell on the Kazan Imperial University, founded in 1804. According to the Charter of 1804, the university had an office for natural and natural history, a mineral cabinet that laid the foundation for modern geological, zoological and botanical museums. In 1815 the formation of a university cabinet of rarities laid the foundations for the organization of ethnographic and archaeological museums. Based on the cabinet of A.M. Butlerov in 1863 was created a museum of the Kazan Chemical School [6].

The Kazan Federal University owns the unique in the universities of Russia a unique combination of diverse museums called the Museum of the History of Kazan University. It includes expositions on the development of mathematics, physics, astronomy, chemistry, biology, medicine, physiology, jurisprudence, philology, history, Oriental studies at Kazan University. Today there are 10 museums in the museum complex of the Kazan Federal University. One of them is the museum of the Kazan Chemical School.

The traditions of the museum of the Kazan School of Chemistry are still in our days. The chemical laboratory, geographically combined with the museum, never stopped its work. The museum hosted lecture classes with students, as well as public lectures, experimental works, scientific discussions. The listeners and collaborators were people of different nations and religions, united by one great educational goal and humane views. Moreover, the territory of the Kazan Academic District covered polyethnic regions of the empire, and the national composition of the students was varied.

The educational process in the museum has its own specifics. Modern university education provides systemic rational knowledge and professional competences [1 - 4], which are formed through the active work of consciousness. The museum educational space attracts emotions and feelings, by immersing in a certain epoch or historical situation. As a result, a person experiences general feelings with a story in the museum or any other museum exhibit.

Thus, the museum educational space allows to preserve and, most importantly, cultivate the patriotic feelings of young people. This sensory process is difficult to implement in an intensive stream of modern visual information.

In the multinational society, the importance of education in the spirit of friendship, tolerance and mutual understanding of people of different nationalities is very important. Tatarstan is a multi-ethnic and multi-confessional republic with centuries-old traditions of peaceful and good-neighborly residence of a large number of nations and nationalities, which greatly increases the sensitivity of the republic's population to any manifestations of xenophobia and ethnic intolerance. An important role in the prevention of xenophobia and ethnic intolerance, the creation of a favorable educational space for students of all ethnic groups, is assigned to the museums of the Kazan Federal University.

Today the museum of the Kazan Chemical School conducts cultural and educational work and is part of the educational process. In the memorial room of the museum there are lectures and practical classes for students in such disciplines as "Development of Chemistry in Kazan", "The Role of Chemistry in the Development of Natural Science Knowledge", "Kazan School of Chemists", "History and Methodology of Chemistry". Round tables, seminars and public lectures of Russian and foreign scientists in the field of chemistry and chemical education are held. With the purpose of forming the need for self-development, the teachers introduce the students to the history of cognitive contradictions in chemistry that served as the impetus for the birth of a new theory. Moreover, students - future teachers of chemistry carry out design work on the history of chemistry, and a number of tasks on chemical disciplines are presented in the form of cases containing historical material.

For example, in case studies on thermodynamics, a historical explosion is described in an attempt to break up a caked mixture of nitrate and ammonium sulfate, and it is proposed to evaluate the measure of the risk of decomposition of ammonium nitrate on the basis of thermodynamic calculations. Another case raises the issue of thermodynamic justification for the possibility of spontaneous decomposition of potassium chlorate. A historical reference is made to the effect of the degree of purity of a given salt (the presence of impurities of combustible substances such as phosphorus and sulfur) on its susceptibility to friction, its use in the history of match production, and the corresponding thought experiment.

Thus, chemistry cases containing historical international material can be considered as a complex integrated system of active learning in the multicultural educational space of the university.

Consequently, the multicultural educational space of the museum contributes to the education of the person's value relationship, the expansion of her sensory experience, the development of patriotism and citizenship, the ability to tolerate social, ethnic, confessional and cultural differences, thereby creating general cultural competencies for university students.

## Conclusion

Museum pedagogy presupposes the inclusion of appropriate elective courses in the content of the university multicultural educational process. The pedagogical potential of the museums of the University is to bring up value orientations through the expansion of sensory experience in the process of communicating with museum objects and creating a favorable educational space for students of all ethnic groups. As a result, in the modern multicultural educational space of the museum, conditions are ensured that ensure the formation of general cultural competencies, protection and support for the development of each student. Consequently, the museums of the university can be regarded as one of the most important means of multicultural education and upbringing of modern youth.

## References

[1] Gilmanshin, I.R., Gilmanshina, S.I. 2016 The formation of students' engineering thinking as a way to create new techniques, technologies, materials *IOP Source: International Scientific-Technical Conference on Innovative Engineering Technologies, Equipment and Materials* Vol 134 **012006**

[2] Gilmanshina S I, Gilmanshin I R 2015 Building axiological competence of graduate students by means of project-based learning *IOP Conference Series: Materials Science and Engineering*, Vol 86 (1) **012029**

[3] Gilmanshina S I, Khalikova F D 2016 Teaching Gifted Adolescents in Terms of the Transforming Natural Sciences Education *Ifte 2016 - 2nd International Forum on Teacher Education* Vol 12 pp 50-54

[4] Gilmanshina S I, Sagitova R N, Gilmanshin I R 2016 Innovative Components for Preparing Graduates and Master's Degree Holders of Natural Sciences, for Education at the University Level *Ifte 2016 - 2nd International Forum on Teacher Education* Vol 12 pp 55-59

[5] Sidorova I B 2012 *Muzei i muzeinoe delo Rossii (dorevolyucionnyi period)* [Museums and museology Russian (pre-revolutionary period)]. Kazan, 299 p

[6] Valitova G F 2014 *Aleksandr Mihailovich Butlerov* [Alexander Butlerov]. Kazan, 431 p

IOP Conf. Series: Materials Science and Engineering **240** (2017) 012051 doi:10.1088/1757-899X/240/1/012051

# Different load types modeling using MatLab

**I S Metelev[1], D V Dvorkin[2] and R G Isakov[1]**

[1]Kazan National Research Technical University named after A.N. Tupolev, Kazan, Russia
[2] Otto-von-Geuricke University, Magdeburg, Germany

ivan0893@mail.ru, dvdvorkin33@mail.ru, ruslanisakov@yandex.ru

**Abstract**. Schemes with balanced, imbalanced and non-sinusoidal loads were modelled in MatLab in this paper. Verification (comparison of calculated values and measured) were done for active power (with difference less than 5%). There was a power phases inequality before transformer and after it.

## 1. General description of scheme

Engineers need to care about electricity quality from the first step of schemes creation. There are numerous of factors, which describe the condition of any scheme in power quality case. In a real schemes different loads (e.g. balanced, imbalanced and non-sinusiodal) can work simultaneously. But for modeling it is useful to model them separately for more accuracy.

General scheme for modeling is shown in the Pic.1. In this scheme: G-generator, Line – transmission line, T-transformer.

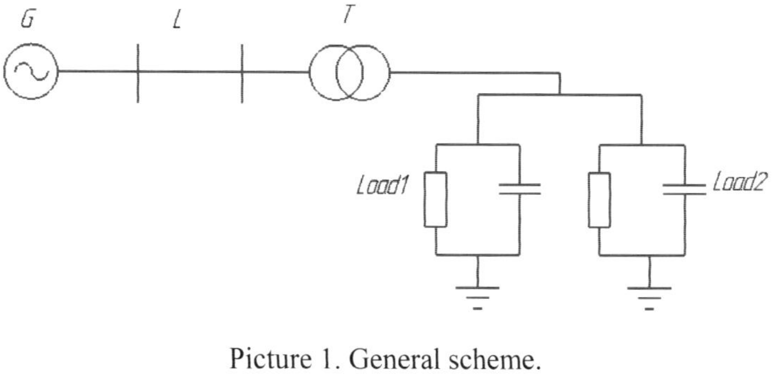


Picture 1. General scheme.

MatLab model for this scheme is shown in the Pic.2.

IOP Conf. Series: Materials Science and Engineering **240** (2017) 012051 doi:10.1088/1757-899X/240/1/012051

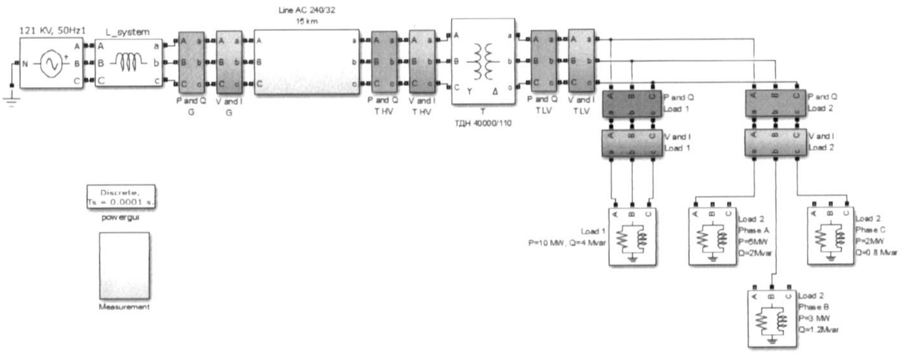

Picture 2. MatLab model.

On this picture Load1 is represented as balanced load, Load2 – as imbalanced.
Location of measurement devices is shown in Pic.3.

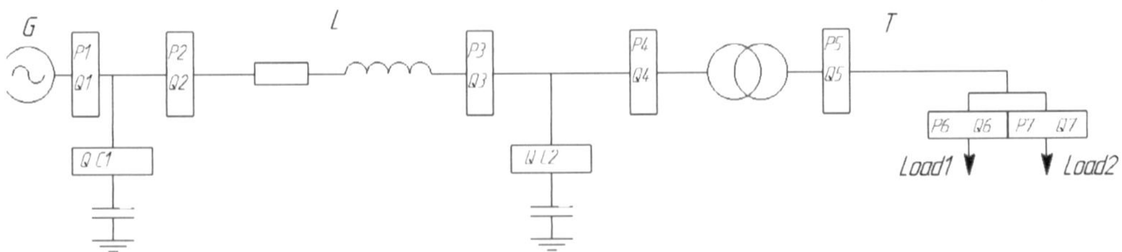


Picture 3. Measurement devices location.

## 2. Balanced load case

For balanced load case scheme, which is shown in Pic.1, consisted of only one load, in which impedances in phases are equal.

Parameters of scheme are next: nominal voltage:110 kV, transmission line length: 15 km, wire type: AC 240/32 (more detailed parameters in [1, Table 3.5, p. 83]), type of transformer:ТДН 40000/110 (parameters are in [1, Table 5.18, p. 250]). Active power of load: 10 MW, reactive power of load: 4 Mvar.

Parameters for MatLab model was calculated using [2].

Measured values are represented in Table 1.

Table 1. Measured power for balance load case

| Number of measurement device/ power | $P_{nom}$ = 10 MW $Q_{nom}$ = 4 Mvar |
|---|---|
| P1, MW | 9.641 |
| P2, MW | 9.641 |
| P3, MW | 9.627 |
| P4, MW | 9.627 |
| P5, MW | 9.553 |

For verification power losses were calculated and compared with measured values. Difference between them should not be more than 5%:

$$\varepsilon = \frac{\Delta P_{net\ meas} - \Delta P_{net\ calc}}{\Delta P_{net\ meas}} \cdot 100\% \tag{1}$$

Equations for calculations are next:

$$\Delta P_{net} = \Delta P_{Line} + \Delta P_{T\,oc} + \Delta P_T \quad (2)$$

Where

$$\Delta P_{Line} = 3 \cdot I_{Line}^2 \cdot R_{Line} \quad (3)$$

$$\Delta P_T = \Delta P_{T\,HV} + \Delta P_{T\,LV}. \quad (4)$$

$$\Delta P_{T\,HV} = 3 \cdot I_{T\,HV}^2 \cdot R_{T\,HV} \quad (5)$$

$$\Delta P_{T\,LV} = 3 \cdot I'^2_{T\,LV} \cdot R_{T\,LV}, \quad (6)$$

Current $I'^2_{T\,LV}$ can be calculated using transformation ratio:

$$k = \frac{U_{T\,HV}}{U_{T\,LV}} \quad (7)$$

$$I'_{T\,LV} = \frac{I_{T\,LV}}{k} \quad (8)$$

Results of calculation are shown in Table 2:

Table 2. Calculated values of losses of power for balanced load case

| | Calculated power losses $\Delta P$, kW |
|---|---|
| $\Delta P_{Line}$ | 0.013 |
| $\Delta P_{T\,HV}$ | 0.011 |
| $\Delta P_{T\,LV}$ | 0.011 |
| $\Delta P_{net}$ | 0.085 |

Error is 3.4%, and that is quite accurate.

## 3. Imbalanced load case

For imbalanced loads case the same scheme with thesame parameters was used, except parameters of load. Imbalanced load can be modelled as a 3 different loads connected in separate phases. Parameters of load are next: $P_a = 5$ MW, $Q_a = 2$ Mvar, $P_b = 3$ MW, $Q_b = 1.2$ Mvar, $P_c = 2$ MW, $Q_c = 0.8$ Mvar.

Procedure of verification is the same in general, but imbalanced load leads to different power floes in phases, and measurements should be made in each phase separately, as well as calculations. Measurement values are shown in Table 3.

Table 3. Measured values for imbalanced load model

| Number of measurement device/ power | Power in phase a $P_a$, MW | Power in phase b $P_b$, MW | Power in phase c $P_c$, MW |
|---|---|---|---|
| P1, MW | 6.456 | 6.266 | 5.169 |
| P2, MW | 6.456 | 6.266 | 5.169 |
| P3, MW | 6.438 | 6.247 | 5.156 |
| P4, MW | 6.438 | 6.247 | 5.156 |
| P5, MW | 5.746 | 5.941 | 6.028 |
| P6, MW | 2.298 | 3.127 | 3.767 |
| P7, MW | 3.448 | 2.814 | 2.26 |

Calculated values are shown in Table 4.

Table 4. Calculated values for imbalanced load model

| Measured quantity/ Value | Phase *a* | Phase *b* | Phase *c* | Total |
|---|---|---|---|---|
| $\Delta P_{Line}$, kW | 0.017 | 0.019 | 0.013 | 0.049 |
| $\Delta P_{T\ HV}$, kW | 0.014 | 0.016 | 0.01 | 0.04 |
| $\Delta P_{T\ LV}$, kW | 0.012 | 0.012 | 0.013 | 0.037 |

There is no difference between calculated and measured values.

From the Table 3 it can be seen that power in phases before transformer (P4) and after it (P5) are different. It is connected with current transformation in transformer. In model transformer has so-called Star/Delta connection (Y/Δ) (Pic.4).

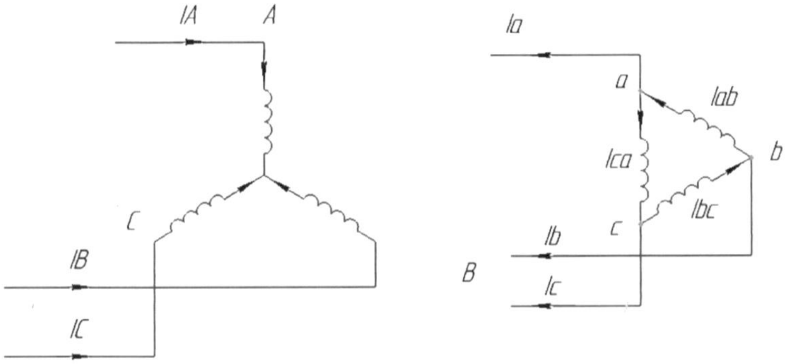


Picture 4. Transformer winding connections: left – Star connection, right – Delta connection.

Current in primary winding are transferred in currents in secondary winding:

$$I_{ab} = k_T(I_B) \tag{9}$$

$$I_{bc} = k_T(I_C) \tag{10}$$

$$I_{ca} = k_T(I_A) \tag{11}$$

From picture next equations can be written:

$$I_a = I_{ab} - I_{ca}, \tag{12}$$

$$I_b = I_{bc} - I_{ab}, \tag{13}$$

$$I_c = I_{ca} - I_{bc}, \tag{14}$$

or, uniting (9) and (12), (10) and (23), (11) and (14), it can be written as:

$$I_a = \frac{k_T(I_B) - k_T(I_A)}{\sqrt{3}} = \frac{k_T(I_B - I_A)}{\sqrt{3}}; \tag{15}$$

$$I_b = \frac{k_T(I_C - I_B)}{\sqrt{3}}; \tag{16}$$

$$I_c = \frac{k_T(I_A - I_C)}{\sqrt{3}} \tag{17}$$

There will be different currents in phases because of different phase loads. If current in phase a after transformer is biggest, current in phase c is smallest and current in phase b is medium between them, then before transformer current in phase B will be biggest one proportional to current ab), current in

IOP Conf. Series: Materials Science and Engineering **240** (2017) 012051 doi:10.1088/1757-899X/240/1/012051

phase A will be approximately medium (proportional to current ca) and current in phase C will be smaller one (proportional to current bc).

Also there is a voltage transformation. In Delta connection there is no neutral point, and that's why it is impossible to calculate phase-to-ground voltages directly. For this purpose artificial neutrals or neutrailer devices are used [3].

Voltages and currents also change the angle (transformer connection group). In modeled case, this group is 11 (angle between primary and secondary voltages is 330 deg).

Therefore, power is multiplication of current and voltage, and according to above, power in phase c after transformer should be bigger than power in phase b, and power in phase a should be the smallest one. Table 3 proves this.

## 4. Scheme with non-sinusoidal load

Non-sinusoidal load is a load, that can create a distortion in a sinusoidal waveform of voltage. Usually sources of this distortions are AC loads, that apply a power from DC sources (invertors).

Any non-sinusiodal signal (voltage or current) can be represented as a sum of the numerous of harmonics (Fourier decomposition):

$$X(\omega t) = A_0 + \sum_{n=1}^{\infty}(A_n\, cos(n\omega t) + B_n\, sin(n\omega t)) \tag{18}$$

In this paper this load is a 6-pulse inverter (Pic. 5).

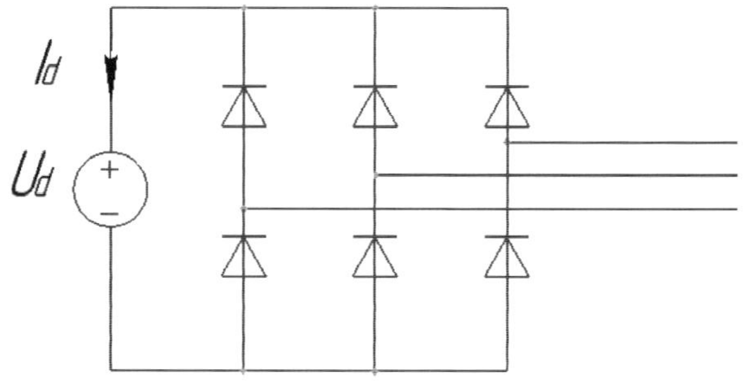


Picture 5. Scheme for 6-pulse inverter

For this device, current can be written (using Fourier decomposition) as:

$$i(\omega t) = \frac{2\sqrt{3}I_d}{\pi}(sin(\omega t) - \frac{1}{5}sin(5\omega t) + \frac{1}{7}sin(7\omega t) - \cdots) \tag{19}$$

It can be seen that there are no even harmonics and harmonics, multiplied for 3:

$$n = 6k \pm 1, \tag{20}$$

where n – number of harmonic, k=1,2,3...

In this work harmonics, which are close to fundamental, will be modelled (5 and 7), as well as harmonic, quite far for fundamental (23), will be modelled. Each of this harmonics exists only in certain frequency:

$$n = 5;\ f_5 = 50\ Hz \cdot 5 = 250\ Hz,$$
$$n = 7;\ f_7 = 50\ Hz \cdot 7 = 350\ Hz,$$
$$n = 23;\ f_{23} = 50\ Hz \cdot 23 = 1150\ Hz$$

Scheme for this case is shown in the Pic. 6.

IOP Conf. Series: Materials Science and Engineering **240** (2017) 012051 doi:10.1088/1757-899X/240/1/012051

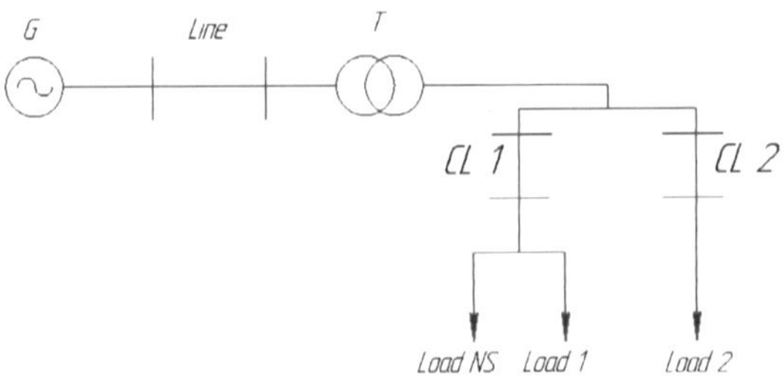


Picture 6.Sheme for non-sinusoidal load case

Om this scheme: G – generator, Line – Transmission Line, T-Transformer, CL1 and CL2 – Cable lines, Load NS – non-sinusoidal load, Load1 and Load2- balanced loads.

Parameters of the scheme are next: nominal voltage: 110 kV, line length: 50 km, type of wires in line: AC 240/32 (more detailed parameters are in [1, Table 3.5, p. 83], transformer: ТДН 40000/110 (see general description, chapter 2).

Parameters of the Cable line 1: length: 50 km, type of wires: ААГУ 120 (more detailed parameters are in [1, Table.3,29, p.106]), voltage – 10 kV.

Parameters of the Cable Line 2: length: 60 km, type of wires: ААГУ 120 (more detailed parameters are in [1, Table.3,29, p.106]), voltage – 10 kV.

Load parameters: Load1: P=8 MW, Q=3Mvar; Load2: P=10 MW, Q=4 Mvar.

Non-sinusoidal load is modeled by the current sources with different frequencies and magnitudes. By the way, current magnitude is decreasing with increasing of harmonic order (Pic.7).

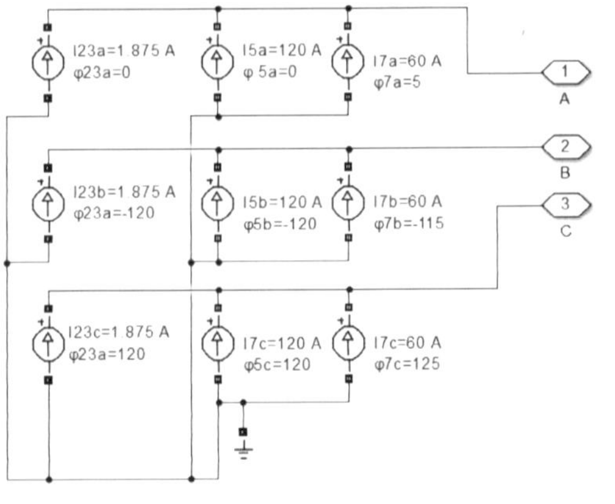


Picture 7. Model of non-sinusoidal load.

Parameters of this model! are next (Table 5):

Table 5. Parameters for non-sinusoidal load model

| **Number of harmonic/ parameter** | **n = 5** | **n = 7** | **n = 23** |
|---|---|---|---|
| Current amplitude in phase a, A | 120 | 60 | 1.875 |
| Angle of phase a, deg | 0 | 5 | 0 |
| Current amplitude in phase b, A | 120 | 60 | 1.875 |
| Angle of phase b, deg | -120 | -115 | -120 |
| Current amplitude in phase c, A | 120 | 60 | 1.875 |
| Angle of phase c, deg | 120 | 125 | 120 |

IOP Conf. Series: Materials Science and Engineering **240** (2017) 012051 doi:10.1088/1757-899X/240/1/012051

Verification was done with the same manner, but for each harmonic. measured values are shown in Table 6.

Table 6. Measured values for non-sinusoidal load case

| **Number of measurement device/ power** | **Power in 1st harmonic, $P_1$, MW** | **Power in 5th harmonic, $P_5$, kW** | **Power in 7th harmonic, $P_7$, kW** | **Power in 23th harmonic, $P_{23}$, kW** |
|---|---|---|---|---|
| P1 | 17.19 | $1.492 \cdot 10^{-4}$ | $-3.672 \cdot 10^{-5}$ | $1.746 \cdot 10^{-5}$ |
| P2 | 17.19 | $0.149 \cdot 10^{-3}$ | $-3.673 \cdot 10^{-5}$ | $1.746 \cdot 10^{-5}$ |
| P3 | 17.05 | -0.522 | -0.1014 | $-9.772 \cdot 10^{-5}$ |
| P4 | 17.04 | -0.5203 | -0.1033 | $-0.119 \cdot 10^{-3}$ |
| P5 | 16.92 | -0.79 | -0.1647 | $-0.132 \cdot 10^{-3}$ |
| P6 | 7.794 | 11.39 | 4.019 | $5.26 \cdot 10^{-3}$ |
| P7 | 6.911 | -3.06 | -8.409 | $-7.596 \cdot 10^{-3}$ |
| P8 | 6.911 | 36.08 | 13.41 | 0.036 |
| P9 | 0 | 66.74 | 21.83 | 0.043 |

Calculated values and error between it and measured values are shown in Table 7.

Table 7. Calculated values for non-sinusoidal scheme model and error

| **Measured quantity/value** | **1st harmonic, kW** | **5th harmonic, kW** | **7th harmonic, W** | **23rd harmonic, W** |
|---|---|---|---|---|
| $\Delta P_{Line}$ | 144.327 | 520.53 | 102.974 | 0.113 |
| $\Delta P_{T\ HV}$ | 37.338 | 119.851 | 22.075 | 0.001752 |
| $\Delta P_{T\ LV}$ | 33.880 | 2414.739 | 460.13 | 0.00632 |
| $\Delta P_{T\ CL1}$ | 813.915 | 16694.647 | 3614.245 | 3.641 |
| $\Delta P_{T\ CL2}$ | 1399.706 | 1773.133 | 623.819 | 0.78 |
| $\sum \Delta P$ | 265.465 | 215.219 | 4823.268 | 4.534 |
| $\varepsilon, \%$ | 0.03 | 1.44 | 1.92 | 4.98 |

Error is raise with increasing of harmonic order. It is connected with so-called skin-effect, where distribution of current on wire is non-linear.

## Conclusion

1. Schemes with different load types were modeled;
2. In this schemes error between calculated and measured values are in normal limits (no more than 5%);
3. Difference in phase power before transformer and after it was explained.

References

[1] Karapetyan I G, Faybisovich D L, Shapiro I M 2009 *Spravochnik po proektirovaniju elektricheskikh setey* [Guide to design of electrical networks], Moscow, ENAS Publ. 392 p

[2] Dvorkin D V 2014 *Opredelenie tipov sredstv kompensatcii iskazhenii sinusoidalnosti i simmetrii napryazhenija i razrabotka metodiki opredelenija mest ich ustanovki.* Master Thesis. [Definition of compensation means of voltage's sinusoidal waveform and symmetry distortion and development of determination of their locations methods. Master thesis]. Moscow 147 pp

IOP Conf. Series: Materials Science and Engineering **240** (2017) 012051 doi:10.1088/1757-899X/240/1/012051

[3] Kaminskiy E A 1973 *Zvezda, treugolnik, zigzag* (Y, Delta, Interstar], Moscow, Energia Publ. 104 pp
[4] Matveev A N 1983 *Elektrichestvo i magnetism* [Electricity and magnetism], Moscow, Vysshaya Shkola Publ. 463 pp
[5] Prokhorov S G and Khusnutdinov R A 2002 *Electicheslije mashiny* [Electrical machines], Kazan, Kazan State University Publ. 140 pp

IOP Conf. Series: Materials Science and Engineering **240** (2017) 012052 doi:10.1088/1757-899X/240/1/012052

# Rise in accuracy of gas stream spectral analysis in mechanical engineering technology

**G F Mukhametzyanova, F G Karikh and I R Mukhametzyanov**

Kazan (Volga region) Federal University, Naberezhnye Chelny Institute (branch of), Department of Materials, Technology and Quality, Russian Federation, Republic of Tatarstan, Naberezhnye Chelny, 423812, Suyumbike Avenue, 10A

gulnara_ineka@mail.ru

**Abstract.** The submitted method of rise in accuracy of gas stream spectral analysis in mechanical engineering technology by controlling the chemical composition of the exhaust gas streams by means of a six-jet electric-arc plasmatron is available.

To provide the quality of products manufacturing operating procedure in mechanical engineering technologythe chemical composition controllingusing various methods such as chemical, X-ray spectral and emission spectral analyses is widely used.One of the most commonly used and short-time is emission spectral analysis. In the world practice the rise in the atomic-emission spectral analysis accuracy with the outgoing gas streams is being achieved by using various types of plasmatrons. Their disadvantage is the low efficiency of particlesthrown away from the high temperature zone penetration into plasma. A high-voltage six-electrode flare discharge lacks thisdisadvantage due to highly effective gas-dynamic capture of the supplied particles. However, it is characterized by insufficiently high excitation temperature - 4 $800^0$C [1 – 3].

The objective of this work is to increase the accuracy of materials atomic emission spectral analysis by means of controlling the chemical composition of the exhaust gas streams generated during the products manufacturing or under the direct chemical composition control.

The use of a new plasma flow generator - a six-jet electric arc plasmatron [4] has been proposed in order toachieve this goal. This one makes it possibleto control the plasma flow temperature in the range of 6 $000^0$C to 12 $000^0$C. For this problem solution a physical model of a six-jet electric arc plasmatron has been developed, that makes it possible to use this temperature interval as a source of thegas flow spectrum excitation under the circumstances of a highly efficient capture of its particles.

The image of a six-jet electric-arc plasmatron is shown in Fig. 1, 2.

Plasma-forming copper heads 1 mounted on dielectric plateaus 2 are rigidly attached to brackets 3 with the ability to move along the axes of heads 1 in a direction perpendicular to tubular struts 4. A

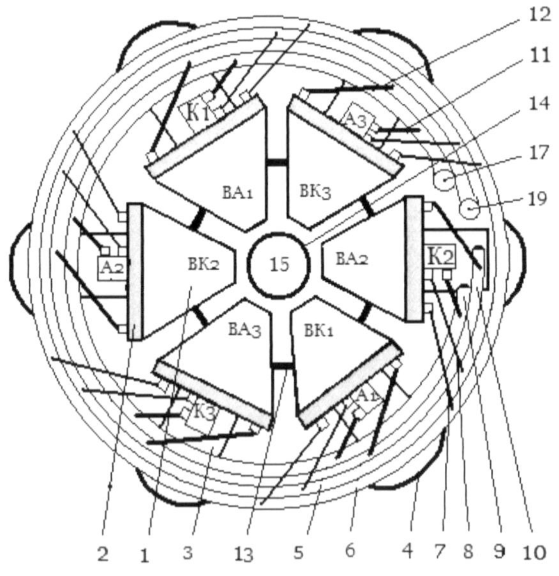


a)

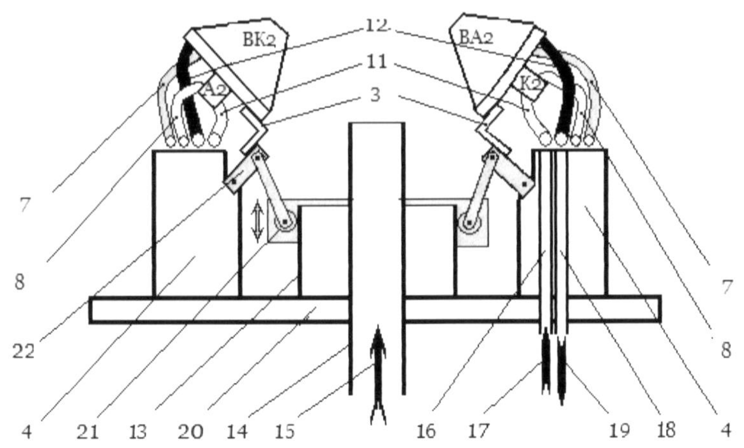


b)

Fig.1 – Scheme of a six-jet electric arc plasmatron a) top view; b) side view: 1 – copper heads; 2 – plateau; 3 – brackets; 4 –tubular posts; 5, 6, 9, 10 – chamber; 7, 8, 11, 12 – hoses; 13, 14 – a branch pipe; 15 – powder material; 16, 17, 19 – channel; 20 – mounting table; 21 – hinged system; 22 – plateau.

tubular chamber 5 for argon supplying to heads which protects the electrodes from oxidation and chamber 6 for distributing the working gas (air) by means of flexible hoses 7 areannularly placed above them. The protective gas is supplied to copper anodes with contacts A1, A2, A3 and tungsten cathodes with contacts K1, K2 and K3 by means of flexible hoses 8.

The cooling water inlet chamber 9 and chamber 10 discharging water into channel 19 are placed above posts 4axially to chambers 5 and 6. The cooling water inlet in the head section is made from the vertical channel 17 (see Fig. 2, which shows the vertical plasmatron section along the flat-bone, passing through the pair of heads axes, powered from one of the phases of the in Use three-phase rectifier, in particular, connected to the anode head with electrode A2 and auxiliary cathode BK2, and a cathode head with cathode K2 and auxiliary anode A2). There is chamber 10 discharging water into channel 19 next to the annular chamber 9. Chambers 9 and 10 connections with the head sections is carried out by means of flexible hoses 11 and 12. The water stream 19 warmed up during the plasmatron operation is directed to be cooled in a radiator (not available in Fig.2), from which the cooled water returns to channel 16. Posts 4 are located on amounting table 20, there is a rigidly mounted branch pipe 14 between, which is forming the analyzed gas stream or the processed powder

Fig. 2 – Six-jet electric-arc plasmatron.

material 15 [5]. Cylinder 13 is placed axially to branch pipe 14 for synchronizing the change in the angle of six heads convergence by means of a hinge system 21, which provides a synchronic change in thegradient angle of heads relatively to thetable plane 20. System 21 comprises a plateau 22 with movable brackets 3, thus providing a change in the magnitude of an interelectrode gap between plasma-forming heads of the plasmatron.

As you can see in Fig.2, the plasma flow is directed vertically upwards during the process of this physical model operation as a source of spectral excitation.

At present, the pilot-industrial plasmatron has been manufactured and is being planned for commissioning at "KamAZ-Metallurgy" PTC in orderto control inreal time the quality of the alloy being prepared [6].

## References

[1] *Metallurgy and steel heat treatment* 1983 [Reference book 3rd edition in 3 volumes]Vol 1Methods of testing and research / Edited by M L Bernstein and A G Rakhshtadt Moscow: Metallurgy p 352

[2] Karikh F G, Mukhametzyanova G F and Petrov D M 2003 A setup for monitoring the chemical composition of gas flows forming during the pyrolysis of solid organic materials *Instruments and Experimental Techniques* Volume 46 Issue 5 pp 717 – 718

[3] Karikh F G, Mukhametzyanova G F and Petrov D M 2003 Continuous control of melting in the composition of waste gases in pyrometallurgy *Machine-building metallurgy* 2 pp 2 – 3
[4] Patent 2529740 Russian Federation 2014 IPC H05H1/26Six-jet electric-arc plasmatron / F G Karich, A F Karich, V S Poprotskey and R V Poprotskey - № 2013129645/07 application 27.06.2013 published 27.09.2014 Bulletin 27
[5] Patent 2375687 Russian Federation 2009 IPC G01J 3/00 G01F 13/00 The method and apparatus for supplying powdered materials to a high-voltage flare discharge / F G Karich, R A Bikulov and R N Abdrakhmanov - № 2008118262/28 application 07.05. 2008 published 10.12.2009 Bulletin 34
[6] Karich F G 2004 *Methodology of metal melting physical and chemical parameters determination on the basis of spectroanalytical data* Thesis for a Doctor of Technical Sciences degree Chelyabinsk: South-Ural State University p 310

IOP Conf. Series: Materials Science and Engineering **240** (2017) 012053 doi:10.1088/1757-899X/240/1/012053

# Process research into metallic pipe wear of hot chamber die casting machines and methods ofincreasing wear resistance

**G F Mukhametzyanova[1], MS Kolesnikov[2], I R Mukhametzyanov[1] and V I Astatshenko[1]**

[1]Kazan (Volga region) Federal University, Naberezhnye Chelny Institute (branch of), Department of Materials, Technology and Quality, Russian Federation, Republic of Tatarstan, Naberezhnye Chelny, 423812, Suyumbike Avenue, 10A
[2]Kazan (Volga region) Federal University, Naberezhnye Chelny Institute (branch of), Department of Mechanical Engineering, Russian Federation, Republic of Tatarstan, Naberezhnye Chelny, 423812, Suyumbike Avenue, 10A

gulnara_ineka@mail.ru

**Abstract.** The kinetics and reasons for metallic pipe wear of hot chamberzinc alloy die casting machines are established.Increasing metallic pipe wear components wear resistance is being achieved by means of die steelДИ – 22 with electroslag remelting modification and electron-beamremelting modification and after the processes of nitriding and boriding besides.

In the process of zinc alloys casting (modifications: Zamak - 4 - 1, Zamak - 10 - 1 and others) the most overloaded pressing unit features of hot chamber die casting machines typesCLT and IDRA appear to be metallic pipes components. Beta testing data carried out with KAMAZ PTCshow that metallic pipes componentsserviceability of IDRA machines made of steel 4Х5МФС doesn'texceed 2,5% ofzinc alloy die casting machinesmolding toolsaverage durability. Metallic pipes are being run at the temperature of zinc meltsequal to723±15K. High physicochemical interaction of zinc melts and steels influences on metallic pipe wire most of all, which depends on solution of metallic pipes componentsinternal channels[1 – 3].Consequently, processes research into metallic pipe wear of hot chamber zinc alloy die casting machines in order to increase wear resistanceis really actual.

To provide a quantitative law of tips internal parts damageability the investigation was held. The test disks had the following measurements: outer diameter - 30mm; internal diameter - 20mm; disk thickness – 5mm.Solutionintensity was estimated from the disk weight loss with regard to surface unit($\Delta P, g/cm^2$) during the period of planned investigations ($n = 1 \div 3\ 000$ cycleswith intervals 100 cycles).The results of tests on the solution kinetics of the metallic pipeactive faces are shown in Fig.1.Trend of a curve$\Delta P = f(n)$ is the evidence ofa compound dependence of steel 4Х5МФСsolution at various stages of operation.Conditionally, it is possible to distinguish 4 characteristic stages of the analyzed process. At the first stage ($n \leq 100$ cycles) the steel solution in melt Zamak - 4 - 1 increases insignificantly. The observed incubation period is associated with the increased resistance to oxide film solution on the metallic pipe surface. At stage *II* ($n > 100$ cycles), a sharp increase in solution rate was established. In the interval n ≈ 500-700 cycles (Stage *III*), the solution rate slows down. At stage *IV* ($n > 3000$ cycles) the intensified solution ofthe metallic pipe active face is again noted.

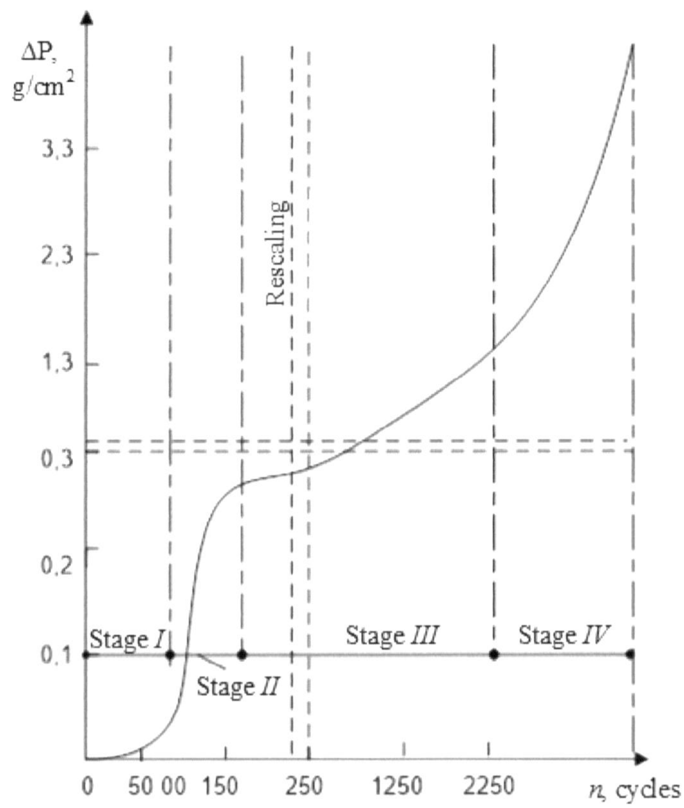


Fig.1 – The solution kinetics of the metallic pipe active faces made from steel 4X5MΦC in melt Zamak - 4 – 1 at 723K.

Metallographic and X-ray diffraction studies performed with DRON-3 installation established the formation of a transition layer on the tip channel surfaces being in contact withthe melt, consisting of the intermetallide phases $FeAl_3$, $Fe_2Al_5$, $Fe_3Zn_{10}$(Stage *II*) [4]. Identification of intermetallide phases was carried out according to characteristic radiation by scanning pulses at oblique cuts of polished sections. The structure of the layer is characterized by the presence of some defects in the form of pores, ruptures and cracks and by a strongsmall hardness $H_\mu$ ~ 1800, (Curve 3, Fig. 2).

As soon as *n* increases up to 1000 cycles, the intermetallide interlayer forms a continuous coating with thickness of ~ 0,08mm. The presence of these intermetallide phases causes an increased resistance to solution, (Stage *III*). Further accumulation of cycles (n>1000) has little effect on the transition intermetallide layer thicknessin the "steel-melt" system. At the same time, there is a slight decrease insmall hardness (Curve 4, Fig. 2), which is associated with the formation of an intermetallide layer with new characteristics ($Fe_5Zn_2$, $FeZn_7$, $FeZn_{13}$). Accordingly, the number of interference lines corresponding to iron-aluminum intermetallide elements such as $FeAl_3$ *and* $Fe_2Al_5$ decreases. At the end of this stage (n ~ 1300 cycles), the thickness of the layer increases, and thesmall hardnessdecreases noticeably (Curve 5, Fig. 2). At the same time there is an increase in the rate of the metallic pipe channel active face solutionalong with the disappearance of iron-aluminum intermetallide elements.

The obtained results make it possible to supplement the existing ideas about the causes for one of the most common types of technical failures in practice (the loss of the channel size and the implied danger of metallic pipes destruction or melt sputtering at the structural parts mating), and also implement a number of applied solutions aimed to increase metallic pipe wear components wear resistance.

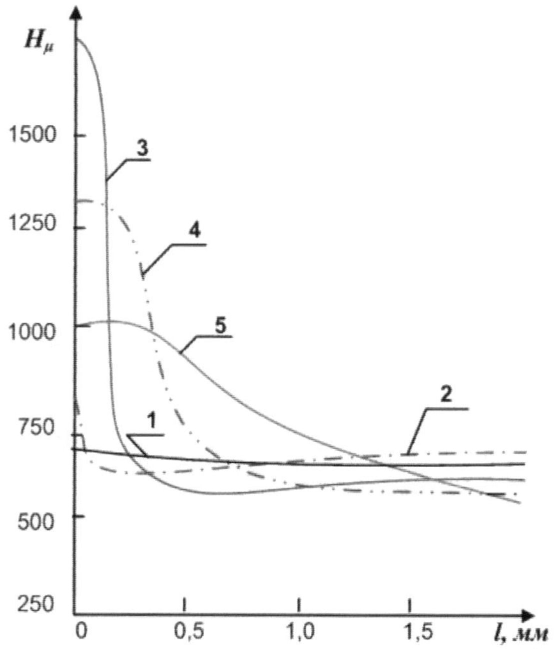


Fig. 2 – Change of small hardness in the surface layer of a metallic pipe material made from steel 4Х5МФС under the condition of interacting with melt Zamak - 4 - 1 at 723K: Curve 1 - initial state; Curves 2,3,4 and 5 after 5, 120, 1000,1300 pressing cycles respectively.

Along with steel 4Х5МФС widely used for metallic pipes manufacturing, such steel grades as 5Х4СВ4МФ, 4Х4ВМФС, 5Х3В3МФС, and steel 4Х4ВМФС (ДИ-22) received with the help of electric-slag and electron-beam remelting, and hardening by chemical heat treatment were subjected to research (Table 1). According to the solubility index (ΔP, $g/cm^2$) steel 4Х4ВМФС (ДИ-22) is promising for the manufacturing of metallic pipe elements of hot chamber zinc alloy die casting machines (from the number of materials studied). In accordance with the analyzed data the components serviceability reserve with the metallic pipes of die casting machines is based in the application of increased metallurgical purity material. The resistance of samples made of steel 4Х4ВМФСreceived with the help of electric-slag and electron-beam remelting to solution inZamak - 4 - 1 is 2.1 and 2.5 times higher on average than after conventional electric arc melting. The use of chemical-thermal treatment (nitriding and boriding) can significantly improve the serviceability of metallic pipes (Table 1).

Table 1 – The resistance oftool steel to solution in melt Zamak - 4 - 1 (melt temperature –723K, the period of planned investigations, $n$ = 3 000 cycles).

| Steel grade | Temper regime | | Hardness, HRC, HB | Strengthening treatment | Hardness, HV | Relative resistance to solution in melt Zamak - 4 – 1 |
|---|---|---|---|---|---|---|
| | $T_{hardening}$, K | $T_{cast}$, K | | | | |
| 4Х5МФС | 1313 | 853 | 50-53 | | | 1 |
| 5Х4СВ4МФ | 1393 | 873 | 52-54 | | | 1,28 |

| | | | | | | |
|---|---|---|---|---|---|---|
| 4Х4ВМФС (ДИ – 22) | 1333 | 853 | 50-53 | | | 1,2 |
| 5Х3В3МФС | 1393 | 873 | 52-54 | | | 1,32 |
| 4Х4МФС (electric-slag remelting) | 1333 | 853 | 50-52 | | | 2,52 |
| 4Х4МФС (electron-beam remelting) | 1333 | 853 | 50-52 | | | 3,0 |
| 4Х4ВМФС | 1333 | 853 | 50-52 | Nitriding, 833K – 8 hours., 803K – 16 hours Layer depth 0,18 – 0,21mm | 860 - 880 | 4,4 |
| 4Х4ВМФС | 1333K (annealing – 2 hours) | | 160-180 HB | Boriding, 30%$B_4C$, 60%$Na_2B_4O_7$, 10%SiC, 1173K – 30 hours. Layerdepth 0,1 – 0,12 mm | 1800 | 4,27 |

## References

[1] Goryunov I I 1973 *Molding tools for die casting* [Reference book] L: Mechanical engineeringp 256
[2] Gini E CH, Zarubin A M and Rybkin V A 2008 *Foundry engineering technology Special types of casting* (Edited by VA Rybkin 3rd edition) M: Academy p 352
[3] Nikitin V I 1967 *Physicochemical parameters in the proses of hard metal exposure to hot metal* Moscow: Atomizdat p 422
[4] Rabkin D M and Lozovskaya A V 1992 *Physical metallurgy of aluminum and its alloys welding* Kiev: Naukova dumka p 160

IOP Conf. Series: Materials Science and Engineering **240** (2017) 012054 doi:10.1088/1757-899X/240/1/012054

# Simulation of parameters of hydraulic drive with volumetric type controller

**V L Mulyukin[1], A V Boldyrev[1], D L Karelin[1], A M Belousov[1]**

[1]Kazan Federal University, Naberezhnye Chelny Institute
Pr. Mira 68/19, Naberezhnye Chelny, Republic of Tatarstan, Russia, 423812

karelindl@mail.ru

**Abstract**. The article presents a mathematical model of volumetric type hydraulic drive controller that allows to calculate the parameters of forward and reverse motion. According to the results of simulation static characteristics of rod's speed and the force of the hydraulic cylinder rod were built and the influence of the angle of swash plate of the controller at the characteristics profile is shown. The results analysis showed that the proposed controller allows steplessly adjust the speed u of hydraulic cylinder's rod motion and the force developed on the rod without the use of flow throttling.

Increase of fuel price and tightening of ecological requirements make the problem of increasing of earthmovers' energy efficiency more and more actual.

Nowadays one of the ways to increase energy efficiency is using the combined power unit with volumetric hydraulic drive. It is characterized by high rate of power density and smooth control of operating parameters.

Traditionally hydraulic drive contains a pump with several hydraulic actuators as well as directional and regulating equipment. Nowadays it is spreading the electric and hydraulic drive with hydraulic actuators controlled by valves with proportional solenoids [1]. The drive is controlled by using of throttle regulator installed in series or in parallel that causes the increase of energy cost.

There is the variant of using the volumetric type controller operating as axial-piston hydraulic machine. In combination with control system it allows to recover energy when lowering the working bodies of the earthmover, improve it's handling and increase efficiency of the entire power unit.

The use of the adjustable volumetric flow divider operating as axial-piston hydraulic machine in hydraulic transmission of mobile machines is considered in the work [2]. The use of gear-type volumetric flow divider is shown in the work [3].

This article is devoted to the modeling of processes in the volumetric hydraulic drive with the volumetric type controller that take place when working body of earthmover is moving.

When the earthmover's working body is lifted by hydraulic cylinder the first and the second sections of the volumetric type controller operate in a pump mode and the third section operates in a hydraulic motor mode. Thereby the pressure generated in the pressure chamber of the hydraulic cylinder is higher than the pressure generated by the pump. When the earthmover's working body is lowered the pump's flow and pressure are not required, so the operating modes are changed allowing to use the hydraulic cylinder as a pump.

For calculation of required pump delivery at a known speed of the output link of hydraulic actuator (direct problem) and calculation of the speed of the output link of hydraulic actuator for a known pump delivery (inverse problem) we use the balance of the input and output flows of the volumetric type controller (fig. 1):

$$Q_n + Q_{cI} = Q_{oII} + Q_{oIII}, \tag{1}$$

where $Q_n$ – pump delivery, m$^3$/s; $Q_{cI}$ – drain flow from the first section, m$^3$/s; $Q_{oII}$ – the second section's inlet flow rate, m$^3$/s; $Q_{oIII}$ – the third section's inlet flow rate, m$^3$/s.

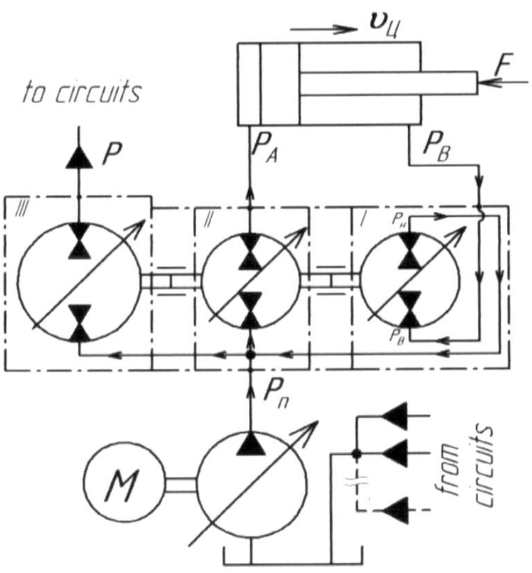


*Figure 1 – Hydraulic drive circuit with volumetric type controller*

We consider that according [4] the average flow rate of axial-piston hydraulic machine is defined as $Q_{cp} = D \cdot tg(\gamma) \cdot S \cdot n \cdot z \cdot \eta_o$ for the section operating in a pump mode and $Q_{cp} = D \cdot tg(\gamma) \cdot S \cdot n \cdot z / \eta_o$ for the section operating in a motor mode and that the shafts of the controller's sections' shafts are rigidly connected so their rotation frequency is the same. Here $D_i$ – the diameter of the pistons' axes in the cylinder block, m; $S_i$ – piston's area, m$^2$; $n$ – controller's shaft rotation frequency, rps; $z$ – number of pistons in the cylinder block; $\gamma_i$ – the angle of the controller's sections swash plates; $\eta_o$ – volumetric efficiency of hydraulic machine (fig. 1).

The resulting system of equations allows to determine the speed of movement of the cylinder rod for various angles of controller's swash plates at a given pump delivery.

For calculation of required pump pressure at a known external force load on hydraulic actuator (direct problem) and calculation of the load at a known pump pressure (inverse problem) we use the balance of forces on the hydraulic actuator, as well as the balance of moments on the shafts of controller's sections operating in a motor mode – $M_I = D_I \cdot tg(\gamma_I) \cdot S_I \cdot z_I \cdot (p - p_n)/\eta_{мI}$ or in a pump mode – $M_{III} = D_{III} \cdot tg(\gamma_{III}) \cdot S_{III} \cdot z_{III} \cdot (p_n - p) \cdot \eta_{мIII}$ [4].

After mathematical transformations provided that the sections I and II have the same dimensions we finally obtain the equations (2-5) representing a mathematical model of the proposed controller:

$$\frac{D_{III} \cdot S_{III} \cdot z_{III}}{D_{II} \cdot S_{II} \cdot z_{II}} \cdot \frac{tg(\gamma_{III})}{tg(\gamma_{II})} \cdot \eta_{мIII} = \frac{\eta_{оц}^2 \cdot \eta_{oI} \cdot \eta_{oII} \cdot \sigma}{\eta_{мII}} + \frac{1}{\eta_{мII}} \left( \frac{p_A}{p_n} - 1 \right) \tag{2}$$

$$F = S_ц \cdot (p_A - \sigma \cdot p_B) \cdot \eta_{мц} \tag{3}$$

$$n = \frac{Q_n}{D_{II} \cdot tg(\gamma_{II}) \cdot S_{II} \cdot z_{II} \cdot \left( \frac{1}{\eta_{oII}} - \sigma \cdot \eta_{oI} \cdot \eta_{oII}^2 \cdot \eta_{цо}^2 \right) + \frac{D_{III} \cdot tg(\gamma_{III}) \cdot S_{III} \cdot z_{III}}{\eta_{oIII}}} \tag{4}$$

$$\upsilon_ц = D_{II} \cdot tg(\gamma_I) \cdot S_{II} \cdot n \cdot z_{II} \cdot \eta_{oII} \cdot \eta_{цо} / S_ц . \tag{5}$$

For the mathematical simulation of changes in the value of the cylinder rod's effort $F_ц$ and it's speed $\upsilon_ц$ at various angles $\gamma$ of controller sections' swash plates we use the software complex « MathCad v.15». We also define the following terms and conditions:

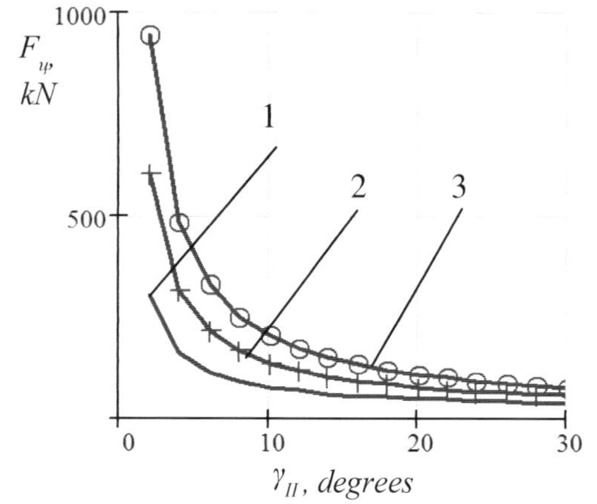


*Figure 2 – Dependence of the force on the cylinder's rod $F_ц$ from the angle $\gamma_I$:*
1- $\gamma_{III}$ = 10°; 2- $\gamma_{III}$ = 20°; 3- $\gamma_{III}$ = 30°.

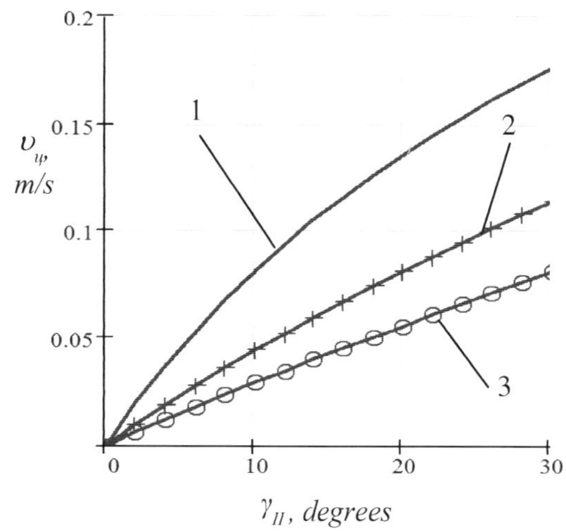


*Figure 3 – Dependence of the cylinder's rod extension speed $\upsilon_ц$ from the angle $\gamma_{II}$:*
1- $\gamma_{III}$ = 10°; 2- $\gamma_{III}$ = 20°; 3- $\gamma_{III}$ = 30°.

- the sections I and II have pistons' area $S_i$ = 3.14 cm$^2$ and the diameters of the pistons' axes in the cylinder block $D_i$ = 100 mm;
- all sections have the same highest control angle $\gamma_i$ = 30°;
- the pressure in the return line of the cylinder $p_B$ and the outlet pressure of section III $p$ are zero;
- pump delivery $Q_n$ = 110 l/min, pump pressure $P_н$ = 6,3 MPa;
- diameter of cylinder's rod $d_ц$ = 80 mm;
- volumetric and mechanical efficiency of hydraulic machines are averaged.

The simulation result is represented as dependence of the force on the cylinder's rod $F_ц$ from the angle $\gamma$ (fig. 2) and dependence of the cylinder's rod extension speed $\upsilon_ц$ from the angle $\gamma$ (fig. 3).

Figure 2 shows that the proposed controller allows to change the force on the cylinder's rod $F_ц$ towards larger values at a constant given pump pressure $P_н$ through the variation of the angle $\gamma$. At small values of the angle of the sections' swash plates $\gamma$ the force $F_ц$ sharply increases so it is necessary to provide the protection of the system from overload.

Dependence on the figure 3 shows that the proposed volumetric type controller allows to steplessly adjust the cylinder's rod speed $\upsilon_ц$ and the force on the cylinder's rod without throttling the flow at a constant given pump delivery.

**Thus**, the use of the proposed volumetric type controller in earthmover's hydraulic drive allows to adjust value of the force generated by hydraulic actuators without increasing the pump pressure and steplessly adjust the speed of the output unit without throttling the flow of hydraulic fluid. This will extend the range of regulation of power and speed characteristics that will significantly improve the earthmover's handling without increasing the power plant capacity.

**References**

[1] Sveshnikov V K 2015 *Machine hydraulic drives* (St. Petersburg: Polytehnika) p 627
[2] Mulyukin V L, Karelin D L and Belousov A M 2016 A mathematical model of the controlled axial flow divider for mobile machines *IOP Conf. Series: Materials Science and Engineering* vol.134 (http://iopscience.iop.org/1757-899X/134/1/012039)
[3] Karelin D L, Gureev V M and Mulyukin V L 2014 Simulation of power redistribution process in volumetric flow divider of multi-circuit cooling system *Bulletin of KSTU named after A. N. Tupolev* vol 4 (Kazan: KNRTU-KAI press) pp 40-44
[4] Bashta T M, Rudnev S S and Nekrasov B B 1982 *Hydraulics, hydraulic machines and hydraulic drives* (Moscow: Mashinostroenie) p 423

IOP Conf. Series: Materials Science and Engineering **240** (2017) 012055 doi:10.1088/1757-899X/240/1/012055

# Improving upset forgingtube billettechnology to avoid the clamp defect on semi-axle housingforging

**D L Pankratov[1], R S Nizamov[2] and I Z Kharisov[2]**

[1]Kazan Federal University, Naberezhnye Chelny Institute, pr. Mira 68/19, Naberezhnye Chelny, Republic of Tatarstan, Russia,423812
[2] Forging Plant of KAMAZ, pr. Avtozavodsky2, Naberezhnye Chelny, Republic of Tatarstan, Russia,423827

pankratovdl@gmail.com, nizamovrs@yandex.ru

**Abstract.** Upset forging with heating on the limited site is simulated. Advantage of type-setting conversion formation method application with a punch in the form of a cone is confirmed.

Technological process of semi-axle housing production in the Forging plant of KAMAZ PJSC represents an upset forging of a tube for 3 conversions. The main problem periodically arising in upset forging is a defect in the form of crack in the inner cavity of the flange portion of the forging. The defect is a clamp on the inner surface of the forging. Modeling process QForm2D V 8.0.4 for basic and experimental technologies are presented in table 1.

On simulation basis the cause of clamping is determined at upset forging of a tube. A method that eliminates the formation of a defect in the semi-finished product by changing the configuration of typesetting conversion was worked out. [1]

Table 1.Comparison of basic and experimental technologies.

| Conversion | Basictechnology | Experimentaltechnology |
|---|---|---|
| 1 conversion  | 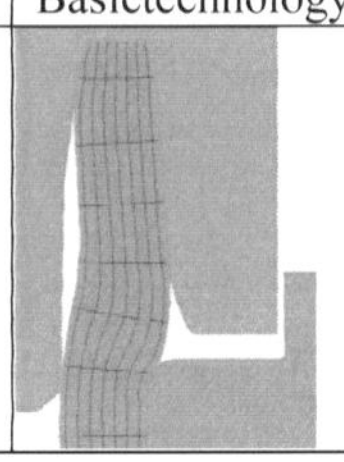 | 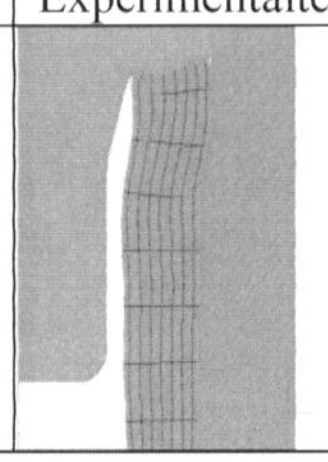 |

| | | | |
|---|---|---|---|
| 2 conversion | 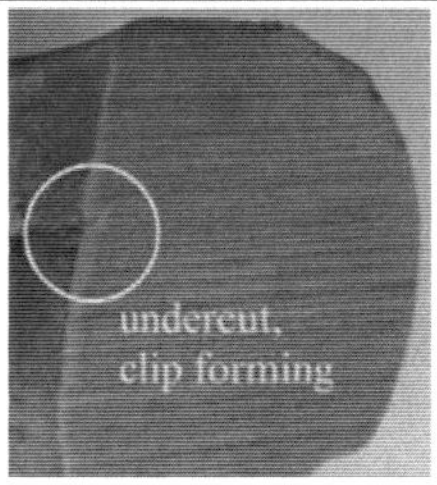  | 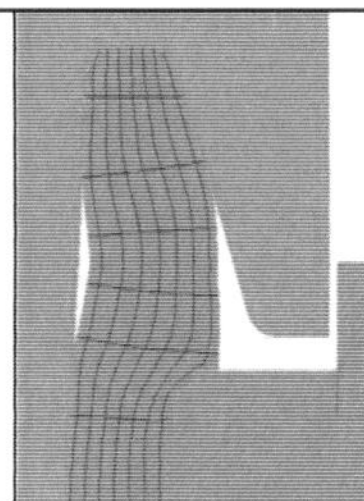 | 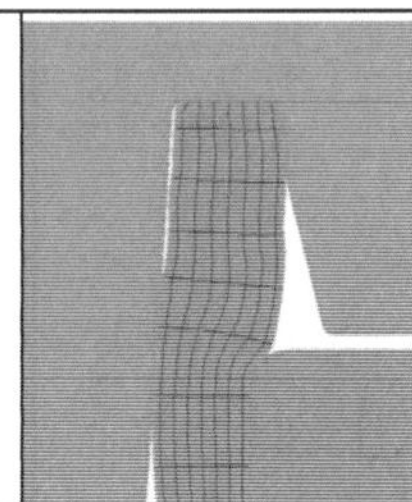 |
| 3 conversion | 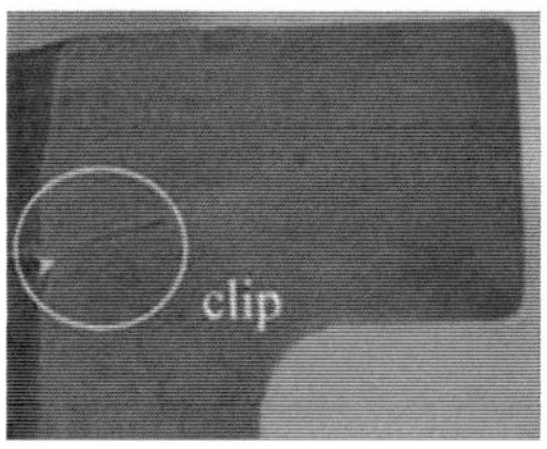  | 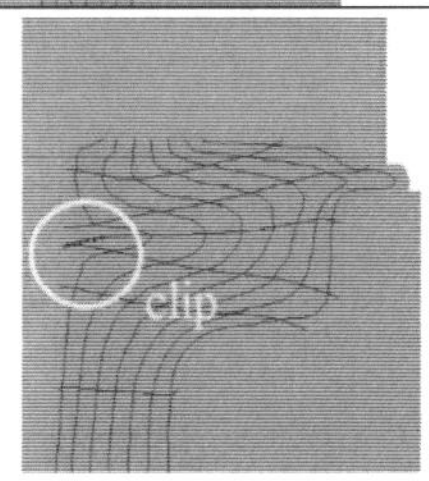  | 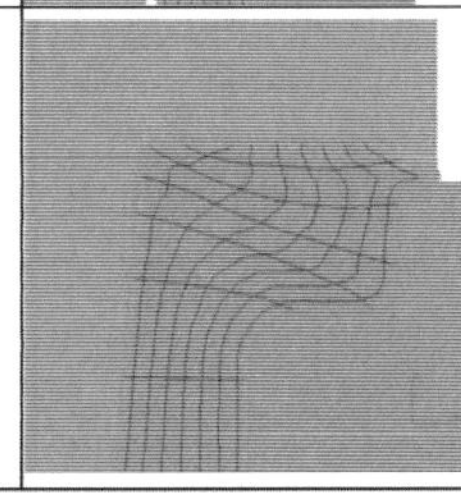 |

However, the stamping of the forging indicates on a significant influence of the billet heating process prior to stamping, as before upset forging a heating is carried out on a limited area. The temperature gradient on the conversion section of the billet is possible to simulate on QForm2D by setting a heat flow for a limited amount of the billet (fig. 1).

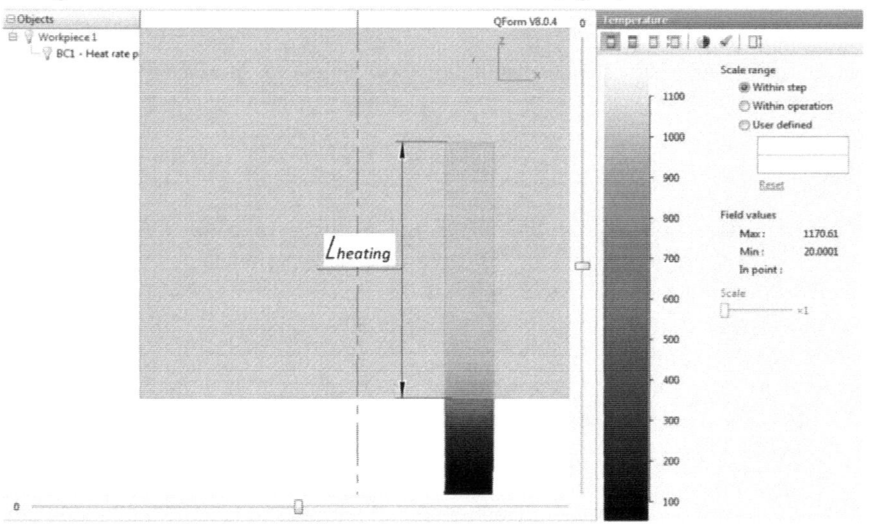


Fig. 1.Simulating heating of a billet section before upset forging.

According Storozhev M.V. recommendations [2, p.127] heating temperature 1180°C before stamping is set. Based on the requirements for end forging temperature of 870°C [2] a simulation of upset forging process for billet heating temperature 1180°C, 1000°C and 900°C was carried out. In general, the heated billet section length is determined from the volume equality of upset billet section and upset forging section [3]. The length of the heated section under the current technological process is 250 mm from the end of the billet and is defined by a QForm "boundary conditions" function. To study the possibility of reducing energy consumption for a preheating,we offer to simulate the process in conditions of heating a section 0,9$L_{heating}$, 0,8$L_{heating}$and 0,7$L_{heating}$, where $L_{heating}$ is the length of the heated section (Figure 1). Previous studies [1] point to the feasibility of calculating the upset forging options by a stamping punch of conical shape with the basis angle of 15° and a stamping punch with a

IOP Conf. Series: Materials Science and Engineering **240** (2017) 012055 doi:10.1088/1757-899X/240/1/012055

flat end. Simulation results of type-setting forged semi-axle housing conversions to the above variants, taking into account the temperature gradient are shown in fig.2.

Simulation with temperature decrease showed that the use of a type-setting stamping punch with flat end increases the likelihood of the clamp defect (fig.2). Besides, on the third conversion for all options considered is a blank area left. Accumulation of scale and technological lubricant in unfilled sections of die bed contour (fig.2) leads to defects as undercut, scale stamping and shooting [4, 5]. In the final conversion, described abnormalities increase the risk of defect stamping in the forging body.

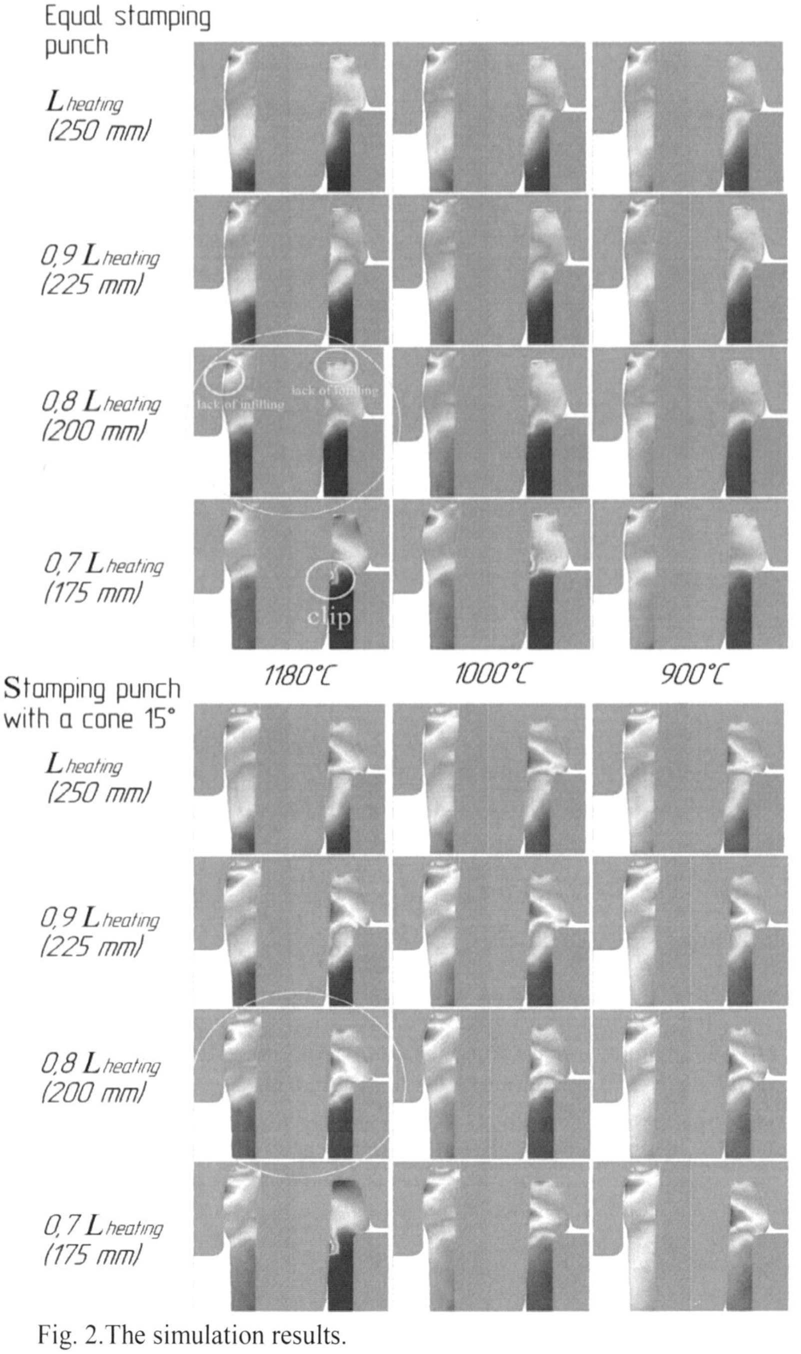


Fig. 2.The simulation results.

The use of a conical stamping punch provides guaranteed filling on conversions at the minimum proposed temperature values of the heated portion length. Due to actual manufacturing conditions it is necessary to use heating temperature 1180°C long and heating length 200 mm respectively (fig.2). The proposed solution will reduce the energy consumption of the heating process before upset forging byapprox. 20%. Comparison of type-setting conversions to the same upset forging conditions ($0{,}8L_{heating}$ = 200mm, 1180°C), represented in the field strains (fig. 3) allows us to conclude that the use of a conical stamping punch allowed to shift shaping in the direction of the unfilled section on the value of Δ. This plastic deformation zone is shifted to the conversion temperature zone on$\Delta_1$. This behavior is due to the law of conservation of volume and due to the decrease in the volume of metal in the upper section of the semi-finished product through the use of a conical stamping punch.

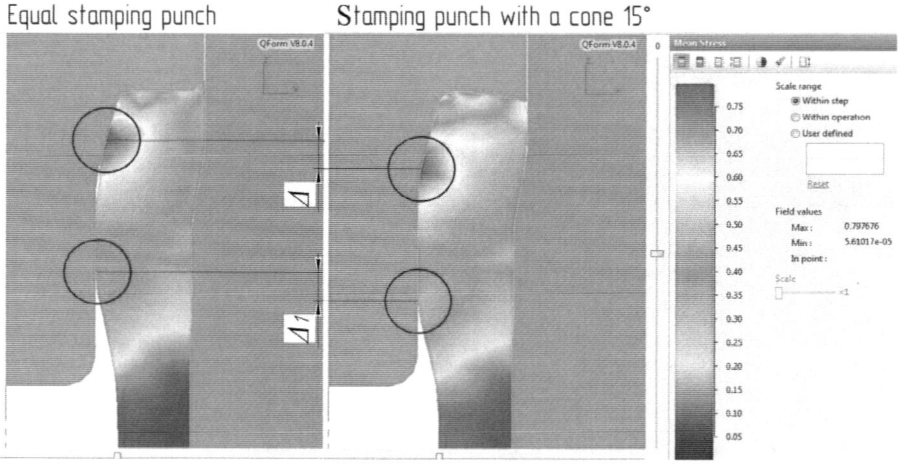


Fig. 3 Type-setting conversion presented in strain field.

Conclusion. Simulation confirms the advantage of the type-setting conversion method with conical stamping punch. Application of the described solutions in practice prevents the formation of latent defects in the finished product. A related effect is the reduction of energy consumption on induction heating of the billet. The proposed method is suitable for the upset forging process of tube billet products. If necessary, it is possible to determine the change in force during upset forging and assess the possibility of cracking [6].

## References

[1] Pankratov D L, Nizamov R S and Kharisov I Z 2016 Investigation of the influence of geometric parameters of stamp engraving type-setting conversions during the upset forging of tube billets to obtain defect-free products *IOP Conf Series:. Materials Science and Engineering* 134 **012047**

[2] Storozhev M V 1969 *Technology Reference for forging and stamping* p 127

[3] Semenov E I 1986 Forging and Stamping. *Directory* Volume 2 p 266

[4] Pankratov D L, Valiev A M 2014 Creation of anisotropic friction conditions in the forging process *IOP Conference Series: Materials Science and Engineering* Volume 69 **012034**

[5] Shaparev A, Savin I 2016 Calculation of the Amount of the Reduction Required for the Formation of Compound Layers during Cold Rolling of Bimetals *Materials Science Forum* Vol 870 pp 328-333

[6] Dmitriev A M, Vorontsov A L 2005 General Theory of upset and the upset forging of cylindrical billets *Vestnik Bauman MSTU. "Engineering" series* p 125

# Alternative evaluation of innovations' effectiveness in mechanical engineering

**A S Puryaev**[1]

[1]Naberezhnye Chelny Institute, Kazan Federal University, 18 Kremlyovskaya street, Kazan, 420008, Russia Federation

aidarp@mail.ru

**Abstract**. The aim of present work is approbation of the developed technique for assessing innovations' effectiveness. We demonstrate an alternative assessment of innovations' effectiveness (innovation projects) in mechanical engineering on illustrative example. It is proposed as an alternative to the traditional method technique based on the value concept and the method of "*Cash flow*".

## 1. Introduction

Along with the necessity to implement the innovation in all areas of human activity, including and in the manufacture of cars, vehicles and other equipment there appears an important question of evaluating the effectiveness of innovative projects. People who think only in economic terms may object and claim about the absence of relevance of such a question as any innovation, in their view, brings a huge effect in expression of monetary terms; innovation is the creation of the new technology, which today will allow to change the lives of consumers and only positively; this is the creation of a new product, which allows to satisfy the tastes and preferences of consumers at the current period of time. But the other side of the coin is left without due attention - the consequences of this innovation in the long term perspective, mainly in non-economic systems of society (national security, ecological sphere, the social system, science and technology, culture and traditions of the country). Notably the consequences of this innovation may be far from positive. Therefore, we have repeatedly raised the question of assessing the effectiveness of investment projects, including innovation projects (innovations) as a relevant and alternative to traditional economic evaluation efficiency of investments. Alternative assessment methodology, based on the author's concept of "*Compramultifactor*", allows taking into account during the process of assessment non-monetary, quantitative and qualitative parameters of the project, especially innovative and aimed at changing the quality of work and life in general.

In the article named *Evaluating of innovative projects' effectiveness at industrial enterprises*, published in conference information package ICIE-2016 ("Industrial Engineering", South Ural State University, 2016) we introduce the general assessment methodology of innovative projects for industrial enterprises. This article is supposed to show its work on hypothetical example of assessing innovative engineering projects.

## 2. Goal

Applying the concept and methodology of "*Compramultifactor*" [2,3,4] to assess the effectiveness of innovative projects presented above (InP) and choose the optimal (efficient) version of the innovative project by the criterion *D→max*.

Decision maker (DM) is necessary to evaluate the effectiveness of three investment projects to introduce innovation (three innovative projects). DM has restrictions on evaluated private evaluation parameters (PEP), as well as their status (strict - strict; the desired level - desirable). Each project has its own meaning for each PEP, i.e. all projects are comparable and identical. Table 1 shows the basic data.

**Table 1.** Meanings of PEP of evaluated innovative projects and restrictions (desired levels) of them

| №№ | Group of critical parameters of effectiveness evaluation | Innovative projects | | | Restrictions (desired levels) | |
|---|---|---|---|---|---|---|
| | | **InP1** | **InP2** | **InP3** | | |
| | The subgroup of high-quality options | | | | | |
| 1 | Compliance with the priority areas of science and technology (8 directions), the list of critical technologies of the Russian Federation (27 technologies) [1] | yes | yes | yes | yes | strict min |
| 2 | The fact of import and (or) uniqueness | yes | yes | yes | yes | strict min |
| 3 | Parameter of uncertainty and risk (UR) in the implementation of the project, points | 2,1 | 2,25 | 2,6 | 2,5 | desirable min |
| 4 | Parameter of project's performance functions' quality (PFQ), points | 2,7 | 3,1 | 3,5 | 3 | desirable min |
| The subgroup of quantitative parameters | | **InP1** | **InP2** | **InP3** | **Restrictions (desired levels)** | |
| 5 | Parameter of the National Security (resource security, RS), % (linguistic variable) | 40 | 45 | 50 | 40 | strict min |
| 6 | Parameter of reducing the risk of manmade disasters and man-made impact (units of reduction of greenhouse gas emissions), $CO_2$e/ year (linguistic variable) | 50000 | 55000 | 75000 | 50000 | strict min |
| 7 | Parameter of GDP growth acceleration (accession rate of added value, ARAV), % (linguistic variable) | 25 | 22,5 | 20 | 20 | strict min |
| | **Group of economic parameters to effectiveness evaluation** | **InP1** | **InP2** | **InP3** | **Restrictions (desired levels)** | |
| 8 | Net Present Value (NPV), thous. of rub. | 1279 | 1554 | 2260 | 1000 | strict min |
| 9 | Internal rate of return (IRR), % | 40 | 32 | 34 | 25 | strict min |
| 10 | Discounted payback period (DPP), years | 3,25 | 4 | 4,5 | 5 | strict max |
| 11 | Investments in the project (IP), thous. of rub. | 5560 | 6240 | 8500 | 10000 | strict max |

IOP Conf. Series: Materials Science and Engineering **240** (2017) 012056 doi:10.1088/1757-899X/240/1/012056

## 3. Decision

Table 2 shows the calculated meanings of the partial functions of desirability (*d*) for all the parameters evaluated, as well as generalized optimization parameters ($D_j$) of evaluated of innovative projects according to the group of critical parameters effectiveness evaluation.

For solving this optimization problem it is offered to use the method of desirability function by Harrington [5], which has the form:

$$d_{ij} = e^{-e^{-y'_{ij}}} \quad (1)$$

where $d_{ij}$ – private desirability function of one-way restriction for *i*-parameter of *j*-innovative project;

$y'_{ij}$ – coded (normalized) meaning of *i*-parameter of *j*-innovative project, transferred into the desirability scale.

Generalized desirability function by Harrington (optimization criterion) of *j*-innovative project ($D_j$) is defined as geometric mean of private desirability's according to formulae:

$$D_j = \sqrt[n]{d_{1j} \cdot d_{2j} \cdot d_{3j} \cdot \ldots \cdot d_{ij} \cdot \ldots \cdot d_{nj}} \quad (2)$$

**Table 2.** Private desirability parameters ($d_{ij}$) and generalized optimization parameters ($D_j$) of evaluated innovative projects according to group of critical parameters of effectiveness evaluation

| №№ | Group of critical parameters of effectiveness evaluation | Innovative projects | | |
|---|---|---|---|---|
| | | **InP1** | **InP2** | **InP3** |
| | The subgroup of high-quality options | | | |
| 1 | Compliance with the priority areas of science and technology (8 directions), the list of critical technologies of the Russian Federation (27 technologies) [1] | 0,37 | 0,37 | 0,37 |
| 2 | The fact of import and (or) uniqueness | 0,37 | 0,37 | 0,37 |
| 3 | Parameter of uncertainty and risk (UR) in the implementation of the project, points | 0,309 | 0,331 | 0,383 |
| 4 | Parameter of project's performance functions' quality (PFQ), points | 0,331 | 0,380 | 0,429 |
| | The subgroup of quantitative parameters | **InP1** | **InP2** | **InP3** |
| 5 | Parameter of the National Security (resource security, RS), % (linguistic variable) | 0,37 | 0,403 | 0,435 |
| 6 | Parameter of reducing the risk of manmade disasters and man-made impact (units of reduction of greenhouse gas emissions), $CO_2$e/ year (linguistic variable) | 0,37 | 0,403 | 0,5 |
| 7 | Parameter of GDP growth acceleration (accession rate of added value, ARAV), % (linguistic variable) | 0,435 | 0,403 | 0,37 |
| | **Generalized parameter of optimization ($D_{1\text{-}7j}$)** | **0,363** | **0,379** | **0,406** |
| | **Rating of innovative project** | **3** | **2** | **1** |

Next, it's needed to evaluate the desirability of PEP economic group for all innovative projects. The calculation results are presented in Table 3.

**Table 3**. Private desirability function parameters of economic groups and generalized parameter of optimization projects on them

| | Group of economic parameters to effectiveness evaluation | InP1 | InP2 | InP3 |
|---|---|---|---|---|
| 8 | Net Present Value (NPV), thous. of rub. | 0,469 | 0,563 | 0,753 |
| 9 | Internal rate of return (IRR), % | 0,578 | 0,470 | 0,498 |
| 10 | Discounted payback period (DPP), years | 0,494 | 0,441 | 0,405 |
| 11 | Investments in the project (IP), thous. of rub. | 0,527 | 0,503 | 0,423 |
| | **Generalized parameter of optimization ($D_{Эj}$)** | **0,515** | **0,492** | **0,503** |
| | **Rating of innovative project** | **1** | **3** | **2** |

It is obvious from calculations, that the first project (InP1) can be considered as prior in using the traditional economic effectiveness' assessment (without restrictions and desired levels of critical parameters' group). It would be correctly to combine the evaluation of critical group and economic as follows. We define a generalized criterion of optimization of innovative projects by the formula (3) and summarize the results in Table 4:

$$D_j = \sqrt[5]{D_{1-7j} \cdot d_{8j} \cdot d_{9j} \cdot d_{10j} \cdot d_{11j}} \tag{3}$$

where $D_{1\text{-}7j}$ – is generalized desirability (parameter) of *j*-innovative project according to group of critical assessment parameters (according to 7 parameters);

$d_{8i}$, $d_{9i}$, $d_{10i}$, $d_{11i}$ – private desirability of economic parameters' group.

**Table 4**

| | Group of economic parameters to effectiveness evaluation *and* $D_{1\text{-}7j}$ | InP1 | InP2 | InP3 |
|---|---|---|---|---|
| 8 | Net Present Value (NPV), thous. of rub. | 0,469 | 0,563 | 0,753 |
| 9 | Internal rate of return (IRR), % | 0,578 | 0,470 | 0,498 |
| 10 | Discounted payback period (DPP), years | 0,494 | 0,441 | 0,405 |
| 11 | Investments in the project (IP), thous. of rub. | 0,527 | 0,503 | 0,423 |
| | **Generalized parameter of optimization ($D_{1\text{-}7j}$)** | **0,363** | **0,379** | **0,406** |
| | **Generalized parameter of optimization ($D_j$)** | **0,481** | **0,467** | **0,482** |
| | **Rating of innovative project** | **2** | **3** | **1** |

## 4. Conclusion

So, using an alternative technique to the traditional assessment of projects' effectiveness, we were able to take into account the constraints and desired levels by predetermined parameters of critical group (group of important and necessary at some point on exact period of time the private parameters' evaluation of diverse physical entity); we managed to combine the assessment of non-economic and economic parameters, presented in a variety of scales (quantitative: clear and fuzzy sets, high-quality, expertly measured) and, as a result get optimum (effective) innovative solution from the set of existing ones..

## References

[1] The list of critical technologies of the Russian Federation 2011 (Retrieved December 1, 2016 from http://kremlin.ru/supplement/988)

[2] Puryaev A S 2015 Private Valuation Parameters of Efficiency the Investment Projects in the Concept "Compramultifactor" *Modern Applied Science* vol 9(11) pp 263-275

[3] Puryaev A S 2015 The Mathematical Apparatus of Compromise of Efficiency Estimation of

Investment Projects *International Business Management* vol 9 pp 856-861
[4] Kharisova A R and Puryaev A S 2014 Competitiveness assessment of engineering products *Source of the Document IOP Conference Series: Materials Science and Engineering* vol 69 issue 1 (article number 012020)
[5] Harrington E C 1965 The desirable function. *Industrial Quality Control* vol 21 no 10 pp 494–498

IOP Conf. Series: Materials Science and Engineering 240 (2017) 012057 doi:10.1088/1757-899X/240/1/012057

# Fuel system for diesel engine with multi-stage heated

**Yu N Ryzhov[1],Yu A Kuznetsov[1], A V Kolomeichenko[1], I S Kuznetsov[1], R Yu Solovyev[2], S N Sharifullin[3]**
[1]Federal State Budgetary Educational Establishment of Higher Education „Orel State Agrarian University named after N.V. Parakhin", Russia
[2]Federal State Budgetary Scientific Institution "Federal Scientific Agroengineering Center of All-Russian Institute of Mechanization", Russia
[3]Kazan Federal University, 18 Kremlyovskaya street, Kazan, 420008, Russian Federation

kentury@rambler.ru

**Abstract.** The article describes a fuel system of a diesel engine with a construction tractor multistage heating, allowing the use of pure rapeseed oil as a diesel engine fuel. The paper identified the kinematic viscosity depending on the temperature and composition of the mixed fuel, supplemented by the existing recommendations on the use of mixed fuels based on vegetable oils and developed the device allowing use as fuel for diesel engines of biofuels based on vegetable oils.

## 1. Introduction.

Despite the development of new deposits of traditional energy sources, the question of their alternatives is extremely sharp. The reason for it is the fact that natural resources are limited, the prices for traditional energy sources constantly increase while the ecological situation on the planet gets worse.

The researches in the field of alternative and renewable fuels have been the prerogative of the energy and resource saving programs of many countries for a long time [1,6,7]. Many scientists of world-wide reputation achieved good results in the sphere of alternative energy sources. Data released in the sphere of atomic energy, solar energy, the usage of wind power generators and tidal energy of seas and oceans, as well as the application of biofuels revealed the positive points of alternative power-engineering, defined tasks and the ways of their solutions. Inspite of high variability of alternative power-engineering, the majority of scientific works with positive results won't get wide manufacturing application in the nearest future for many reasons. Therefore, many European scientists and engineers think that "atom", "hydrogen" and recuperative system will be applied in manufacturing only in 40-50 years. This time will take a transitional period, when biofuel will be used as a fuel for already existing combustion engine models. By now biofuel gained trust and confidence of many countries. It is used in its pure state (alcohols replacing gasoline; biodiesel replacing traditional mineral diesel fuel; biogas in gas generator units) as mixtures with mineral fuels (E5, E15, E80, E85, B10, B20, etc.) and as a fuel additives [2,7].

In the context of agricultural enterprises the most suitable alternative for diesel fuel nowadays is bleached vegetable oil (VO), meeting the requirements of the German standard V 51605. The use of pure VO as a fuel for diesel engines results in the following negative effects: loss of power, increased fuel consumption, shortened life of the fuel equipment, impair starting in cold weather, increased incrustration and lacquer formation, etc. [1,3,8].

The presence of negative moments in the operation of diesel engines due to VO, is primarily connected with the increased values of the kinematic viscosity compared to diesel fuel (DF). In terms of agricultural enterprises heating is the most effective way to reduce the viscosity of the pure VO and its mixtures with diesel fuel with a high content of VO.

## 2. Materials and methods.

Based on the temperature characteristics and the seasonal field work, we investigated the dependence of the kinematic viscosity of the pure RO and its mixtures with diesel fuel on the temperature and the content of the RO in the mixture with diesel fuel.

As the RO we used a sample obtained by hot pressing followed by filtration by settling of these breeds of rape: Northerner, Salsa, Ahat, Luned, Rohan, Krauser, Solar KL, Mobil KL, Lighthouse, Zorn, Proxima, Capital, Leader, Winner, Sfint and Heros. The DF applied in the research met the requirements of GOST R 52368 - 2005 " The National Standard of the Russian Federation. Diesel fuel Euro. Specifications were in line with quality D, class 2.

In the process of research we determined the kinematic viscosity of pure VO, as well as its mixtures with DF, where DF content was gradually increased to 100% in 10% increments. The temperature range of viscosity determination was from 20 to 90°C. The determination of kinematic viscosity was carried out according to GOST 33-2000 "Oil products. Transparent and opaque liquids. Determination of kinematic viscosity and calculation of dynamic viscosity". The mixture heating was conducted in a water-bath, the instruments for measuring kinematic viscosity were capillary viscometers type IWF-1m. The experimental methods were treated with the methods of interpolation and extrapolation

## 3. The results of the research.

The determination of dependences of kinematic viscosity on temperature and composition of fuel mix (Fig. 1 and 2) was the result of the researches, first of all.

The more VO the mixture contains, the lower the content of harmful substances in the exhaust gases and the fuel price and the higher energy independence of the enterprise (on condition of intrafarm VO production), however the kinematic viscosity of the mixture is higher. At a temperature of 0°C the VO viscosity is by 36,67 … 22,003 times higher than of the DF. Gradual temperature increase of the fuel mixtures to 40°C gives sharp decrease in viscosity. For example, at pure RO the kinematic viscosity of the testing sample while heating to 40°C decreases by 4,5 times in comparison with value at 0°C and attains 24,503 cSt. Further temperature increase entailed to slower decrease in viscosity. And at a temperature of 90 °C the viscosity of pure RO attains the value of about 7,384 cSt that allows to minimize the above-mentioned negative moments at operation of diesel engines of a domestic production on this fuel.

For temperature range from 0 to 40 °C for all the fuel mixtures sharp decrease in viscosity is typical. The raised content of DF in them (from 40% and more) also gives decrease in fuel mix viscosity.

In general the analysis of the results of researches with testing samples confirmed the possibility of use as fuel for diesel engines fuel mixtures containing less than 30% of VO at an ambient temperature 20°C. However, the use of fuel mixtures with the low content of VO is ineffective from the point of view of ecology and economy, and the use of fuel mixtures with the high content of VO without heating is impossible. The heating of VO up to the temperature of 70 … 90°C provides decrease in kinematic viscosity, increase in the fineness of injection, reduce of the flame length, i.e. promotes the improvement of the carburetion and combustion processes [5].

We offered a design (Fig. 3) of a dual-fuel system of the tractor diesel with VO multistage heating, which provides the use of vegetable-based oil. It contains a mineral fuel tank 1, a vegetable-based fuel tank 2, where are a heat exchanger 6 and a temperature sensor 21 of the vegetable-based fuel set,

IOP Conf. Series: Materials Science and Engineering **240** (2017) 012057 doi:10.1088/1757-899X/240/1/012057

mineral fuel intake line 3, vegetable-based fuel intake line 7, electro hydro dispenser 11 located in front of the fuel fine filter 12, ultrasonic filter 13 located in front of the injection pump 14, electric heaters 16 located in front of the nozzles 15, a fuel drain line 17 from the nozzles 15 and a fuel drain line 18 from the injection pump 14, an electronic control unit (ECU) 19, a position sensor 20 of the fuel pump rail, located in the fuel pump regulator body. Line 3 of the intake mineral fuels contains first -stage fuel filter 4 and an electric booster pump 5 with a pressure relief valve. Line 7 of the vegetable-based fuel intake, in its turn, contains first-stage fuel filter 8, fuel pump 9 of a standard power supply system and posistor heater 10 [4].

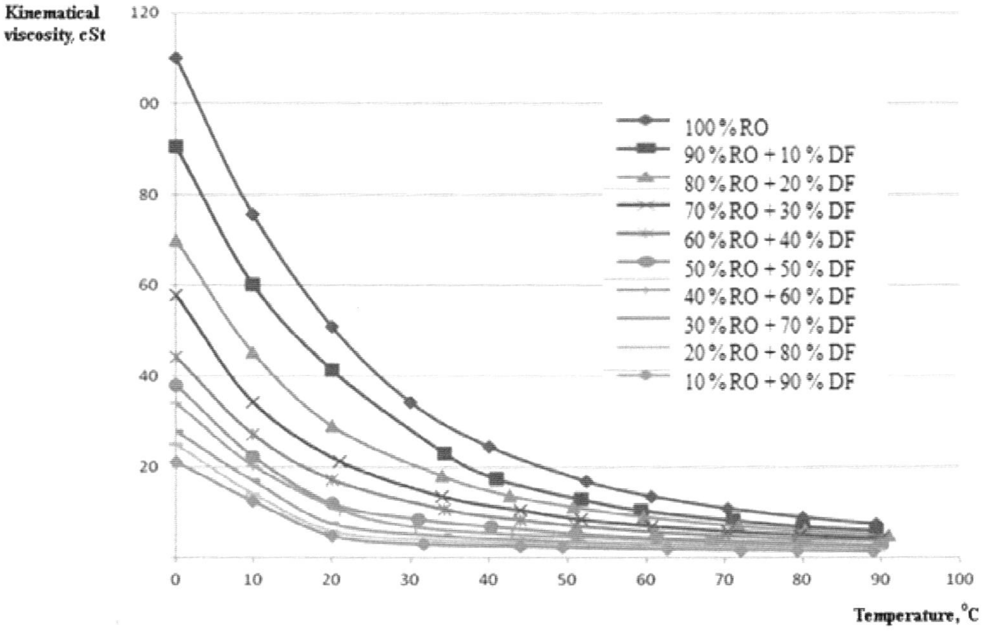


**Figure 1.** Dependence of the kinematic viscosity of the fuel mixture temperature

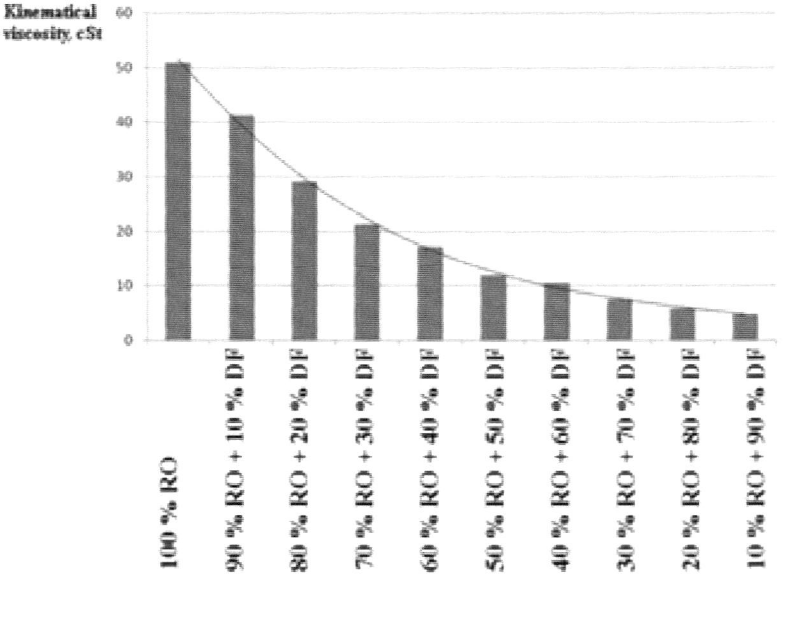

**Figure 2.** The dependence of kinematic viscosity on the component composition at 20 ° C

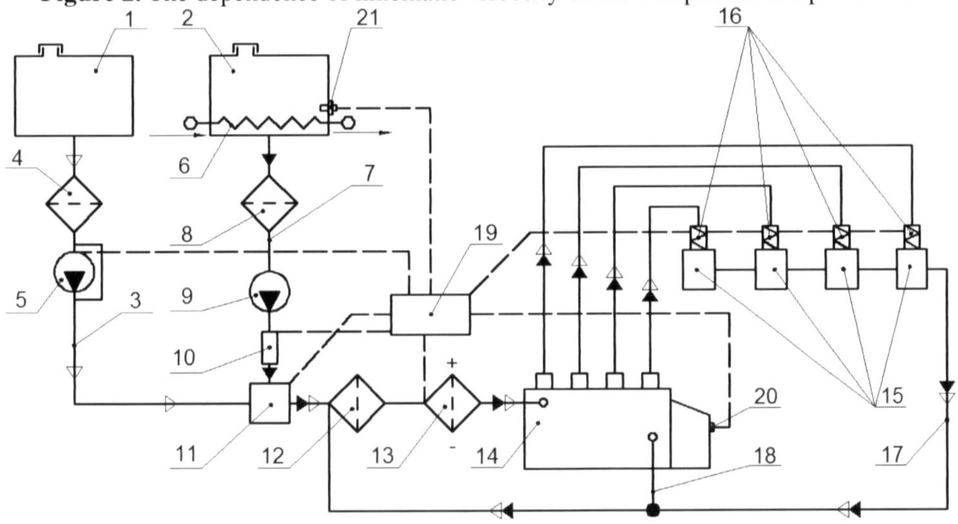


**Figure 3.** The diagram of the dual-fuel system of the tractor diesel with multistage heating

The first heating stage includes heat exchanger 6, connected to the engine cooling system and heating the vegetable-based oil in tank 2 to the temperature of 20 ... 30°C, that ensures its pumpability even at low ambient temperatures.

The second heating stage is represented by posistor heater 10, which is self-regulated and serves to heat the vegetable-based oil to a temperature of 60 ...70°C, which ensures its good pumpability through the fine filter 12 and ultrasonic filter 13.

The third stage of heating is realized as electric heaters 16 mounted directly in front of the nozzles 15. The vegetable-based oil at this stage is heated to a temperature of 80 ... 90°C. Electric heaters 16 contain posistors, which stand for control units, so they are self-regulated. The heating of the vegetable-based oil to a temperature of 80 ... 90 °C during injection ensures less flame length and a finer spray compared with lower preheating temperatures.

The dual-fuel system of a tractor diesel with multistage heating operates in the following way:

The start and warming up of the engine is carried out on diesel fuel. At the same time it is supplied by the electric boost pump 5 from the tank through the first- stage fuel filter 4, electro hydro dispenser 11, the fine filter and ultrasonic filter 13 in the high pressure pump 14. Then the diesel fuel through the electric heaters 16 is supplied to the nozzles 15. When the engine is warming up and during the operation with diesel fuel ultrasonic filter 13 and electrical heaters 16 are switched off. The excess of diesel fuel from the nozzles 15 and the fuel pump 14 through the drain lines 17 and 18 are supplied to the fine filter 12.

When heating and during the engine operation the coolant of its system circulates through the heat exchanger 6. Once the temperature of the vegetable-based oil reaches the level of 20 ... 30 °C in the tank 2 of the electric control unit 19, perceiving this option by the temperature sensor 21, switches the electro hydro dispenser 11 in the position of supplying vegetable-based oil and also starts the work of heaters 10 and 16 as well as ultrasonic filter 13. The main function of the ultrasonic filter 13 is further reduction of the kinematic viscosity of the vegetable-based oil.

After switching the electro hydro dispenser 11 in the position of supplying vegetable-based oil, the oil supply stops because the ECU 19 turns the electric boost pump 5 off (spool electro hydro dispenser at the signal, the ECU 19 blocks the access of diesel fuel in the system), the fuel boost pump 9 of the standard power system, delivers warmed to 20 ... 30°C vegetable-based oil from tank 2 through a first-stage fuel filter 8 into the posistor heater 10 where it is heated to 60 ... 70 °C. Then the oil is sent by the electro hydro dispenser 11 through the fine filter 12 in an ultrasonic fuel filter 13, where the vegetable-based oil is subjected to cavitation treatment, whereby its viscosity is further reduced, and it is supplied into the high pressure pump 14 and then to the electric heaters 16, where it is heated up to 80 ... 90 °C and the injection nozzles 15 spray it into the combustion chamber. The excess of vegetable-based oil from the nozzles 15 and the high pressure pump 14 through the lines 17 and 18 are sent to the fine filter 12.

Through the use of PTC resistors in heaters 10 and 16, the latter automatically maintain the temperature of vegetable-based oil in the fuel system. In the PTC heater 10, they have two functions: they are heating elements and at the same time they maintain the oil temperature in the range of 60 ... 70°C. The heating elements in the electric heaters 16 are spirals, while posistors serve as elements supporting the temperature in the range of 80 ... 90°C.

At steady state conditions of the minimum stable speed and regimes of part-load engine (to 35%), the latter runs on diesel fuel. [4] When working on the above-mentioned modes the rail position sensor 20 of the fuel pump sends a signal to the ECU 19, which puts the system on diesel fuel. The ECU 19, used in the dual-fuel system, allows a person to switch from one fuel to another one manually.

Before stopping the engine, the fuel system is converted to pure diesel fuel and the engine should run on it for about 5 minutes (depending on operating mode).

The usage of the offered dual-fuel system of the tractor diesel engine will allow to apply vegetable-based oil as a fuel in cold season and improve the overall efficiency of the engine on this type of fuel.

## 4. Conclusion.

1. The dependences of kinematic viscosity on temperature and content of fuel mix are defined.
2. The existing recommendations about application of the fuel mixtures on the basis of vegetable-based oils are added.
3. The devices allowing to apply biofuel on the basis of VO as fuel to diesel engines are developed.

## References.

[1] Devyanin S N, Markov V A, Semenov V G 2007 *Vegetable-based oils and fuels for diesel engines* Moscow State Agrotechnical University Pp 400

[2] Government of the Russian Federation 2012 *Comprehensive Biotechnology Development Program of the Russian Federation for the period until 2020* dd 24 No 1853-p

[3] Kurochkin A A, Josan A A, Ryzhov Y N, Golovin S I 2013 Heated rapeseed oil as a way to improve the efficiency of its use as a fuel *Herald Eagle GAC* No 1 (40) pp 209-12

[4] Kuznetsov Yu A, Ryzhov Yu N, Kurochkin A A, Mikhaylova Yu L, Mikhaylov M R 2014 Dual-fuel system of the diesel with multi-stage heating, running on compounded (plant and mineral) fuels in a climate of russian federation *Tractors and power machines* No 2 pp 36-40

[5] Ryzhov Y N, Josan A A, Kurochkin A A 2012 The peculiarities of the injection and combustion of rapeseed oil and diesel fuel in modern engines *Tractors and farm machinery* No 6 pp 19-20

[6] Ryzhov Y N, Inshakov A P, Kurochkin A A 2014 Dual-fuel system of the tractor diesel with multistage heating *Tractors and farm machinery* No 6 pp 11-13

[7] Savelyev G S 2010 *Technologies and means of automotive equipment adaptation to the work on alternative fuels* Dis.Doc. Engineering Sciences: 05.20.01; Moscow State Research Institution All-Russian Scientific Research Institute of Farm Mechanization pp 304

[8] Ukhanov A P, Ukhanov D A, Rachkin V A, Ivanov V A 2015 *Three-fuel system tractor diesel* Pat 2403431 RF, IPC C 1 F02M43 / 00,. No 2009116953 / 06

IOP Conf. Series: Materials Science and Engineering **240** (2017) 012058 doi:10.1088/1757-899X/240/1/012058

# Increase of reliability of contact networks of electric transport, due to increase of strength of the joint unit of pipes of different diameters

**L S Sabitov[1], N F Kashapov[1], I R Gilmanshin[1,4], I Z Gatiyatov [2] and I L Kuznetsov [3]**

[1] Kazan Federal University, 18 Kremlyovskaya street, Kazan, 420008, Russian Federation
[2] Kazan State Power Engineering University
[3] Kazan State University of Architecture and Civil Engineering
[4] Kazan National Research Technical University named after A N Tupolev, Kazan, Russian Federation

sabitov-kgasu@mail.ru

**Abstract.** The feature of the stress state of the supports of the contact networks is the presence of a joint of pipes of different diameters, the ultimate state of which is determined, as a rule, the strength of the weld. The proposed unit allows to increase the reliability and strength of the connection and also exclude the presence of a weld bead on the outer surface of the pipe of smaller diameter in the place of its attachment to the upper end of the support ring.
Key words: Reliability, transport, contact network, pipe connection

One of the main elements of the contact network are the supports that provide the given position of the contact suspension in the plan and profile above the roadway, due to which the transmission of electricity to the rolling stock and the reliability of the current collection are realized. Because of their lack of redundancy, they directly affect the continuity of the technological process of transportation, and in case of destruction they create a threat to violations of traffic safety and people's lives.

The analysis of damage to the supports of the contact networks of urban electric transport made it possible to identify two characteristic problems: metal supports - damages in the junction of pipes of different sections; Reinforced concrete supports - corrosion of concrete and reinforcement.

To connect the links of the two link steel pillars to each other, a smaller diameter pipe is welded with six wedges. In a pipe with a larger diameter, holes are made under the places of the wedges. Then the lower link is pushed with force to the upper, after which welded joints connect a pipe of larger diameter with wedges. The pipe joint is closed from above by a flange, which is welded with a discontinuous suture.

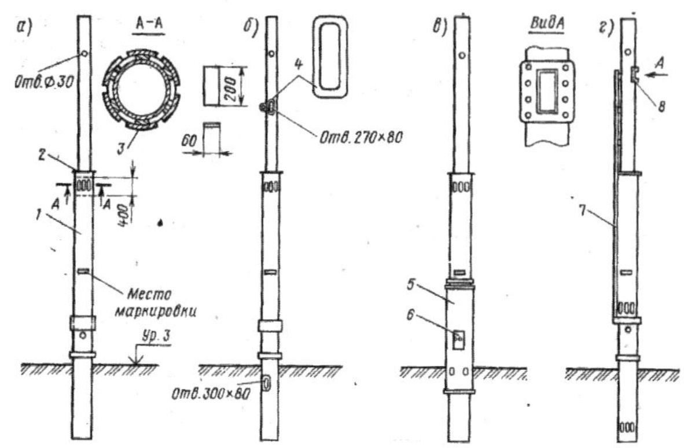

Fig.1. Steel standard bearings:
*а*- type of OS; б -type OS-2B; *в* - type OSC; *г* - type of OSG;
1- body of support; 2-flange; 3-wedge; 4-overlay, δ = 8; 5- base; 6 - the door;
7- pipe Ø 45x4; 8 - overlay δ = 10

The connection of pipes of different diameters is carried out through the end face (Fig. 2, a).

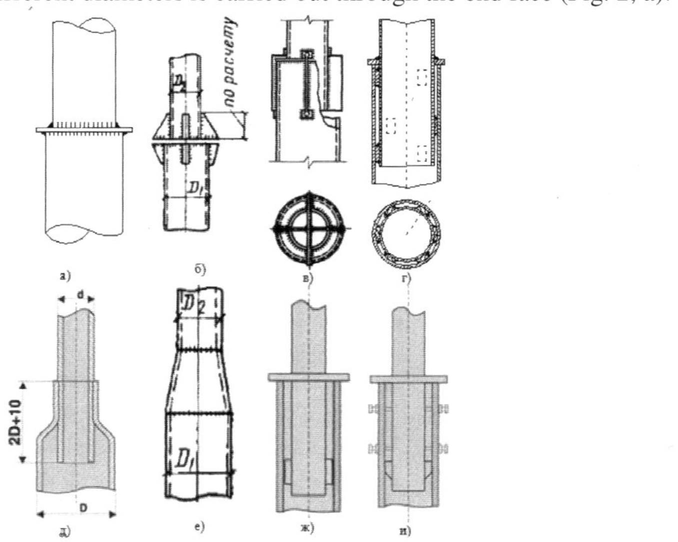

Fig.2. Types of connection of tubular rods of different diameters
a) butt by means of a face piece, б) connection through a profile with ribs,
в) pipe in the pipe with cross mortise cutters, г) joint through intermediate plates, д) connection with gradual transition to a smaller diameter (cold rolling), e) with a conical liner, g) welded combined connection with a flange, e) combined bolted compound.

For the supports of contact nets, the essential question is the weldability of the slice with the main rods. The peculiarity of the stressed state of supports with such connections is the presence of a joint of pipes of different diameters, the limiting state of which is determined, as a rule, by the strength of the welded seam. Even with the maximum allowable weld joints, the strength condition is not met. Which leads either to an increase in the cross section, or the introduction of additional elements in the form of plates, edges to increase the strength of the nodal section. With this increase in the cross section, the cross section naturally leads to over-consumption of steel, given the massive construction of the supports. And the introduction of additional elements in the unit on one side increases the laboriousness, on the other, worsens the aesthetic qualities of the supports (Figure 3).

Fig. 3. The actual state of the supports and their assemblies

The pipe connection assembly of different diameters includes the ends of pipes of larger and smaller diameters on which the respective connecting parts are connected, connected by tightening bolts, the connecting piece of the larger diameter pipe $D_0$ is an end plug with holes for passage of the coupling bolts connected to the end of a larger diameter pipe, The connecting part of a pipe of smaller diameter $d_0$ is made in the form of a support ring whose internal diameter corresponds to the diameter $d_0$, the base of the support ring has a cylindrical belt with an outer diameter $D_0$ which, to the height L from its upper boundary to the upper end of the support ring, decreases to a diameter $d$, the base outside is planted on the end of a pipe with a smaller diameter, hot-fit flush with its end, which is welded to the base of the support ring by a welded seam with edge cutting; holes for a spanner with holes coaxial to the holes in the end plug are formed along the contour of the outer surface of the base of the support ring, The support ring and the end plug are interconnected by tightening bolts inserted from the inside of the end plug into its holes, the heads of which are welded to it, and for connecting pipes of larger and smaller diameters $D_0$ and $d_0$, respectively 325 and 219 mm, their walls are respectively 10 and 9 mm, and the height $L$ of the support ring and the outer diameter of its upper end $d$, respectively 200 and 225 mm, the outer diameter of the support ring $D(x)$ is determined, depending on the current coordinate $x$, measured from the upper boundary of the cylindrical base belt of reference ring, from the table.

| $x$, mm | 0 | 5 | 10 | 15 | 20 | 25 | 30 | 35 | 40 | 50 | 100 | 200 |
|---|---|---|---|---|---|---|---|---|---|---|---|---|
| $D(x)$, mm | 325,0 | 289,8 | 271,3 | 259,8 | 252,1 | 246,5 | 242,4 | 239,2 | 236,7 | 233,2 | 228,7 | 225,0 |

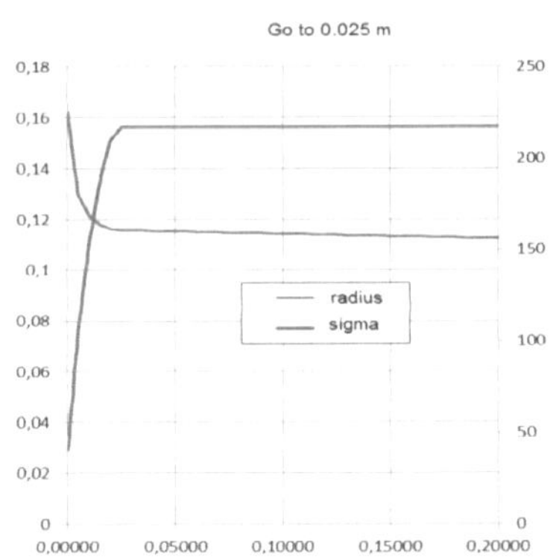


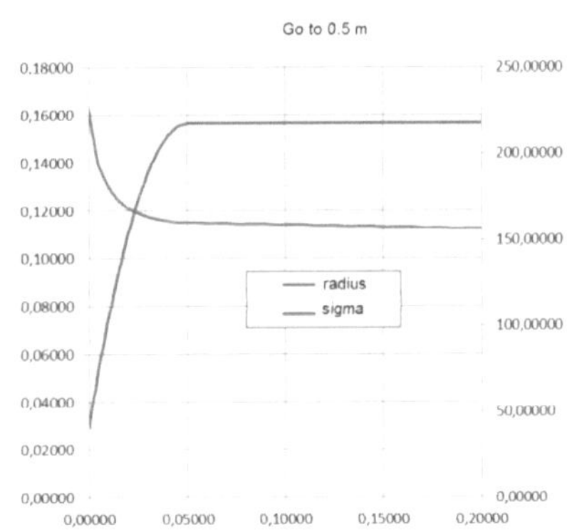

IOP Conf. Series: Materials Science and Engineering **240** (2017) 012058 doi:10.1088/1757-899X/240/1/012058

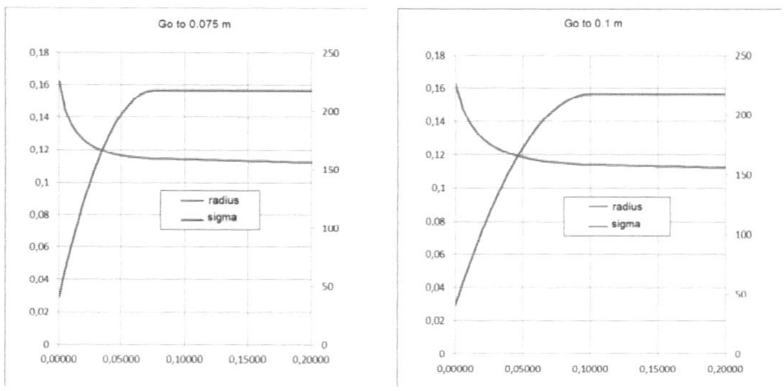


Fig. 4.

The proposed construction makes it possible to increase the reliability of the connection and its strength by increasing the moment of resistance of a pipe of smaller diameter, with a bearing ring mounted on it, in the area of its attachment to a pipe of larger diameter, providing a smooth transition from diameter $D_0$ to diameter $d_0$, as well as exclusion on the outer surface of a pipe of smaller diameter in the place of its attachment by the upper end of the support ring.

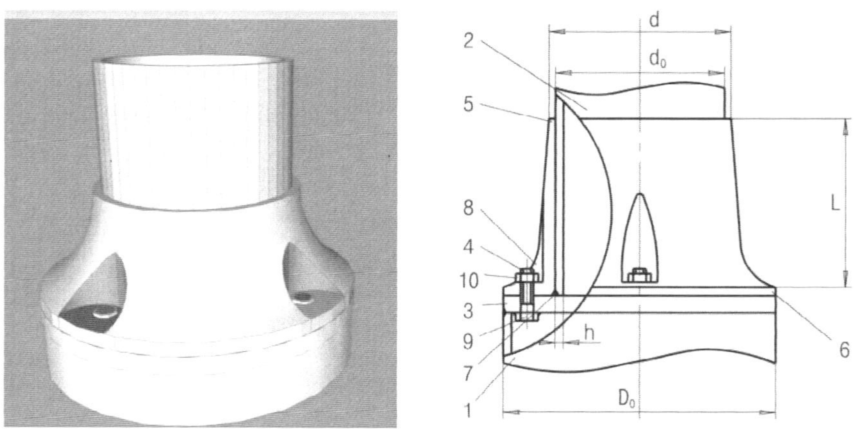


Fig. 5.

The shape of the outer surface of the support ring that surrounds the end of the smaller diameter pipe is selected so that in all sections of the connection the bending stresses do not exceed the allowable value $\sigma_{max}$ and should be as uniform as possible while providing a smooth transition from diameter $D_0$ to diameter $d$. Based on these conditions, the most complete use of the support ring material is carried out and the local stress concentration associated with the abrupt change in the cross-section of the base of the support ring at the place of its rigid embedding in the end plug of a larger diameter pipe is eliminated.

The calculation of the pipe joint for bending of pipes according to the ANSYS program showed that the flexural stresses at the welding site of a pipe with a diameter $d_0$ = 219 mm to the flange, with the same loading scheme as for the proposed sample of the junction unit, is 359.4 MPa, which is 32.8 %, which is greater than the value obtained in the calculation for the proposed connection variant.

Conclusions: The proposed connection option allows to increase the reliability of the connection and its strength, and also to exclude the weld on the outer surface of a pipe of smaller diameter at the place of its attachment with the upper end of the support ring.

IOP Conf. Series: Materials Science and Engineering **240** (2017) 012058 doi:10.1088/1757-899X/240/1/012058

**References**

[1] Afanasyev A S 1988 *Contact networks of tram and trolleybus: A textbook for SPTU* Moscow: Transport 264 p
[2] Kuznetsov I L, Sabitov L S, Isaev A V 2012 *Structures with joints of steel pipes of different diameters* Kazan: KGASU 123 p
[3] Sabitov L.S. Development and investigation of joints of pipes of different diameters, Izvestiya KazGASU. 2008. № 1 (9). Pp. 102-105.
[4] Sabitov L S, Kuznetsov I L, Gatiyatov I Z 2014 Experimental studies of joints of pipes of different diameters in the supports of contact networks of electric transport *Bulletin of Civil Engineers* No 6 (47) pp 90-95
[5] Kuznetsov I L, Sabitov L S *Node of connecting pipes of different diameters* Patent for the invention of the Russian Federation No 2365805
[6] Kuznetsov I L, Sabitov L S *Method for manufacturing a junction of pipes of different diameters* Patent for the invention of the Russian Federation No 2382266

# Method of the steel aluminum wires protection of air transmission lines from fast vibration damage at the output from the support clamp

**L S Sabitov[1], N F Kashapov[1], Yu M Strelkov[2] and I L Kuznetsov[3]**

[1] Kazan Federal University, 18 Kremlyovskaya street, Kazan, 420008, Russian Federation
[2] Kazan CHPP-1, Engineering Center "Energoprogress"
[3] Kazan State University of Architecture and Civil Engineering

sabitov-kgasu@mail.ru

**Abstract.** Conducted theoretical research of the stress-strain state of flexible mechanical connections, the varieties of which are wires of overhead transmission lines, cable cable mounting structures, strings frequency sensors of force and small displacements at the nodes of a rigid fixing. Chosen form of nodules, a flexible mechanical connection at the nodes of the rigid fixing of the conditions of increasing resource durability, and also reduce the maximum total stresses from stretching and bending at the nodes of a rigid fixing.

Investigation of the stress-strain state of a flexible mechanical connection in the nodes of a rigid embedding.

In technology, there are several varieties of flexible mechanical communication with the nodes of rigid embedding, which include the wires of overhead power lines, cable-stay cable fastenings of structures, strings of frequency sensors and small displacements. All these elements, despite the different areas and purposes of application, are similar in that they have a high length-to-diameter ratio are subjected to significant tensile loads and transverse vibrations. The result of these actions may be damage to the flexible mechanical connection in the nodes of their rigid sealing, where the stresses from stretching and from bending occur.

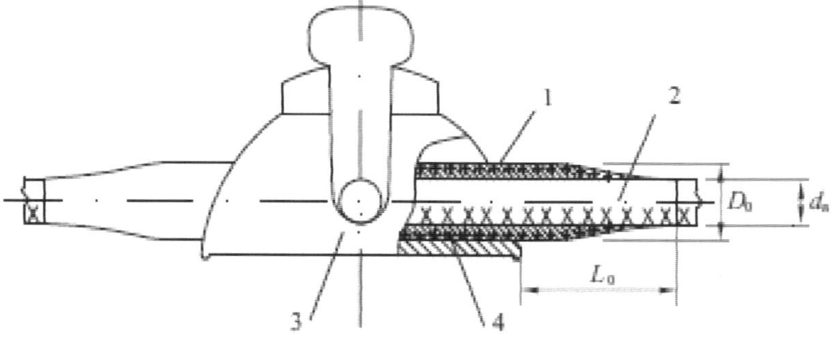

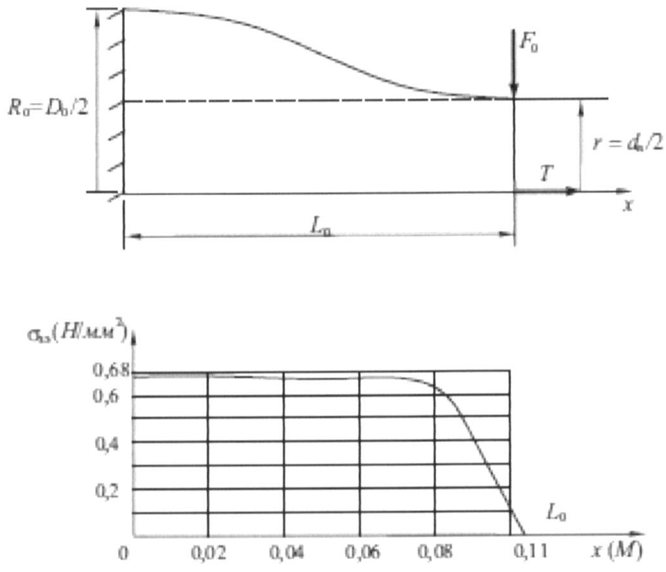


**Fig.1 Method of protection of VLEP wires**

For example, in the typical nodes of rigid termination of strings to reduce these stresses by increasing the diameter of the ends of the string, they are milled in round nipples, sealing in split cones, and also making the string in one piece with a conical thickening of the ends [1,2].

However, in the mentioned fastening options, local overvoltages are possible due to string crushing, concentration of stresses in the places of transition of the conical part of the string to cylindrical, which leads to a decrease in the strength of the string at the nodes of rigid sealing and worsens the stability of the oscillation frequency.

The wires of overhead transmission lines are to some extent subject to fluctuations caused by wind, similar to the impact of cables and cable ties, therefore during operation, their fatigue vibration damages in the nodes of rigid sealing are possible. To protect the wires from these damages at the output of the support clamp, which is a wire end clamping unit, it is recommended to use spiral searchlights wound on the wire at the installation site of the supporting clamp. [3]. In this case, the wire with the searchlight has insufficient rigidity and flexural strength that Does not allow to provide effective protection of a wire from fatigue damages.

In order to solve the above problems, it is proposed in this paper to form a thickening of a material having similar mechanical characteristics to the initial diameter of the thickening $D_0$ at the point of embedding equal to several diameters $d_п$ of a flexible mechanical coupling, smoothly passing in its cantilever part to Diameter d. In this case, the shape of the cantilever part of the thickening is selected from the condition of increasing the resource stability of the flexible mechanical connection, as well as reducing the maximum total stress from stretching and bending in the rigid embedding unit.

For the wire of the overhead power line and the cable-stayed cable, the formation of a thickening of the proposed shape is ensured by welding them to the rigidity nodes of the termite by sealing the security element with the given profile of the console part. In the case of a string of a frequency sensor, it is rational to produce it in one piece with the thickening of its ends in accordance with the proposed form.

The procedure for selecting the profile of the thickening formed in the node of the rigid termination of the flexible mechanical connection, as well as evaluating the efficiency of the proposed method for increasing its life-endurance, is given in the example of an overhead power transmission line.

The ability of a wire with a protective element to withstand without damaging vibrational vibrations is primarily determined by the level of cyclic flexural stresses at the point of its exit from the supporting clamp, and also in the section at the outlet of the wire from the protective element. Proceeding from these conditions, the shape of the cantilever part of the protective element is chosen so that in all sections along the console length the flexural stresses do not exceed the permissible value $\sigma_{max}$ and, if possible, they are the same, in order to use the material of the protective element as completely as possible, providing a smooth transition from the diameter $D_0$ to the diameter $d_П$ of the wire. For an arbitrary section of the console, which is spaced apart $x$ from its fixing point, the conditions for ensuring the required strength have the form

$$\frac{M_{изг}(x)}{W(x)} \le \sigma_{max}, \tag{1}$$

where $M_{изг}(x)$ bending moment, $M_{изг}(x) = F_0(L_0 - x)$;

$W(x)$ oscillation resistance moment, $W(x) = \pi D^3 / 32$, where $D = D(x)$ - diameter of a circle in a section $x$, and $D(0) = 2R_0$, $D(L_0) = 2r.$

The condition of equal bending resistance for the console will be

$$F_0(L_0 - x) = \sigma_{max} \cdot \frac{\pi D^3}{32}, \tag{2}$$

or

$$R^3 = \frac{4F_0}{\pi\sigma_{max}} \cdot (L_0 - x), \tag{3}$$

where $R = R(x)$

Rewrite this equation in the form

$$R^3 = \beta(L_0 - x), \tag{4}$$

where $\beta = \frac{4F_0}{\pi\sigma_{max}} = \frac{R_0^3}{L_0}$.

A smooth transition from the diameter $D_0$ of the protective element to the diameter $d_П$ of the wire is ensured when

$$\left.\frac{dR}{dx}\right|_{x=L_0} = 0. \tag{5}$$

Look for $y \equiv R^3$ in the form

$$y = a + bx + cx^2 + ex^3 \tag{6}$$

of the following conditions

1) $x = 0$, $y = R_0^3$;

2) $x = L_0$, $y = r^3$;

3) $x = L_0$, $y' = 0$,

This condition is equivalent to equality (5);

4) $\min_e I = \int_0^{L_0} [\beta \cdot (L_0 - x) - y]^2 dx, \quad (7)$

when this condition is satisfied, the actual longitudinal profile of the console is minimally deviated from the theoretical one, which is a cubic parabola, which is caused by the need to perform a smooth transition from the diameter $D_0$ of the protective element to the diameter of the wire $d_П$

from condition1) follows that $a = R_0^3$,

from condition 2) $r^3 = R_0^3 + bL_0 + cL_0^2 + eL_0^3$,

from condition 3) $b + 2cL_0 + 3eL_0^2 = 0$, or $b = -2cL_0 - 3eL_0^2$.

then

$$r^3 = R_0^3 - 2cL_0^{\ 2} - 3eL_0^{\ 3} + cL_0^{\ 2} + eL_0^{\ 3} = R_0^3 - cL_0^{\ 2} - 2eL_0^{\ 3}$$

$$\text{и } b = -2\frac{R_0^3 - r^3}{L_0} + eL^2, \tag{8}$$

$$c = \frac{R_0^3 - r^3}{L_0^{\ 2}} - 2eL. \tag{9}$$

In this way

$$y = R_0^3 + \left(eL_0^{\ 2} - 2\frac{R_0^3 - r^3}{L_0}\right)x + \left(\frac{R_0^3 - r^3}{L_0^{\ 2}} - 2eL_0\right)x^2 + ex^3. \tag{10}$$

The minimum condition for the integral I (7)

$$\frac{\partial I}{\partial e} = 0 \text{ or } \frac{\partial}{\partial e}\int_0^{L_0}\left[\beta\cdot(L_0 - x) - y\right]^2 dx = 2\int_0^{L_0}\left[\beta\cdot(L_0 - x) - y\right]\cdot\frac{\partial y}{\partial e}dx = 0,$$

With allowance for (10) leads to

$$\int_0^{L_0}\left[\beta\cdot(L_0 - x) - y\right]x(L_0 - x)^2 \mathrm{d}x = 0 \tag{11}$$

or $$\int_0^{L_0}\left[\beta x\cdot(L_0 - x)^3 - x(L_0 - x)^2 y\right]\mathrm{d}x = 0. \tag{12}$$

After calculating the integral (12), we obtain an expression for determining the coefficient $e$

$$e = \frac{21}{4}\frac{\beta}{L_0^2} - \frac{35}{4}\frac{R_0^3}{L_0^3} - \frac{7}{2}\frac{b}{L_0^2} - \frac{7}{4}\frac{c}{L_0}. \tag{13}$$

Substituting in (13) the value of the coefficients β from (4), b from (8) and c from (9), we have

$$e = \frac{7}{4}\frac{R_0^3 - 3r^3}{L_0^3}. \tag{14}$$

Thus, in equation (6), all the coefficients are defined:

$$a = R_0^3;\ b = -2\frac{R_0^3 - r^3}{L_0} + eL_0^{\ 2};\ c = \frac{R_0^3 - r^3}{L_0^{\ 2}} - 2eL;\ e = \frac{7}{4}\frac{R_0^3 - 3r^3}{L_0^3}.$$

After transformation (6), we obtain an equation for determining the values of the diameter of the cantilever parts of the protective element $D(x)$ in places after exiting the supporting clamp, depending on the current coordinate $x$, measured from the extremities of the supporting clamp

$$D(x) = 2\sqrt[3]{a + bx + cx^2 + ex^3}, \tag{15}$$

где $a = \frac{D_0^3}{8};\ b = -\frac{D_0^3 - d_П^{\ 3}}{4L_0} + e\cdot L_0^2;\ c = \frac{D_0^3 - d_П^{\ 3}}{8L_0^2} - 2e\cdot L_0;\ e = \frac{7}{32}\cdot\frac{D_0^3 - d_П^{\ 3}}{L_0^3}.$

Along with increasing the resource resistance of the wire, the proposed method makes it possible to prevent power line failures due to mechanical overloads of wires in the event of a dance of wires caused by a combination of considerable wind loads with ice deposits. In the most dangerous section of the wire, at the point of its exit from the supporting clamp, there is a total tension from the wire stretching and from its bending. During the dance of wires, the tension of the wire T substantially increases due to the fact that the specific loads on the wire from the weight of the wire covered with ice and the wind pressure on the ice-covered wire can be several times higher than the load of the own weight of the wire (for AC 120/19 wire more than in 5 times [8]). In addition, when the air flow is applied to the wire at a speed of 30 m / sec, an alternating bending moment occurs in this section. In this case, the effectiveness of the proposed method of protection against wire damage in this case is determined by a significant reduction in the dangerous section of the wire, with a coaxially welded protective element, of a maximum voltage σΣ equal to the sum of the tensile stresses $\sigma_p = 4T / \pi D_0^2$ and from bend $\sigma_{изг} = 32M_{изг\max} / \pi D_0^3$. This is due to the fact that the sectional area of the wire with the protective element and its moment of resistance is significant, respectively, 4 and 8 times higher than the specified values for the wire itself.

## References

[1] Novatsky P V, Knorring V G, Gutnikov V S 1970 *Digital devices with frequency sensors* L., "Energy" 424 p
[2] Osadchenko E P 1979 *Designing sensors for measuring mechanical quantities* M., "Mechanical Engineering" 480 p
[3] Kolosov V G, Ryzhov S V 2005 Increase of the resource resistance of overhead lines during vibration by installing spiral projectors in the boats of the supporting clamps *Electro* No 6 pp 45-51

IOP Conf. Series: Materials Science and Engineering **240** (2017) 012060 doi:10.1088/1757-899X/240/1/012060

# Algorithm of calculation of energy consumption on the basis of differential model of the production task performed on machines with computer numeric control (CNC)

**D T Safarov, A G Kondrashov, G F Glinina, L R Safarova**

Russia, Republic of Tatarstan, Naberezhnye Chelny institute (branch) of the Kazan (Volga) Federal University, 423812, Naberezhnye Chelny, Mira Ave., 68/19

safarov-dt@mail.ru, leila-r.85@mail.ru

**Abstract.** The calculation algorithm, power consumption of all consumers involved in the operation and production tasks developed by the example of workplaces equipped with CNC machines is developed. The algorithm takes into account the actual status, operating modes and switching sequence of all electricity consumers.

## 1. Introduction

Severe conditions of enterprises competition for a big share of the market at realization of products require the simultaneous solution of several tasks: improvement of products quality, decrease of resources consumption at products production , reduction of wastes at certain operations of a shaping process.

Due to the developing import substitution for the enterprises of mechanical engineering the organization of preparation of new products production becomes an important task. Most machine-building enterprises are equipped with machines with the numerical software that allows to realize a set of options of performing technological cutting passes of processing one detail. At the same time the existing CAD packages allow to generate trajectories and a program code of the driver for machine drive gears, and also to give preliminary estimate to processing duration. But enterprises need more – they need to have information on expediency of this or that option from the point of view of its cost.

Modern requirements to machine-building productions include minimization of all types of resources. Processing by cutting was and remains the main technology of a shaping of working elements of car details.

The significant item of expenditure in the process of cutting is the expense on electricity consumption. This article tells about cutting-down of energy resources consumption.

Economic methods of calculation define energy consumption with no details, as the work of rated power of electric motors on process duration. At such calculation of an expense of the electric power its size doesn't depend on the actual content of technological operation in any way. Traditionally at technological design the maximum size of forces of cutting only for check of drive gears of the main movement and supply is defined for providing a necessary power reserve. This method of calculation of electricity consumption within separate operation and then in production task doesn't receive further development.

## 2. Theoretical part

IOP Conf. Series: Materials Science and Engineering **240** (2017) 012060 doi:10.1088/1757-899X/240/1/012060

The main source of the electric power loss is the process of overcoming resistance to cutting forces, in the course of removal of an allowance of cutting, friction forces in mobile knots – guides, bearings, ball-screw couples, dispersion of current in the form of heating of engines at the idle running of electric motors, and on various modes of operation, etc. Besides in the process of tasks performance other consumers of the electric power also act. Cutting fluid supply pumps turn on, electric motors of lubrication of mobile parts, conveyors on removal of shaving, loading and transport gears work.

Thus, to calculate the actual consumption of energy and to build it in the process of technological design, it is necessary to consider performance of tasks in a workplace in the form of the following technological components (subprocesses) enclosed by each other: production task, setup and adjustment cycle, operation, workpiece-setting block, tool-setting block, cutting pass, cutting stroke, separate step [1, 2]. The sequence of performance and duration of each elementary process is formed as a temporary chain [3].

Calculation of electricity consumption is carried out not at once, but consistently developing into the final sum at each hierarchical level, beginning with the simplest element – a separate step (fig. 1) before a separate operation and further from operation to production task. The calculation procedure described in the article is harmoniously built in a more general technique of design of technologies of by-products and waste lifecycle in the course of preparation for release of car components [4].

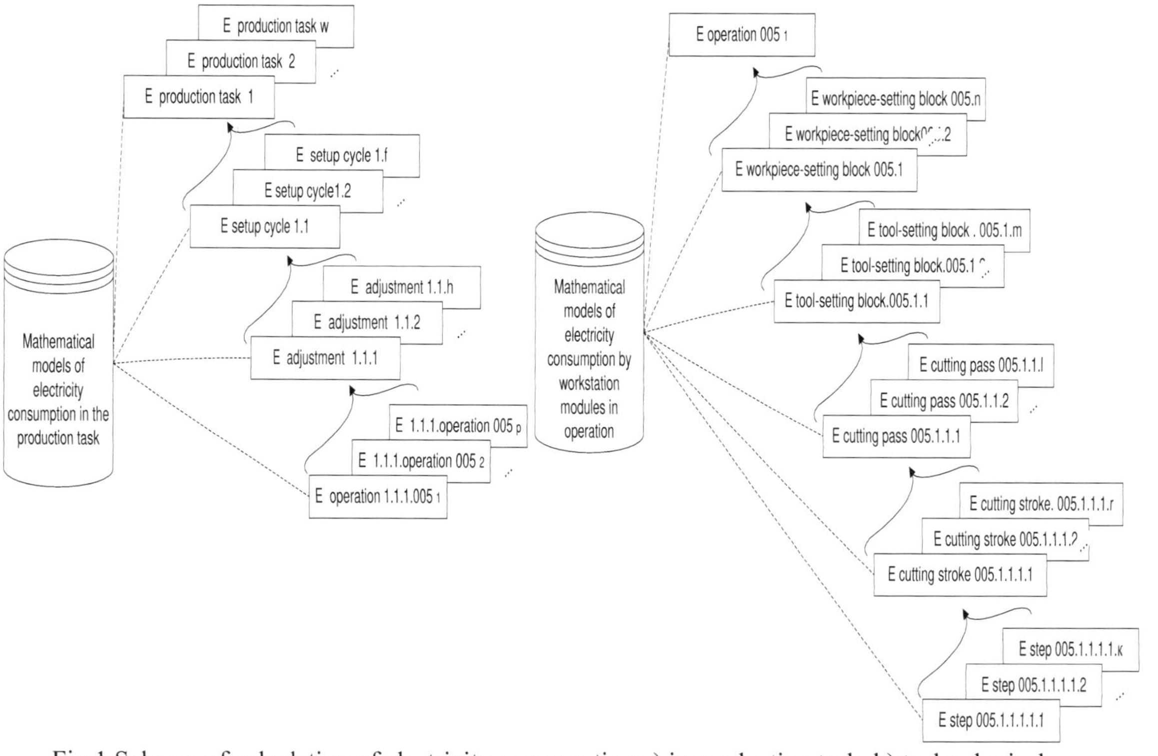


Fig.1 Scheme of calculation of electricity consumption a) in production task, b) technological operation

Basic data for carrying out calculation is the text of the driver of the machine with CNC. The program code by which hierarchical components of the process, and also operation modes of drive gears –number of spindle turns, size of running supply, accelerated supply, coordinates of points of modules conveyances are analyzed.

After that the calculation with application of the developed algorithm which initial fragment is given (fig. 2) is carried out.

In a step, which is the simplest structural element of operation, the shaping of certain sites of preparation with removal of an allowance is carried out. At this time the electric power is consumed by drive gears of the machine, and also by a spindle in the process of overcoming the moment of cutting, the main, radial, axial component of forces of cutting.

Estepi = (Mcut i*ϖi + Pyi*Vsyi + Pxi*Vsxi + Pzi*Vszi) *$t_{stepi}$ [W]

where Mcuti – the cutting torque [H*m];

ϖ – the angular speed of a spindle [turn/sec];

Pyi, Pxi, Pzi – components of the cutting force [H];

Vsyi, Vsxi, Vszi – feederate along the coordinate axis's [m/sec];

$t_{stepi}$– duration of a step i [sec].

Further electricity consumption is summarized on each step of the driving program and total electricity consumption ΣEc.s. in cutting strokes. Then, in each cutting pass to the calculated size of electricity consumption processes of electricity consumption of drive gears of supply in the accelerated supplies and off takes of calipers on each axis of coordinates, a spindle block in the processes of start-up, electromagnetic braking of a spindle, electricity consumption at pumping of a cutting compound cooling the tool in a cutting zone are added.

Ec.p.i = N a.s.t.x.i * t a.s.t.x.i + N a.s.t.y.i * t a.s.t.y.i + N a.s.t.z.i * t a.s.t.z.i + Nsp.start * t sp.start.i + Σ Ec.s. + N a.o.t.x.i * t a.o.t.x.i +N a.o.t.y.i * t a.o.t.y.i +N a.o.t.z.i * t a.o.t.z.i + N sp.stop * t sp.stop.i + $N_{CCF}$ * tc.p.i

where N a.s.t.x.i, N a.s.t.y.i, N a.s.t.z.i, N a.o.t.x.i, N a.o.t.y.i, N a.o.t.z.i – the power of electric motors for the accelerated supplies and off takes of the tool on coordinates of X, Y, Z [W];

t a.s.t.x.i; t a.s.t.y.i, t a.s.t.z.i, t a.o.t.x.i, t a.o.t.y.i, t a.o.t.z.i – duration of processes of the accelerated conveyances [sec.];

Nsp.start –the spindle block electric motor power spent in the course of start-up of a spindle and its untwisting to working rotary speed [W];

ΣEc.s. – total electricity consumption in the cutting strokes [W*sec.];

Nsp.stop – the power consumed by the electric motor in the course of braking of a spindle block [W];

t sp.stop.i – duration of process of braking [sec];

$N_{CCF}$ – the power of the electric motor, system of coolant-cutting fluid supply spent for its pumping [W];

tc.p.i – duration of the cutting pass;

Then energy consumption is summarized and there is a total electricity consumption in cutting passes ΣEc.p. Further summation is carried out on a similar algorithm, with the processes, inherent to each hierarchical level, and consumers.

## 3. Conditions for better efficiency of the offered technique.

The technique is effective only in the presence of mathematical models of consumption by all consumers on all modes of operation. This information can be obtained as a result of direct measurements of electricity consumption by machine modules with CNC and other auxiliary systems (transport, cleaning of shaving, supply of a cutting compound, lubricants) as a result of carrying out the planned production experiment according to the schedule. The experiment can be local, on specification of consumption at the demanded combination of supply, rotary speed, depth of cutting and brand of material, or much wider, in which other consumers and the modes of their work are considered.

Monitoring of power of all consumers of the electric power involved in production of products can effectively solve the following problems:

- control of technical condition of separate drive gears of machine systems with further planning of operations of service or repair of the details, solved now directly by measurements of the detail or the machine parameters [5-14],

- control of wear of the tool for its timely change, traditionally solved by introduction in production of statistic regulation of the process [10];
- decrease of electricity consumption by optimization of work of drive gears in the course of a shaping at a stage of technological design of operations, which is now not solved fully by the existing systems of the automated technological design.

The developed technique is tested at a stage of technological design at the choice of options of operations performance with the smallest electricity consumption on grinding machines with CNC and showed the efficiency when sharpening of the cutting edges of the special cutting tool [15-17]. Convergence of settlement and actual consumption didn't exceed 5%.

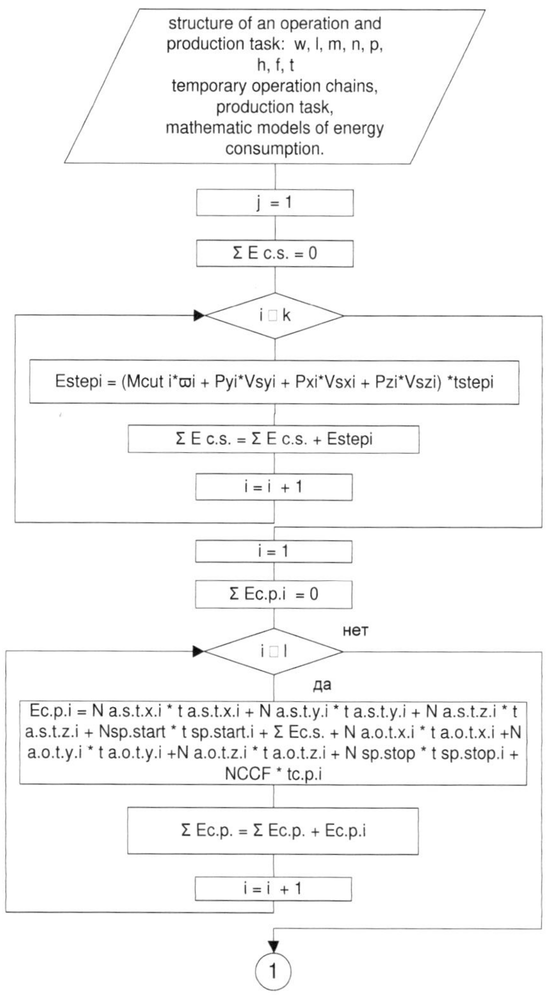


Fig. 2. Algorithm of calculation of need for the electric power in structural elements of technological operation: step, cutting stroke, cutting pass.

## References

[1] Safarov D T, Fedorova K A and Ilyasova A 2015 Sistematization of content of intra operational processes as basis of modern technological documentation *Collection of scientific works of the International scientific and technical conference* Pp 315-17
[2] Safarov D T 2010 Model of accomplishment of a single task on the example of handling of spiral drills *Collection of works of international scientific and technical conference* Pp 254-57

[3] Safarov L R, Glinina G F 2013 Time expence as a basis of calculation of amounts of by-products, waste and emissions in shaping operations *Bulletin of Saratov State Engineering University* No 2 (71) pp 91-93
[4] Kasyanov S V, Safarova L R 2016 The concept of designing of technologies of lifecycle of by-products and waste in the course of preparation for a new car component production *Materials of the International scientific and technical conference "Innovative Machine-building Technologies, the Equipment and Materials — 2016"* pp 309-12
[5] Kondrashov A G, Kasyanov S V, Safarov D T and Kuznetsova A V 2013 Diagnostic measurements of geometrical parameters of spatial and difficult details of autocomponents by one-coordinate altimeter *Control. Diagnostics* No 8 pp 60-64
[6] Kasyanov S V, Safarov D T 2004 Diagnosis of technical state of equipment and tools according to indices of technological accuracy *Avtomobil'naya Promyshlennost* Issue 5 pp 24-28
[7] Kondrashov A G, Safarov D T 2014 Prediction of accuracy when handling by cutting / News of higher educational institutions *Mechanical engineering* No 12 pp 63-69
[8] Balabanov I P, Balabanova O N, Groshev A V 2015 Formation of initial data of the workpiece batch in simulation modeling precision forming *IOP Conference Series: Materials Science and Engineering* Volume 86 Issue 1
[9] Khusainov R M, Belov S F, Chukhontseva O V 2014 Diagnosis of CNC machine tools in terms of circular interpolation's accuracy figure *IOP Conference Series: Materials Science and Engineering* Volume 69 Issue 1
[10] Safarov D T, Fedorova K A and Ilyasova A I 2016 Algorithms of development of making special techniques in APQP manufacturing process of automotive components *IOP Conference Series: Materials Science and Engineering* Volume 134 Issue 1 **012036**
[11] Kasyanov S V, Safarov D T and Kondrashov A G Patent for the invention *A way of diagnosing of a relative position and rigidity of the tool equipment in boring operations on indicators of accuracy of the processed details* No 2496611 bulletin No 30
[12] Kondrashov A G, Kasyanov S V, Safarov D T, Minnekhanova T A, Kuznetsova A V *A way of measurement of coordinates of the center and radius of cylindrical sites of details* Patent No 2581384 bulletin no 11
[13] Kasyanov S V, Safarov D T, Kondrashov A G *The automated complex for diagnosing of turning technological system on indicators of accuracy of the processed details* Patent for useful model No 133040 the bulletin No 28
[14] Kasyanov S V, Safarov D T, Kondrashov A G *A complex for diagnosing of turning technological system on indicators of accuracy of the processed detail*s Patent for useful model No 133039 the bulletin No 28
[15] Balabanov I P, Kondrashov A G 2014 Shaping of cutting part of angle milling cutters with nonzero geometry *World Applied Sciences Journal* Volume 30 Issue 12 pp 1731-34
[16] Chemborisov N A, A G Kondrashov 2008 End Machining of Teeth by a Single-Turn Worm Mill *Russian Engineering Research* Vol 28 No 8 pp 809-11
[17] Kondrashov A G, Safarov D T and Davletshina G K 2015 Bit for the Core Drilling of Nonmetals *Russian Engineering Research* Vol 35 No 8 pp 617–18

# System Analysis and Decision-Making During Synthesis of High-Performance Hybrid Boilers

**T R Safin, I A Konakhina and G R Khamidullina**

Kazan Federal University, 18 Kremlyovskaya street, Kazan, 420008, Russian Federation

rr-088@mail.ru

**Abstract.** The decision-making analysis for synthesis of high-performance hybrid boiler plants is based on current philosophy of system analysis and synthesis of combined heat and power plants. Energetic and exergetic utilization is used as performance criteria.

A generalized description of the system analysis approach for complex heat and power plants is referred to in [1]. However, as each new heat and power plant differs in structure and modes of operation, implementation of this approach requires basic methodological provisions of the system analysis, synthesis and optimization to be applied to the approach itself. A high-performance hybrid boiler plant would be an option in this case.

The boiler plant is a combination of conventional and alternative power sources. The conventional source generates heat energy for external consumers in the form of supplying hot water for heating, ventilation and tap hot water supply. The auxiliary system of the boiler plant required for heating up raw water, deaerator tank and demineralized water is facilitated by an alternative source – the solar trap [2] and the heat from an outside flue gas heat regenerator [3].

The boiler plant is also being discussed to operate as a mini-CHPP with gas reciprocating engines to generate the power for auxiliary systems of the boiler plant and to supply the power to third parties. Partially, the power is generated by an alternative source – solar cell batteries [2].

Therefore, the boiler plant is a combined source of heat and power energy generated by both the conventional and alternative sources, i.e. solar traps and solar cell batteries.

Based on the available philosophy of the system analysis, synthesis and optimization of heat and power complexes, the optimum structure and parameters of the hybrid boiler plant should be picked as follows [1].

1. External system parameters related to load connections and releases to atmosphere will be set.

2. The hybrid boiler plant will be structurally described to identify chains of dependent components (circuits) available or may be integrated after upgrading.

Structural modelling will define the number of equations $n_{\mathrm{E}}$ required for calculation of one or another component using simultaneous equations [4]

$$\left.\begin{aligned} n_{\mathrm{I}} &= n_{\mathrm{P}}' + n_{\mathrm{AC}} - n_{\mathrm{uncert.}} \\ n_{\mathrm{E}} &= n_{\mathrm{P}} - n_{\mathrm{I}} \end{aligned}\right\}, \tag{1}$$

where $n_P^{'}$ is the total parameters of incoming connections; $n_{AC}$ is the number of conditions applied to the parameters of incoming connections; $n_{uncert.}$ is the number of uncertainties for the parameters of outgoing connections; $n_I$ is the number of independent parameters; $n_P$ is the total number of parameters.

3. Further, a mathematical model will be constructed using balance equations and inequalities of imposed parametric limitations. The decisions on how to make islanding the plant, identify the components of circuits, chains of components to be calculated in sequence are followed, i.e. the results of structural analysis are in use.

The modelling will find a best parameter combination for the facility under process of synthesis based on a single or multiple criteria. In this case, it is the energetic and exergetic performance.

The problem can be defined as follows: To find the largest value of the function [5]

$$F\left(\tilde{O},\overline{X}\right)_{\sigma} \quad \text{when} \quad \left(\tilde{O},\overline{X}\right)\in R, \tag{2}$$

where $R$ is the acceptable region defined by

$$\begin{aligned} &\left(f_p\right)_{\min} \le f_p\left(X,\overline{X}\right) \le \left(f_p\right)_{\max}, \qquad p=\overline{1,a} \\ &\overline{X}_{\min} \le \overline{X} \le \overline{X}_{\max}, \quad \overline{X} \in \boldsymbol{\sigma}_s; \\ &X \in L_t \end{aligned} \tag{3}$$

The functions $F\left(\overline{X}\right)$ and $f_p\left(\overline{X}\right)$ are assumed to be differentiable if the values $X$ are fixed.

Here $X$ is the total discretely variable parameters of the boiler plant; $\overline{X}$ is the total independent parameters that continuously vary (seasonal parameters, weather conditions etc.); $\boldsymbol{\sigma}$ is the total external factors characteristics; $f=\left\{f_1,f_2,\ldots,f_a\right\}$ is the sum of operational characteristics of the boiler plant under the process of synthesis that are used for setting the limiting conditions; $\boldsymbol{\sigma}_S$ is the spatial dimension; $L_t$ is the finite aggregate of discrete elements of dimension $t$; "min" and "max" means a minimum and a maximum value, respectively.

Energy efficiency has been considered as an energy-related effectiveness criterion [1]:

$$\eta_E = \frac{\sum \Delta Q_i^{\text{useful}}}{\sum \Delta Q_i^{\text{applied}}}, \tag{4}$$

The formula (4) uses the following features: $\Delta Q_i^{\text{applied}}$ is the heat value for the energy consumed by the plant with $i$-flow, kW; $\Delta Q_i^{\text{useful}}$ is the heat value for the useful effect of $i$-flow energy perception, kW.

Exergy efficiency has been considered as a thermodynamic effectiveness criterion:

$$\eta e = \frac{\sum \Delta E_i^{\text{useful}}}{\sum \Delta E_i^{\text{applied}}}, \tag{5}$$

The formula (5) uses the following features: $\Delta E_i^{\text{applied}}$ is the heat value for the exergy consumed by

IOP Conf. Series: Materials Science and Engineering **240** (2017) 012061 doi:10.1088/1757-899X/240/1/012061

the plant with *i*-flow, kW; $\Delta E_i^{useful}$ is the heat value for the useful effect of *i*-flow exergy perception, kW.

4. An optimum solution is going to be found for the entire range of parameters by the gradient methods to have a combination of parameters that allow for maximum performance of both the energy and exergy efficiency.

The provisions of the system analysis have been implemented in a real facility – the boiler plant at Gazprom-dobycha Nadym, LLC taking weather conditions of Nadym into consideration [3]. The effect of upgrading the boiler plant is estimated at 1,320 tons of oil equivalent per annum that corresponds to efficiency rise approximately by 10%.

## References

[1] Nazmeev Yu G and Konakhina I A 2002 *Heat and energy systems and power balances of production plants* (M.: Publishers of Moscow power engineering institute) p 407

[2] Khusnutdinova E M, Konakhina I A and Khamidullina G R 2015 Proceedings of the international scientific and technical conference *Innovative engineering technologies, machinery, materials – 2015* pp 338-34

[3] Safin T R and Konakhina I A 2014 *Optimization of thermal conditions of hot-water boiler with economizer* Proceedings of the international scientific and technical conference «Innovative engineering technologies, machinery, materials – 2014 pp 264-61

[4] Sazanov B V and Sitas V I 1990 *Heat and energy systems of production plants* (M.: Energoatomizdat) p 304

[5] Konakhina I A 2004 *Generalized problem definition related to optimization of operating conditions during synthesis of combined energy production systems* Izvestiya vuzov: Power Engineering Issues pp18-13

IOP Conf. Series: Materials Science and Engineering **240** (2017) 012062 doi:10.1088/1757-899X/240/1/012062

# Production of thermally conductive carbon foams and their application in automobile transport

**V M Samoylov[1], E A Danilov[1], E R Galimov[2], V L Fedyaev[2,3], N Ya Galimova[2] and M A Orlov[4]**

[1] JSC NIIgrafit, Moscow, Russian Federation
[2] Institute of Mechanics and Mechanical Engineering of Kazan Scientific Centre of Russian Academy of Science, Kazan, Russian Federation
[3] Kazan National Research Technical University named after A N Tupolev, Kazan, Russian Federation
[4] Bauman Moscow State Technical University, Moscow, Russian Federation

vsamoylov@niigrafit.org, morenko@imm.knc.ru, maksim.orlov@emtc.ru

**Abstract.** The report examines the possible methods of synthesis and properties of a new class of carbon materials – thermally conductive foams based on mesophase pitches and their application in automobiles to control the heat flow.

## 1. Introduction

Currently, in many foreign countries research aimed at the study of such materials as foams based on polymers (foam plastics), inorganic materials including carbon are carried out. The range of application of these materials is extremely wide [1]. The heat-insulating carbon foams on the basis of phenolic resins developed by the Union Carbide company (USA) in the early 60s' of the 20$^{th}$ century [2] are well-known. One of the examples of a new class of carbon materials is thermally conductive foam based on mesophase pitches [1, 3].

## 2. Research part

For the first time materials of this class were obtained by Klett in the USA in the year 2000 [3]. Currently, industrial production of thermally conductive carbon foams based on the mesophase pitches is performed by ORNL and Poco Graphite companies. The properties of these foams are shown in table 1.

**Table 1.** Properties of carbon foams.

| Properties | ORNL | POCO |
|---|---|---|
| Density, g/sm$^3$ | 0.25 – 0.65 | 0.56 – 0.90 |
| Porosity, % | 78 – 88 | 61 – 70 |
| Coefficient of thermal expansion, $10^{-5}$ K$^{-1}$ | 2.0 | 0.7 – 0.8 |
| Compressive strength, MPa | 1.0 – 3.5 | 3.0 – 2.7 |
| Tensile strength, MPa | 0.7 – 1.6 | 0.6 – 1.9 |

Physical-mechanical properties of carbon foams are sufficiently high. Even at a density of about 0.27 g/sm$^3$ compressive strength for heat-insulating carbon foams based on polyurethane or phenolformaldehyde resin is 10-15 MPa at thermal conductivity less than 10 W/m · K [4-6]. The strength of thermally conductive foams based on mesophase pitches varies from 1 to 20 MPa at a density of 0.25 to 0.65 g/sm$^3$. This strength is sufficient to meet not only operational requirements but also to allow the production of large-scale products up to 3 m in length and up to 150 mm thick.

A useful feature of carbon foams is coefficient of thermal expansion which remains constant in a sufficiently wide variation range of temperatures from 200 to 600 °C. Coefficient of thermal expansion at 300 K for the heat-insulating carbon foam based on phenolformaldehyde resin is from 5 to 6 $\cdot 10^{-6}$ K$^{-1}$, for thermally conductive foam on the basis of mesophase pitches from 1.2 to 2.6 $\cdot 10^{-6}$ K$^{-1}$ [4-6].

The thermal conductivity of graphitized carbon foams based on mesophase pitches significantly changes depending on the nature of the raw pitch and the peculiarities of technological process. As it was shown in work [6] with an increase in the density of thermally conductive carbon foams based on mesophase pitches from 0.25 to 0.65 g/sm$^3$ thermal conductivity increases from 40 to 210 W/m · K.

Furthermore, carbon foams possess an improved fire resistance [1] and withstand even prolonged exposure to the torch flame in the air [5, 6]. The maximum operating temperature in the air for the thermally conductive POCO carbon foam is 400°C [6].

It is clear that there are significant differences in the structure of carbon foams caused by both the nature of the initial raw material and the applied technologies [1, 4, 6]. First of all we are talking about the microstructure (graphitizability) of carbon foams and their porosity.

The viscosity of the pitches and resins with an increase of temperature at carbonization varies quite widely. The amount of volatile matter dicharged with increasing of the processing temperature also varies [1, 4]. Formation of thermally conductive carbon foam from mesophase pitches has a number of significant peculiarities [1, 4]. When carbonizing the pitches at the first stage of heating (up to 350°C) melt viscosity continuously decreases the escaping gases originally form intruding bubbles. If pressure is applied when carbonizing the bubbles of volatile matter begin to inflate the pitch staying within the volume of the carbonized mass.

The size of these bubbles is determined not only by the nature of the pitch but the temperature range of carbonization under pressure. Increasing the temperature and pressure in the lower viscosity range increases the sizes of the formed bubbles. At the final stage (470-530°C) the initial structure of the carbon foam is formed. After the external pressure release the bubbles breakthrough resulting in foam with a structure similar to that shown in figure 1. For comparison, the same figure shows the structure of the heat-insulating carbon foams based on the phenolformaldehyde resin.

It should be kept in mind that in besides the external differences there is a difference in the microstructure of the carbon material itself. For heat-insulating foams it is non-graphitisable carbon material consisting of polycrystals with the sizes less than 5 nm and an interlayer distance more than 0.342 mm. In contrast, for thermally conductive foam it is a material with high capacity for graphitization having the crystal sizes of 100 nm or more and interlayer distance close to the monocrystal – 0.335-0.336 nm.

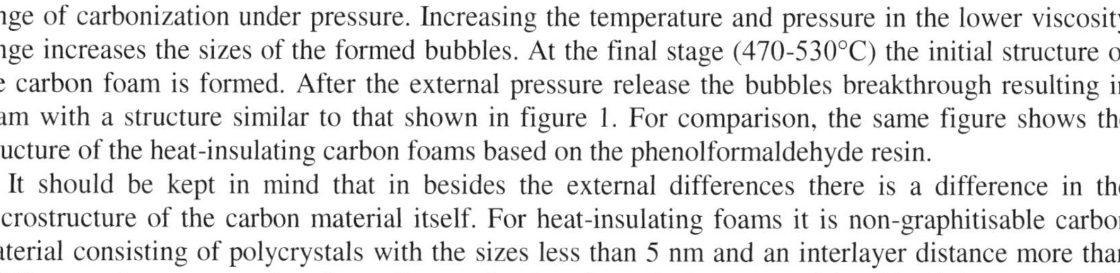


**Figure 1.** SEM images of heat-insulating carbon foams of non-graphitisable resins (a, b) and thermally

conductive carbon foam based on mesophase pitches (c) obtained using different technologies.

The basis of the thermally conductive foam technology (figure 2) is carbonization under pressure of inert gas followed by stabilization of the structure of the foam due to the thermal oxidation or carbonization [1, 4].

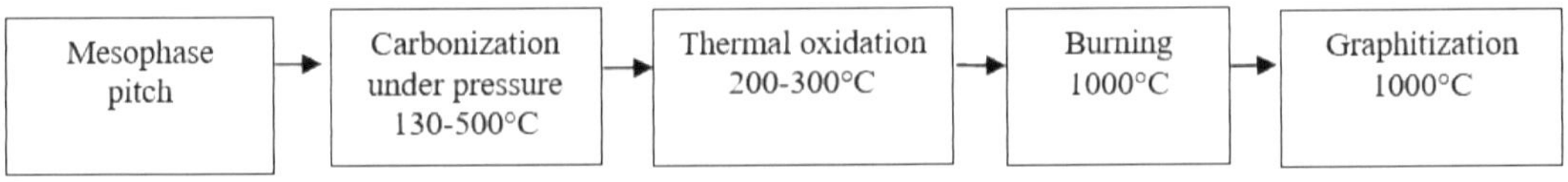


**Figure 2.** Production stages of thermally conductive foam.

Application of thermally conductive foam to control the heat flows is due to the following: thermal conductivity of carbon foams is 4 times higher than that of copper and 6 times higher than that of aluminium whereas the mass is not more than 20% from the mass of these metals for the surface varying from 5000 to 50000 $m^2/m^3$ [4-6]. The carbon foam accumulates heat on 65% less than cooper dissipates heat 15 times faster than copper [4-6].

## 3. Conclusions

As these papers show the application of foam allows reducing the sizes of the conventional automotive type radiator and making it lighter and more efficient. In this case the radiator does not have to be placed in the front part of the vehicle it will significantly improve its aerodynamics and the released heat can be used, for example, for heating of the passenger compartment. Currently, racing teams of formula A show great interest in such radiators. When producing the armored personnel carriers and armoured vehicles these radiators reduce the vulnerability of the machinery. Application variants of carbon foam for cooling the breaks (the local overheating is eliminated) and also in the cooling systems of oil and transmission fluid are considered.

It is also promising to use carbon foams as cathodes of lithium-ion current supplies and in the acid accumulators as well as in the power supply systems of hybrid vehicles as part of the batteries [7, 8].

## References

[1] Michio I, Jieshan Q and Quangui G 2015 *Review Carbon foam: Preparation and application. Carbon* (vol 87) pp.128 –152.

[2] Ford, Walter 1964 *Method of Making Cellular Refractory. Thermal insulating Material* US Patent 3,121,050

[3] Klett J 2000 *Process for Making Carbon Foam.* US Patent 6,033,506

[4] Klett J 2002 *International Seminar on Advanced Applications for Carbon Materials* (Jeju Island, Korea)

[5] Nidia C G and James W K 2002 *International Seminar on Advanced Applications for Carbon Materials* (Jeju Island, Korea)

[6] Klett J, Ott R and A McMillan 2000 *Heat exchangers for heavy vehicles utilizing high thermal conductivity graphite foams* (SAE Technical Paper 2000-01-2207)

[7] *The only carbon-foam agm battery is now available for marine applications* 2016 (http://www.bruceschwab.com/uploads/2016-firefly-handout-2pg.pdf)

[8] Jurchenko D, Kelley K 2007 *Carbon foam liberates lead-acid batteries for emerging military missions Defense Electronics*, (706RFDEF3.indd http://www.rfdesign.com)

This work was performed with financial support from Ministry of Education and Science of the Russian Federation in the framework of the Federal Target Program “Research and development in priority development fields of science and technology sector of Russia for 2014-2020" the agreement No. 14.583.21.0057 on 28th of July 2016 (Project ID RFMEFI58316X0057).

IOP Conf. Series: Materials Science and Engineering **240** (2017) 012063 doi:10.1088/1757-899X/240/1/012063

# Investigation of physical-mechanical and performance properties of glass fiber reinforced polyurethane materials used in the machine building industry

**L N Shafigullin[1], V I Astashchenko[1], N V Romanova[1], A R Ibragimov[1], G R Shafigullina[1] and A N Shafigullina[1]**

[1]Naberezhnye Chelny Institute (branch), Kazan Federal University
Sjujumbike street 10a, Naberezhnye Chelny, 423800 Tatarstan, Russia

misharin_82@mail.ru

**Abstract.** The paper presents the investigation of physical-mechanical and performance properties of the glass fiber reinforced polyurethane (PUR) materials made using the long fiber injection process. It was found that glass fiber reinforced polyurethane could be used to manufacture interior parts with different surface textures (instrument panels, door trim, armrests).

## 1. Introduction

Today hundreds of elastic and rigid gas-expanded material brands are manufactured from synthetic resins, natural rubber and high molecular compounds. These materials include polyurethanes (PUR) which are used in all the manufacturing and construction sectors [1]. The present-day PURs have low production cost, low toxicity, allow operation in a wide temperature range. Due to a large number of PUR components their properties may vary widely [2].

Polyurethanes are used as noise- and heat-insulating materials, to manufacture vehicle interior and exterior parts, decorative features in construction. Reinforced polyurethanes are used as structural materials [3].

Glass fiber reinforced polyurethanes make it possible to manufacture large-sized components with a multilayer structure. It consists of the various combinations of hard and porous polyurethane, short and long glass fibers to meet different performance and aesthetic requirements in a variety of applications. Glass fiber reinforced PURs are the analogues of fiberglass, ABS sheets (acrylonitrile-butadiene-styrene copolymer) and acryl (polymethyl methacrylate) which are currently in use. Today removal of styrene monomers and other similar solvents from high-volume manufacturing is not only an economic advantage but also important factor in terms of occupational safety and health, and environmental security.

## 2. Body text

The investigations were made on the products which consist of polyurethane film (surface layer), glass fiber reinforced rigid polyurethane (reinforced layer (base)), semi-rigid PUR (intermediate (bonding layer)). The products were manufactured using long fiber injection technology (LFI) [6].

The table shows the results of the investigation of mechanical and performance properties. The tests were performed at a temperature of 21$^{\circ}$C, air humidity of 55%.

The test results revealed the high values of physical-mechanical and performance properties of glass fiber reinforced PURs. The obtained results allow us to recommend glass fiber reinforced PURs as an alternative for conventional fiberglass materials in the production of interior parts with various surface textures (instrument panels, door trim, armrests) [6].

Table. Physical-mechanical and performance properties of products made of glass fiber reinforced PUR

| Property name | Test results | | | Test methods |
|---|---|---|---|---|
| | Product 1 | Product 2 | Product 3 | |
| Tensile strength of the surface layer, kgf<br>Longitudinal direction<br>Transverse direction | <br><br>45<br>43 | <br><br>45<br>43 | <br><br>45<br>43 | GOST 17316 [7] |
| Ultimate elongation of the surface layer, %<br>Longitudinal direction<br>Transverse direction | <br><br>237<br>193 | <br><br>192<br>180 | <br><br>187<br>193 | |
| Strength of bond between the surface layer, semi-rigid PUR and reinforced layer | No cracks on the surface, no layer separation in the part | – | No cracks on the surface, no layer separation in the part | GOST 17317 [8] |
| Temperature resistance:<br>+100 °C for 12 hours<br>-45 °C for 16 hours | No separation of the surface layer, no changes in appearance, no deformation, no layer separation | No separation of the surface layer, no changes in appearance, no deformation, no layer separation | No separation of the surface layer, no changes in appearance, no deformation, no layer separation | TTM 37.102.0026 – 2007 procedure p.1 [9] |
| Cold resistance<br>-45 °C for 6 hours | No changes in appearance | No changes in appearance | No changes in appearance | TTM 37.102.0026 – 2007 procedure p.3 [9] |
| Heat resistance (change in part geometry, %) | 0.47<br>No separation and swelling of the surface layer | 0.24<br>No separation and swelling of the surface layer | 1.01<br>No separation and swelling of the surface layer | TTM 37.102.0026 – 2007 procedure p.2. [9] |
| Shrinkage, % | 0.22 | – | 0.16 | TTM 37.102.0026 – 2007 procedure p.5. [9] |
| Moisture resistance, % wt. | 0.35 (no separation and swelling of the surface layer) | – | 0.35 (no separation and swelling of the surface layer ) | TTM 37.102.0026 – 2007 procedure p.4. [9] |

| | | | | |
|---|---|---|---|---|
| Odor intensity, points, maximum | 1.0 (allowed for use) | 2 (allowed for use) | 1.0 (allowed for use) | TTM 37.102.0026 – 2007 procedure p.6. [9] |

## 3. Conclusions

LFI technology helps to increase the production flexibility and quality, reduce the production cost of products, especially large-sized ones, thanks to a high quality of manufacturing and a lower amount of production waste.

## References

[1] Internet resource: Gas-expanded plastic materials (expanded polystyrene). Available at http://chem21.info/info/895542/ (accessed 13.11.2015)

[2] Terenteva N N, Danilov V A, Kuzmin M V and Verkhunov S M 2007 Laboratornyi praktikum po distsipline khimiia poliuretanov [Polyurethane chemistry laboratory practicum] (Cheboksary: Ulianov Chuvash State University Publ.) p 97

[3] Internet resource: Avtotekhnik OOO. Available at http://автотехник.net/ru/news/44-osvoeno-novoe-izdelie-2014.html (accessed 13.11.2015)

[4] Internet resource: http://www.dow-izolan.com/ru/products/ (accessed 06.07.2016)

[5] Internet resource: http://poly-tex.ru/fiberglass/glass-roving/ (accessed 06.07.2016)

[6] Internet resource: http://www.nccrussia.com/ru/lfi (accessed 14.10.2016)

[7] GOST 17316–71 1972 Artificial soft leather. Measuring method of tearing load and elongation of break (Moscow, Izdatelstvo Standartov Publ.)

[8] GOST 17317–88 1988 Artificial leather. Method for determination of interlayer bonding strength (Moscow, Izdatelstvo Standartov Publ.)

[9] Material specification No. 37.102.0026. 2007.

IOP Conf. Series: Materials Science and Engineering **240** (2017) 012064 doi:10.1088/1757-899X/240/1/012064

# Ensuring uniformity of metals and alloys structure by multiple extrusion

**V G Shibakov, D L Pankratov, A P Andreev and R F Utyaganov**

Kazan Federal University, Naberezhnye Chelny Institute, Naberezhnye Chelny, Republic of Tatarstan, Russia

pankratovdl@gmail.com

**Abstract.** Key parameters influencing the deformation force, intensity ofplastic deformation and its distribution at IPDmultipleextrusion are considered, recommendations about a choice of rational parameters are made.

One of the methods for obtaining submicrocrystalline or nanocrystalline structure in metals and alloys is aintense plastic deformation (IPD) [1].

The main process steps of the proposed method are shown in Figure 1 [2, 3].

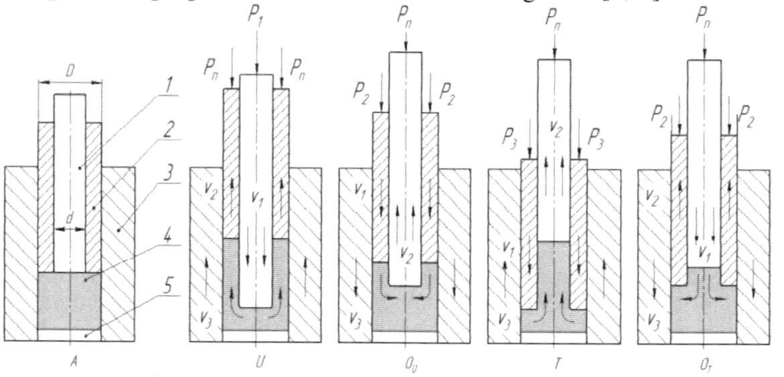


Figure 1 Scheme SPD extrusion process:

A - the starting position; U - reverse extrusion "cup"; $O_U$ -education original form radial extrusion of the "cup"; T -back squeezing bar with thickening; $O_T$ - the formation of the original form of the radial extrusion rod with thickening; 1 - solid punch; 2 - hollow punch; 3 - the matrix; 4 - billet; 5 - fasonny punch; $P_1$ – $P_3$- extrusion force at different stages; $p_c$ - the power of the counterpressure ($0 \le P_c < P_2$, $0 \le P_c < P_3$); D, d - diameter punches; $v_1$, $v_2$, $v_3$- speed tool motion

On the basis of the circuit shown in Figure 1 is formed a number of technology options for multiple IPD extrusion:

1. $A+U_1 \rightarrow O_к$
2. $A+T_1 \rightarrow O_к$
3. $A+U_1+O_1+T_2 \rightarrow O_к$

IOP Conf. Series: Materials Science and Engineering **240** (2017) 012064 doi:10.1088/1757-899X/240/1/012064

4. $A+T_1+O_1+U_2 \rightarrow O_к$
5. $A+U_1+O_1+K+U_2 \rightarrow O_к$
6. $A+T_1+O_1+K+T_2 \rightarrow O_к$
7. $A+U_1+O_1+K+T_2 \rightarrow O_к$
8. $A+T_1+O_1+K+U_2 \rightarrow O_к$,

where K - tilting the semi-finished product prior to extrusion, $O_к$ - form semi-final return to the original

The main factors affecting the reliability of the process and the formation of a homogeneous deformation, and thus the structural state [4] semi-finished metal product $O_к$ are flow sheetIPD and geometric parameters of the tool and the billet.

To assess the influence of geometrical parameters of the tool and definition of rational modes of IPD process multiple extrusion was simulated by varying the following parameters:

- Radius of curvature of solid and hollow punches;
- Geometry of the matrix, and the edges of the hollow punch;
- The scale factor by a proportional increase in the billet, tooling and equipment options.

The study was conducted by simulating the various options in the software package QForm 2D. Simulation parameters are shown in Table 1. For ease of comparison, the results obtained when the plastic deformation range of different embodiments, the same extrusion and has a gradation from 0 to 20 (Figure 2)

Table 1. Initial simulation parameters during extrusion IPD

| Material billet | The initial temperature of the billet, C° | The initial instrument temperature,C° | Lubricant | Deformationspeed, mm/s |
|---|---|---|---|---|
| BT-8 | 650 | 650 | Liquid glass | 1 |

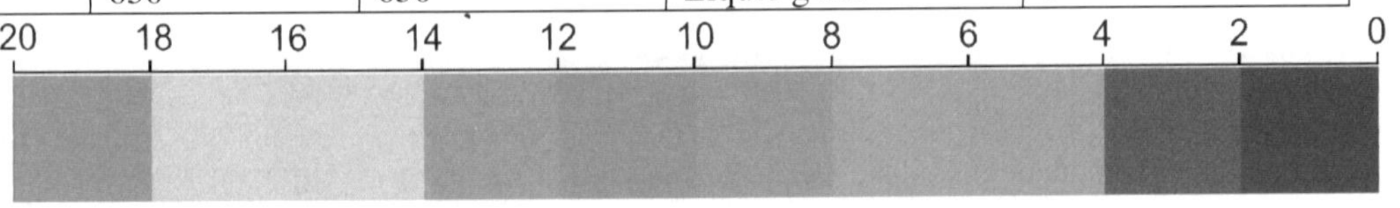


Figure 2.Plasticdeformationscale

Rational parameters of the edges of the matrix and the punch (Figure 3) were determined by simulation while varying according to the data in Table 2.

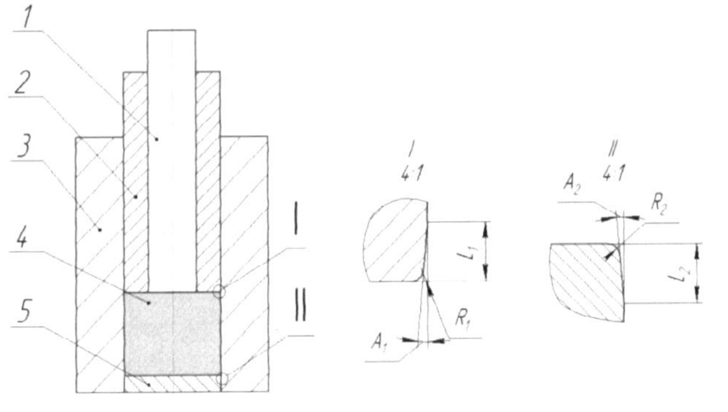


Figure 3. The geometrical parameters edges of the die and punch,

where $R_1$ - radius of curvature of the hollow punch, $R_2$ - Radius of curvature of the matrix; $L_1$ - length of slope on the hollow punch; $L_2$ - length of slope on the matrix; $A_1$ - the angle of inclination on the hollow punch; $A_2$ - the angle of inclination on the matrix; 1 - solid punch; 2 - hollow punch; 3 - the matrix; 4 - billet; 5 - shaped punch;

Table 2 Possible values of the parameters of the matrix edges and hollow Poisson

| Variant№ | $R_1$, mm | $R_2$, mm | $L_1$, mm | $L_2$, mm | $A_1$, degrees | $A_2$, degrees |
|---|---|---|---|---|---|---|
| 1 | 1 | 1 | 1,5 | 1,5 | 10 | 10 |
| 2 | 1 | 1 | 2,5 | 2,5 | 10 | 10 |
| 3 | 1 | 1 | 3,5 | 3,5 | 10 | 10 |
| 4 | 1 | 1 | 4,5 | 4,5 | 10 | 10 |
| 5 | 0,5 | 0,5 | 1,5 | 1,5 | 10 | 10 |
| 6 | 0,5 | 0,5 | 2,5 | 2,5 | 10 | 10 |
| 7 | 0,5 | 0,5 | 3,5 | 3,5 | 10 | 10 |
| 8 | 0,5 | 0,5 | 4,5 | 4,5 | 10 | 10 |
| 9 | 1 | 1 | 1,5 | 1,5 | 5 | 5 |
| 10 | 1 | 1 | 2,5 | 2,5 | 5 | 5 |
| 11 | 1 | 1 | 3,5 | 3,5 | 5 | 5 |
| 12 | 1 | 1 | 4,5 | 4,5 | 5 | 5 |

Based on the simulation results, we can conclude the feasibility of using the following parameters of the matrix edges and hollow punch with $L_1$, $L_2$is not less than 1 mm and not more than 2.5 mm, the angle A is not more 5-7,5°degrees and value of the radius R of not more than 1mm due to the possibility of increasing clip at a higher value of these parameters (Figure 4).

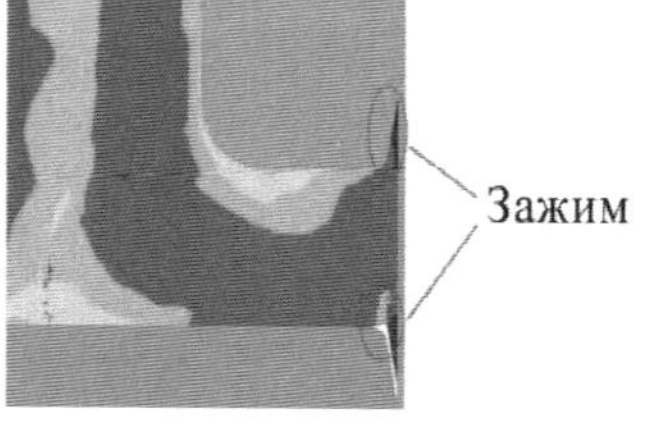


Figure 4.Displayoftheclipduringthedeformation

The scale factor, ie proportional increase in the size of the billet and the IPD parameter has little effect on the rate of plastic deformation and its distribution. Revealed insignificant differences (Figure 5) are associated with a decrease in the friction influence.

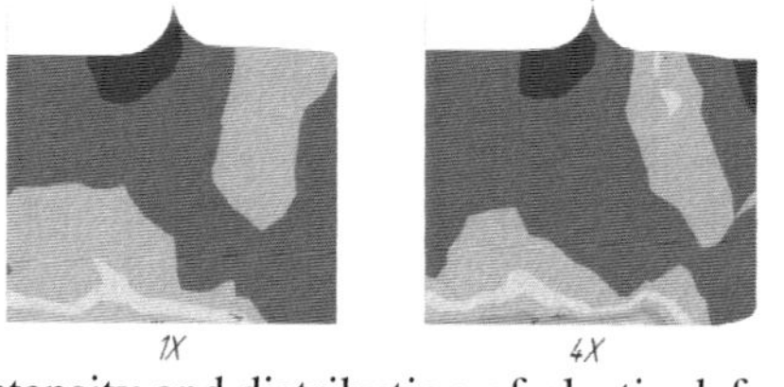


Figure 5. Changes in the intensity and distribution of plastic deformation in a proportionate increase in the billet and IPD settings to four times (4X)

**Conclusions:**

1. The factors influencing the effect of deformation, plastic deformation intensity and its distribution in the IPD multiple extrusion are defined.
2. Recommendations for the choice of rational parameters IPD repeated extrusion process are given.

## References

[1] Bernstein M L, Zaymovsky V A 1979*Mechanical properties of metals* M, Metallurgy 495 pp
[2] Andreev A P, Shibakov V G, Pankratov D L and Andreeva S I 2015 Prediction of samples failure during severe plastic deformation by multiple extrusion IOP *Conference Series:*

*Materials Science and Engineering* Volume 86 Number 1 pp 12036-4
[3] The Patent №2189883 of Russian Federation, MKI 7 21 J 5/00, 13/02, C 21 D 7/02. The process of plastic structure formation of metals during severe plastic deformation and a device for its implementation

IOP Conf. Series: Materials Science and Engineering **240** (2017) 012065 doi:10.1088/1757-899X/240/1/012065

# Algorithms of walking and stability for an anthropomorphic robot

**R T Sirazetdinov, V M Devaev, D V Nikitina, A Y Fadeev and A R Kamalov**

Kazan Federal University, 18 Kremlyovskaya street, Kazan, 420008, Russian Federation

rif-kat@inbox.ru, vdevaev@yandex.ru, daria.92rus@gmail.com, Andrjuhafad@yandex.ru , arttambel@gmail.com

**Abstract.** Autonomous movement of an anthropomorphic robot is considered as a superposition of a set of typical elements of movement - so-called patterns, each of which can be considered as an agent of some multi-agent system [ 1 ]. To control the AP-601 robot, an information and communication infrastructure has been created that represents some multi-agent system that allows the development of algorithms for individual patterns of moving and run them in the system as a set of independently executed and interacting agents. The algorithms of lateral movement of the anthropomorphic robot AP-601 series with active stability due to the stability pattern are presented.

**Introduction.** The problem is considered to realize the lateral step of an anthropomorphic robot AR-601 series produced by "Android Technics" with ensuring its stability [1, 6]. To manage the robot, an information and communication infrastructure is developed, which is a multi-agent system that allows running algorithms for executing individual moving patterns as independently functioning and interacting agents. Agents are located on one or more controllers and interact with the main processor of the robot through an Ethernet type network. In this paper, algorithms for lateral moving of an anthropomorphic robot AP-601 series with active resistance are presented. In this case, the mathematical models and algorithms of the robot step patterns in the side are separately considered and stability of the robot in its longitudinal symmetry plane is ensured. When performing the work, the experience of modeling in related fields of science is taken into account [7, 8].

**Algorithm of stability.**

Consider the robot stability pattern in the longitudinal plane. Figure 1 shows the construction of the foot of AP-601 robot. The foot is connected to the shin by means of an electric drive, which sets the angle between them. A sole of the foot is attached to the foot using a spring. There are force sensors, which allow to obtain the values of the moments of forces that arise in the articulation of the foot with the shin. Assumed that the body of the robot, hip and shin are fixed relative to each other, stability is achieved by changing the inclination of the shin relative to the foot.

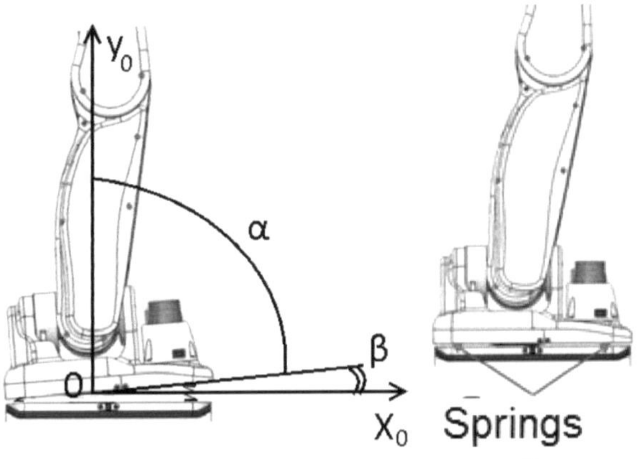


Fig. 1 Scheme of the node shin robot AR-601

In Figure 1, α - is the angle of rotation the shin to the foot, β - is the angle of tilt of the foot relative to the sole (floor).
Then $\gamma = \alpha + \beta$ – tilt of the foot relative to the floor.
Let $M_g = l * m * g * \sin M_g = l m g \sin(\gamma)$ – the torque created by the weight of the robot, $M_\beta = K_\beta * \beta$ $M_\beta = K_\beta \beta$ - the torque created by springs in the soles of the robot, l –the distance from the foot to the center of mass of the robot, g – acceleration of gravity, mg – Weight of the robot,$K_\beta$ – coefficient of elasticity of the sole.
The equation of dynamics tilt of the robot is described by the following differential equation:

$$I_m \acute{\gamma} = l m g \sin(\gamma) - K_\beta \beta \quad (1)$$

Assuming the value of γ to be small, replace sin γ by γ:

$$I_m \acute{\gamma} = l m g \gamma - K_\beta \beta \quad (2)$$

It is required to ensure the stability of the γ angle.
Let the required dynamics be described by the equation:

$$I_m \acute{\gamma} = K_1 \gamma + K_2 \acute{\gamma} \quad (3)$$

$K_1$ and $K_2$ Coefficients of elasticity of the left and right soles of the robot.

To reduce the equation to the following form:

$$K_1 \gamma + K_2 \acute{\gamma} = l m g \gamma - K_\beta \beta \quad (4)$$

We deduce β by α: $\beta = \gamma - \alpha$
Then:

$$K_1 \gamma - l m g \gamma + K_\beta \gamma + K_2 \acute{\gamma} = K_\beta \alpha \quad (5)$$

$$\alpha = \frac{K_1 + K_\beta - l m g}{K_\beta} \gamma + \frac{K_2}{K_\beta} \acute{\gamma} \quad (6)$$

If $\alpha = \alpha(t)$ is given by the formula (6), then the dynamics of the robot oscillations on the feet will correspond to the expression (3). $K_1$ and $K_2$ Are determined from the desired dynamics,l, m and $K_\beta$ are known.

The algorithm for controlling foots to ensure stability: from the moment sensors in the feet of the robot, the indicators $\upsilon_\gamma(t_i)$ at the moment $t_i$are taken and the values $\gamma(t_i)$, $\acute{\gamma}$ $(t_i)$ are calculated. Calculates $\alpha = \alpha$ (t) according to the formula (6) and is sent for execution in the foot drive.

**Step algorithm.**

We consider the step of the robot in the side as a typical element of walking [1]. It is assumed that the rotational movement of the robot around the vertical axis, which occurs when the free leg is moved forward, is compensated by rotations of the robot body. This makes it possible to divide the motion in the longitudinal (sagittal) plane from motion in the transverse plane. Therefore, here we consider the dynamics of the lateral (in the transverse plane) movement of the robot that occurs when walking.

Step is carried out in the following order. The robot crouch's on one knee, then the knee is bent and the robot is given an impulse, which results in movement on one leg according to the law of motion of the pendulum. If the center of mass does not intersect the vertical plane passing through the point of support of the foot, then after a while the robot will return to the initial position with support on both legs. The time for which the robot pushed off and stood on both feet is the step time. The task is to develop an algorithm for straightening the jogging leg, which provides a given time for the step.

The time of the step is determined by the speed of the center of mass of the robot acquired at the moment of tearing off the leg, because force influence on the robot at the moment of detachment of the jogging leg ends [1]. We believe that we can provide a horizontal position of the robot's torso. Consider the equations for calculating the distance from the fulcrum to the center of mass of the robot [2] (Figure 2).

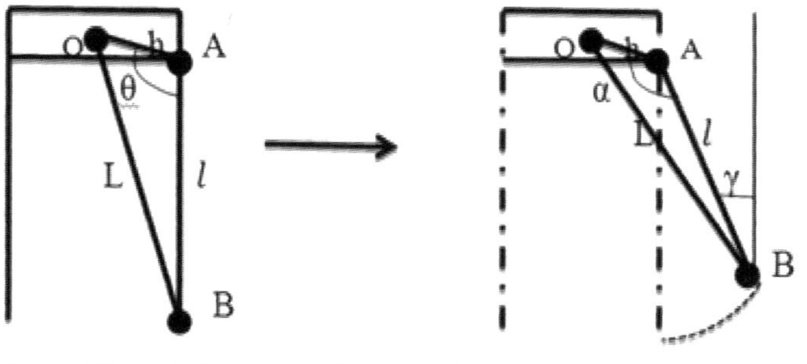


Fig. 2 The position of the center of mass of the robot

L – distance from the point of support of the foot to the center of mass of the robot,
l – current length of the leg, from the fulcrum to the attachment of the leg to the thigh
h – length from the point of attachment of the leg to the thigh to the center of mass of the robot, const;
θ – an angle between the center of mass of the robot and the initial position of the support leg, const;
γ – current angle formed after the transfer of the leg.
α - angle between the new position of the foot and the center of mass.
$\alpha = \theta + \gamma$;

By the cosine theorem [5], we calculate the length L:

$$L^2 = h^2 + l^2 + 2hl\cos\alpha. \quad (7)$$

Consider robot movements on one leg according to the law of a physical pendulum, inverted 180 degrees:

$$J\ddot{\varphi} = mgL \cdot \sin(\varphi)$$

m - robot mass,
φ – Angle of tilt vertically,

g – acceleration of gravity (9,81 м/с$^2$),
L - length from the point of support of the foot to the center of mass of the robot,
J – moment of inertia about the axis passing through the point of support, where

$J = J_0 + mL^2$
$J_0$ – Moment of inertia relative to the center of mass of the robot.
Let the angle $\varphi$remain small all the time (less than 10 degrees), we can replace it $\varphi = \sin(\varphi)$.
Then the equation takes the following form:

$$J\ddot{\varphi} - mgL\varphi = 0 \quad (8)$$

The solution of the differential equation:

$$\varphi(t) = C_1 e^{S_1 t} + C_2 e^{S_2 t} \quad (9)$$

$$\dot{\varphi}(t) = S_1 C_1 e^{S_1 t} + S_2 C_2 e^{S_2 t} \quad (10)$$

where,

$$S_{1,2} = \pm\sqrt{\frac{mgL}{J}} \quad (11)$$

Under the initial conditions $t_0 = 0, \varphi_0 = C_1 + C_2$

$$C_1 = \varphi_0 - C_2 \quad (12)$$

$$C_2 = \frac{\dot{\varphi}_0 - \acute{S}_2\varphi_0}{S_1 - S_2} \quad (13)$$

By specifying $\varphi_0$, $\dot{\varphi}_0$ we calculate the behavior of the function.
We calculate the necessary time for the pendulum to reach the upper point($t_{max}$), at zero initial velocity
($\dot{\varphi}_0 = 0$, that$\varphi_{max} = 0$. (Figure 3)

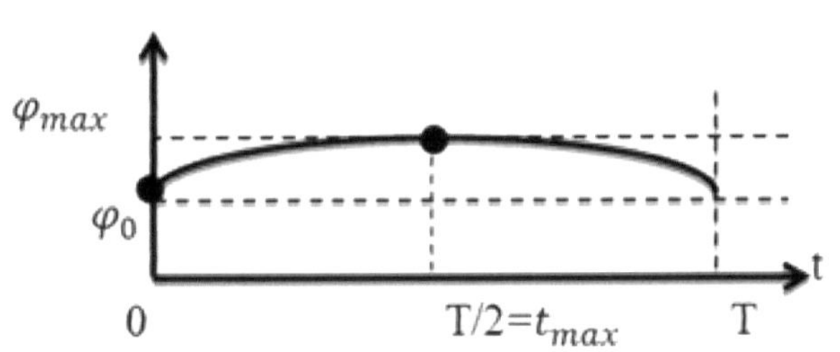


Fig. 3 Time to reach "zone" 0"

$\varphi_{max}$ – angle of reaching the top position of the pendulum, at. $t_{max}$.
We substitute in (10) the initial condition ($\dot{\varphi}_0 = 0$:

$$\dot{\varphi}_0(t_{max}) = S_1 C_1 e^{S_1 t} + S_2 C_2 e^{S_2 t} = 0$$

$$S_1 C_1 e^{S_1 t} = -S_2 C_2 e^{S_2 t}$$

Proceeding from (11) $S_1 = -S_2 = S$:

$$e^{St} = \sqrt{\frac{C_1}{C_2}}$$

Substituting formulas (12) and (13)

$$e^{St} = \sqrt{\frac{\varphi_0 S - \dot{\varphi}_0}{\varphi_0 S + \dot{\varphi}_0}} \tag{14}$$

$t = t_{max}$

$$t_{max} = \frac{1}{2S}\left(\ln(\varphi_0 S - \dot{\varphi}_0) - \ln(\varphi_0 S + \dot{\varphi}_0)\right) \tag{15}$$

The time step T = $2t_{max}$, substituting (14) in (9)

$$\varphi_{max} = \sqrt{\left(\varphi_0^2 - \frac{\dot{\varphi}_0^2}{S^2}\right)} \tag{16}$$

Calculating the time and angle at which the physical pendulum and the center of mass of the robot are in the "zone "0" of the vertical plane, using equations (15) and (16), we find the maximum permissible initial velocity $\dot{\varphi}_{max}$so that the robot does not fall and is in the "zone "0".

"Zone" 0" is a short-term equilibrium position on the supporting leg, when the leg is lifted from the floor. In this case, the robot is in the vertical position when the angle of rotation of the foot of the support leg is in the zero position, or in a position close to 0.

Set $\varphi_0$, it is necessary to find$\dot{\varphi}_0$, that $\varphi_{max} = 0$

It follows from (17):

$$\dot{\varphi}_{max} = -S\varphi_0 \tag{17}$$

Thus, to consider the behavior of the robot on one leg, we regard it as a physical pendulum rotated 180 degrees, where the initial angle of deviation from the vertical $\varphi_0$is known, h, J, m, is calculated L. We set this initial velocity $\dot{\varphi}_0$so that it is not higher then $\dot{\varphi}_{max}$ .

Consider the pattern of repelling the leg from the support surface. The range of permissible speed is known $0 \le \varphi_0 \le \dot{\varphi}_{max}, \forall \varphi_0$. It is necessary to bend the knee of the jogging leg by a certain angle β to ensure $\dot{\varphi}_0$that it is at the given angle $\varphi_0$.

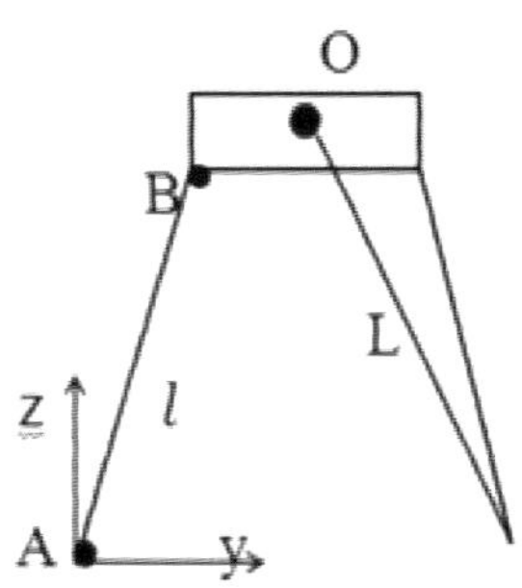


Fig. 4.1 the position of the legs after the first step.

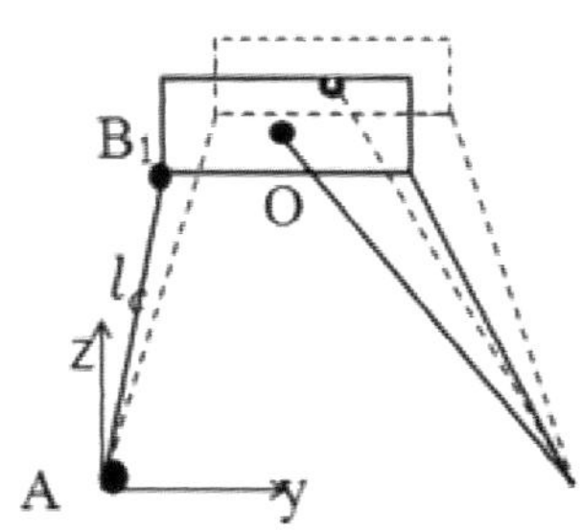


Fig. 4.2 squatting on the jogging leg.

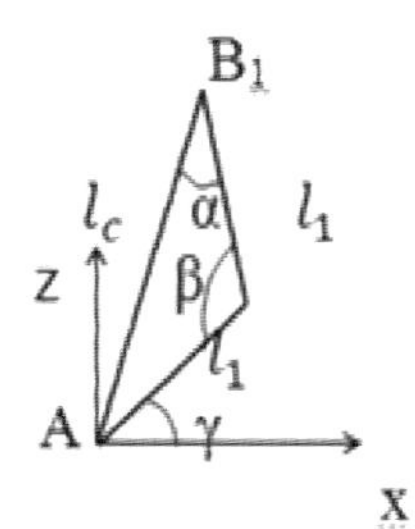


Fig. 4.3 squat the knee. Side view

$l_c$ – Length of leg with squat knee,

$l_1$ – Length of the leg and thigh of the robot's leg,

$\alpha$ – Angle of bend in the plane Oxy,
$\beta$ – Squatting angle of the knee in the plane Oxy,
$\gamma$ - Angle of bend in the plane Oxy.

At the moment of repulsion, when straightening there will be a sharp slowing of the speed, because the weight of the carried body will be equal to the total weight of the robot. At the moment of repulsion from the reference surface the initial velocity should be equal $\varphi_0$, it is necessary to sit down at angle $\beta_0$.

The algorithm was tested on a robot model in the Gazebo simulator, also tested on an anthropomorphic robot AR-601, whose control by ROS. [4]

**Conclusion:** When starting the pattern of active stability, we can calculate such a $\beta$ (knee squatting angle in the Oxy plane) and the knee straightening speed so that, given the initial conditions and the robot's position, it was possible to bring the robot to "zone 0", for further controlled "fall" of the robot and exercising a walking robot.

**References**

[1] Devaev V M, Nikitina D V, Fadeev A Y 2016 Balancing of the anthropomorphous robot walking *IOP Conference Series: Materials Science and Engineering* Past 2 p 22-26
[2] Sirazetdinov R, Kamalov A, Nikitina D, Katsevman E 2016 AR-601 anthropomorphic robot modeling and virtualization toolset for research and education purposes *IOP Conference Series: Materials Science and Engineering* 134(1) **012048**
[3] NPO Androidnaya technika: http://npo-at.com/products/ar-600e/
[4] ROS documentation: http://wiki.ros.org
[5] Ponarin Ya P 2004 *Theorema cosinusov: Elementary geometry* In 2 volumes Moscow: MCNMO pp 84-85
[6] Bogdanov A, Kutlubaev I, Permyakov A and Sichkov V 2015 Building an anthropomorphic robot including interaction control. Irreversible processes in nature and technology *Digest of Eighth-Russian Conference* pp 228-29
*[7]* Kashapov N F, Saifutdinov A I, Fadeev S A 2014 The model of the positive column of a glow discharge with the influence of the acoustic oscillations *Journal of Physics: Conference Series* 567(1) **012004**
[8] Saifutdinov A I, Saifutdinova A A, Kashapov N F 2016 Modeling the dielectric barrier micro-discharge in argon at atmospheric pressure *Journal of Physics: Conference Series* 669 (1) **012044**
[9] Sirazetdinov R, Devaev V, Kamalov A, Katsevman E 2015 Program complex of modeling and virtualization anthropomorphic robot AR-601 bases on ROS and GAZEBO systems *IMMOD-2015* pp 328-31

IOP Conf. Series: Materials Science and Engineering **240** (2017) 012066 doi:10.1088/1757-899X/240/1/012066

# Noise mapping inside a car

**V A Strekalov, R R Shaimuhametov**

Kazan Federal University, 18 Kremlyovskaya street, Kazan, 420008, Russian Federation

ramil.shai@mail.ru

**Abstract.** We present new wireless sensor telemetry system designed to be mounted on different nodes of the car and record acoustic noises. The proposed system is mapping noise inside a car. The noise field is recorded at the ten control regions simultaneously.

The research of the noise inside a car has several aspects:
1. The noise as an environmental factor for the environment;
2. The noise affecting the performance of the staff affected by them.
The amplitude and frequency characteristics of the noise are considered international and Russian standards, technical specifications and other regulations (for example, "Rules №51» UNECE GOST R 41.51 - 2004 "UNIFORM PROVISIONS CONCERNING THE CERTIFICATION OF VEHICLES HAVING AT LEAST FOUR WHEELS WITH REGARD TO NOISE ", etc.).
Automotive companies conduct numerous tests to study the causes of noise and methods to combat them.

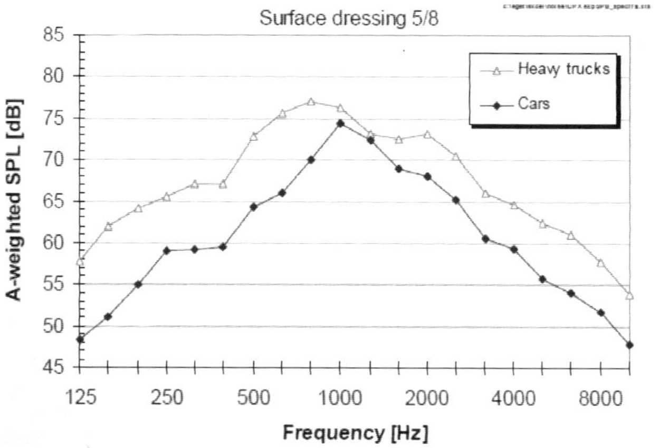


Fig. 1. The noise of the vehicle.

Fig. 1 shows the spectra of the movement of noise in dB for the frequency band from 125 to 8000 Hz to 80 km / h speed (cars), 70 km / h (trucks) [1].

IOP Conf. Series: Materials Science and Engineering **240** (2017) 012066 doi:10.1088/1757-899X/240/1/012066

The equipment used to measure noise in tasks, can be divided into two groups:
1. Measuring noise sound volume level (in dB);
2. The noise spectrum measurement systems (frequency response).

The second group of equipment allows, inter alia, create a "map" to locate the source of the noise, the noise detection reasons drafting a "model" of the spectrum individual node noise.
In the Engineering Institute of the Kazan Federal University to an end the work on the "map" of the vehicle noise. In the original equipment developed with 10 microphones installed in different nodes of the machine removed noise spectra [2].

Fig. 2 shows the arrangement of microphones and noise of individual nodes.

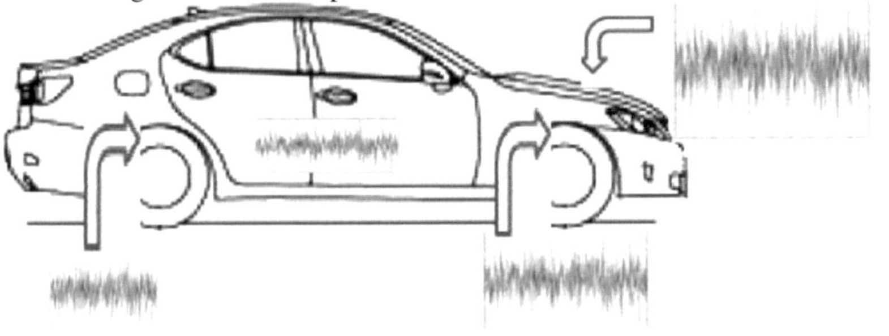

Fig. 2. The spectra of noise on the individual nodes of the vehicle (speed of 70 km / h, asphalt).

The developed equipment allows:

- Carry out spectral and correlation analysis of audio signals with one hundred sensors in the range of 50 Hz - 15 kHz. (ADC sensor to the microcontroller 30 kHz).
- Transfer the data from the sensors to the computer for the unlicensed 433 MHz. transmission channel provides reliable transmission of all the sensors from the vehicle compartment.
- The system-on-chip CC1111 [3] allows to perform advanced signal processing: filtering, spectral analysis.
- To conduct a full analysis on the computer, which may be located far away (100 m) from the object with sensors. The software allows to perform spectral and correlation analysis of acoustic signals.

A low-power system-on-chip (C1111, Texas Instruments, USA) based wireless body area network was developed to achieve parallel recording of acoustic signals. The sensors [4] units are incorporated with 433-MHz RF transceiver, programmable gain amplifier, and microphone to collect the acoustic data and send it to base station through wireless transmission.

## References

[1] Steven H, Küppers D, Blokland G J, Houten, M H, Loon R 2000 International Validation Test for the 'Close Proximity Method' (CPX). *Raadgevende ingenieurs bv, Bruistensingel* 232, NL-5232 AD 's-Hertogenbosch, the Netherlands
[2] Knudsen K, Sjøj S M N, Jacobsen F, Haddad K, Hald J 2012 Noise mapping inside a car cabin *Paper presented at Joint Baltic-Nordic Acoustics Meeting*
[3] C1111Sub 1GHz System On Chip with MCU, Texas Instruments, User Guide
[4] Cheer J and Elliott S 2015 Multichannel control systems for the attenuation of interior road noise in vehicles *Mechanical Systems and Signal Processing* vol 60 pp 753–69

# Designing physically realizable state observer for estimating the kinematic parameters of the road train

**V G Volkov[1,2] and D N Demyanov[1]**
[1]Naberezhnye Chelny Institute, Kazan Federal University, Syuyumbike 10A, Naberezhnye Chelny, Russian Federation
[2]Scientific and Technical Center of PJSC "KAMAZ", Transportnyi proezd 70, Naberezhnye Chelny, Russian Federation

demyanovdn@mail.ru

**Abstract**. In this paper the observability of state variables of the linear dynamic system, describing truck and semitrailer lateral motion on a high constant longitudinal velocity and minor rotation angles of steered wheels, is analyzed. The synthesis of a physically implementable reduced state observer for estimation of immeasurable state variables on the lateral displacement magnitude of the vehicle's frontal part is realized.

## 1. Introduction

Currently, one of the most effective approaches to transportation of bulky and heavy loads is the use of road trains of various types, for example, truck tractors with semi-trailers and trucks with trailers on a rigid coupling. Due to the large kinematic turning radius, considerably big weight and dimensions of the above-described vehicles particularly the development of effective active driver assistance systems that enhance driving stability and controllability is getting an acute problem [1].

As a rule, for its operation, such systems require the fullest possible information about the current parameters of the kinematic motion (state variables), such as the lateral component of the velocity vector, the angular velocity and the folding train angle. Obtaining this information through direct measurements is significantly difficult due to the complexity and high cost of measurement equipment suitable for the use under the road speed conditions of high vibration load. The use of monitoring devices to retrieve information about the unobservable (invisible) part of the state vector is more appropriate.

Thus, the aim of this study is to assess the observability of the linearized model of road trains lateral movement, to develop physically realizable reduced observing device for it and to study its efficiency.

## 2. A description of the model used

We consider the linear model of road train lateral movement on absolutely flat surface, resulting in the study [2]. This model written in a matrix form is a system of equations:

$$\begin{bmatrix}\dot{v}_y\\ \dot{\omega}_{z1}\\ \dot{\omega}_{z2}\\ \dot{Y}\\ \dot{\varphi}_1\\ \dot{\varphi}_2\end{bmatrix} = M^{-1}\left[A_1 \quad A_2 \quad A_3 \quad A_4 \quad A_5 \quad A_6\right]\begin{bmatrix}v_y\\ \omega_{z1}\\ \omega_{z2}\\ Y\\ \varphi_1\\ \varphi_2\end{bmatrix} + M^{-1}\begin{bmatrix}k_1-\delta_1 N_1\\ \left(k_1-\delta_1 N_1\right)l_1\\ 0\\ 0\\ 0\\ 0\end{bmatrix}\theta, \qquad (1)$$

$$M = \begin{bmatrix} m_1+m_2 & m_1 d_1 & -m_2 d_2 & 0 & 0 & 0\\ m_1 d_1 & J_1+m_1 d_1^2 & 0 & 0 & 0 & 0\\ -m_2 d_2 & 0 & J_2+m_2 d_2^2 & 0 & 0 & 0\\ 0 & 0 & 0 & 1 & 0 & 0\\ 0 & 0 & 0 & 0 & 1 & 0\\ 0 & 0 & 0 & 0 & 0 & 1\end{bmatrix}, A_1 = \begin{bmatrix} -\dfrac{k_1+k_2+k_3+k_4+k_5}{v_x}\\ (\sigma-1)(m_1+m_2)v_x-\dfrac{k_1 l_1-k_2 l_2}{v_x}\\ \dfrac{k_3 l_3+k_4 l_4+k_5 l_5}{v_x}\\ 1\\ 0\\ 0\end{bmatrix},$$

$$A_2 = \begin{bmatrix} -\sigma(m_1+m_2)v_x-\dfrac{k_1 l_1-k_2 l_2}{v_x}\\ -\dfrac{k_1 l_1^2+k_2 l_2^2}{v_x}-m_1 d_1 v_x\\ m_2 d_2 v_x\\ 0\\ 1\\ 0\end{bmatrix}, A_3 = \begin{bmatrix} \dfrac{k_3 l_3+k_4 l_4+k_5 l_5}{v_x}\\ 0\\ -\dfrac{k_3 l_3^2+k_4 l_4^2+k_5 l_5^2}{v_x}\\ 0\\ 0\\ 1\end{bmatrix}, A_4 = \begin{bmatrix}0\\0\\0\\0\\0\\0\end{bmatrix},$$

$$A_5 = \begin{bmatrix} \delta_3 N_3+\delta_4 N_4+\delta_5 N_5-\left(k_3+k_4+k_5\right)\\ 0\\ \dfrac{k_3 l_3+k_4 l_4+k_5 l_5}{v_x}\\ 0\\ 0\end{bmatrix}, \ A_6 = \begin{bmatrix} k_3+k_4+k_5-\left(\delta_3 N_3+\delta_4 N_4+\delta_5 N_5\right)\\ 0\\ -\left(k_3 l_3+k_4 l_4+k_5 l_5\right)\\ 0\\ 0\\ 0\end{bmatrix}.$$

Here $v_y$ – the lateral speed of the tractor truck in the moving coordinate system associated with the body of the tractor; $\omega_{z1}$, $\omega_{z2}$ – angular speed of the tractor truck and semi-trailer; $\varphi_1$, $\varphi_2$ – rotation angles of tractor truck and semi-trailer bodies respectively; $Y$ – the value of lateral displacement of the fifth wheel (saddle-coupling unit) in the absolute coordinate system associated with the road, $\theta$ – the turning angle of the steering wheel.

Mass coefficients and moments of inertia – $m_1$, $m_2$, $J_1$, $J_2$, $d_1$, $d_2$ , as well as of the distance from fifth wheel to the centers of mass of the tractor truck and semi-trailer; $l_i$, $k_i$, $\delta_i$, $N_i$ – the distance from the rotation axis to the center of mass of the corresponding level, the automobile drag coefficient, the rolling resistance coefficient, and the normal reaction of the road for the $i$-th train truck axle; $v_x$ – longitudinal component of velocity in the moving coordinate system; $\sigma$ – factor taking into account the rotating masses.

## 3. Investigation of the system observability

Since the system under the study consists of six state variables, each of which can be measured or observed, totally there are 64 variants of output vector, two of which, corresponding to the cases when the system is fully measurable and not measurable, may be excluded from the consideration.

To reduce the order of the observer and, as a consequence, reduce the requirements for computing resources of on-board information and control system, we estimate the fundamental possibility of constructing a reduced observer, restoring not the entire state vector, but only its unmeasured part [3]. For this purpose we consider the sequence of each of the possible combinations of state variables that determine the measurable output signal. After that, by a non-degenerate transformation of the state vector we present the system (1) in block form:

$$\begin{bmatrix} \dot{y} \\ \dot{z} \end{bmatrix} = \begin{bmatrix} A_{11} & A_{12} \\ A_{21} & A_{22} \end{bmatrix} \begin{bmatrix} y \\ z \end{bmatrix} + \begin{bmatrix} B_1 \\ B_2 \end{bmatrix} \theta, \qquad y = \begin{bmatrix} I & 0 \end{bmatrix} \begin{bmatrix} y \\ z \end{bmatrix}. \tag{2}$$

Here $y$ and $z$ is measurable and immeasurable part of the state vector, respectively, in new coordinates.

Observability of the system (2) for each case we define analyzing the pair of matrices ($A_{22}$, $A_{12}$) on the observability per Kalman [3]. The analysis results show that the reduced observer can be constructed for any set of measurable state variables, provided that it contains the value of lateral displacement of the fifth wheel coupling.

## 4. Observer Synthesis

Considering the previous results, it can be assumed that the most simple and least expensive way to determine the state vector of the model (1) is a measurement of $Y$ and all other variables using a reduced observer of the 5 order. However, the direct measurement of $Y$ is associated with considerable difficulties in the physical realization, so it seems appropriate to consider alternatives.

The analysis of existing and future hardware configurations of automotive control systems [4] allowed proposing a sufficiently effective and simple solution to the problem. We will use the front part of the lateral displacement of the tractor as measured variable value, measurement of which can be carried out, for example, with the help of the subsystem used to track road lanes.

The reduced observer resulting equation looks like:

$$\dot{\hat{z}} = (A_{22} - LA_{12})\hat{z} + (A_{21} - LA_{11} + (A_{22} - LA_{12})L)y + (B_2 - LB_1)u, \tag{3}$$

where $L$ is the matrix of observation.

The value of $L$ can be determined using known methods [3], by specifying requirements for the quality indicators of evaluation process:

$$L = \begin{bmatrix} -20.3967 & 39.4035 & -144.4212 & 12.8957 & -8.3916 \end{bmatrix}^T.$$

Figure 1 presents the results of numerical modeling in the environment of MATLAB / Simulink process of the observer work (3) during the typical road train truck maneuver.

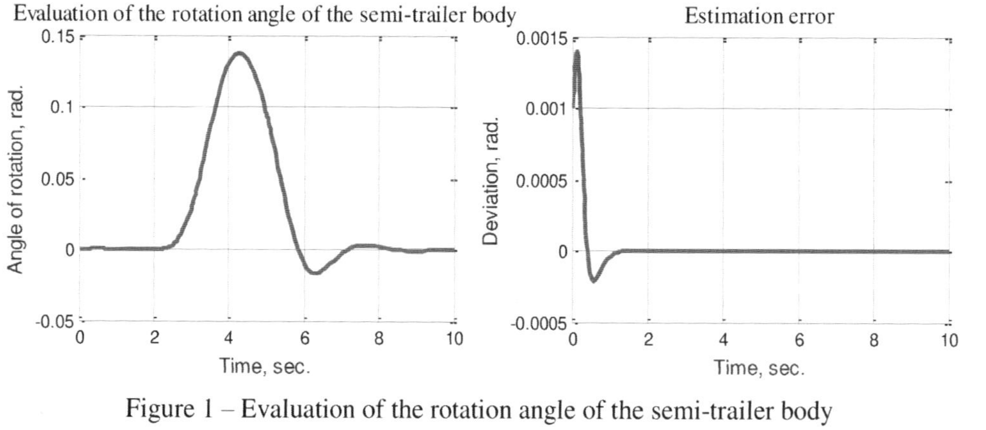


Figure 1 – Evaluation of the rotation angle of the semi-trailer body

It can be seen that even with non-zero initial conditions the estimation error over time asymptotically approaches zero.

## 5. Conclusions

The research of observability of a dynamic system, describing the lateral movement of the train truck, has been carried out. The magnitude of lateral displacement of the tractor truck front part, determined by using the image processing algorithms, has been used as a measurable variable for constructing the reduced observer. Results of the study can be used in practice in the development of active driver assistance systems.

## 6. Acknowledgment

This work was supported by RFBR (grant № 16-38-00042).

## References

[1] Smirnov G A 1990 *Wheeled Vehicles Motion Theory* (Moscow: Mechanical Engineering) 352 p
[2] Dem'yanov D N 2014 Development of the linearized model of the motion train truck *Proceedings of the final scientific conference* pp 375–80
[3] Kuzovkov N T 1976 *Modal control and watching devices* (Moscow: Mechanical Engineering) 184 p
[4] Belousov B N, Naumov S V and Ksenevich T I 2015. Trends in the development of automotive control systems *Automotive Industry* No 1 pp 33–36

IOP Conf. Series: Materials Science and Engineering **240** (2017) 012068 doi:10.1088/1757-899X/240/1/012068

# Application of reverse engineering in the production of individual dental abutments.

**A V Yunusov[1], R N Kashapov[1,2], L N Kashapov[1], E O Statsenko**
[1] Kazan Federal University, 18 Kremlyovskaya street, Kazan, 420008, Russian Federation
[2] Laboratory of Radiation Physics, Kazan Physical-Technical Institute of the Kazan Scientific Center of the Russian Academy of Sciences, Kazan, 420029, Russia

adely911@gmail.com

**Abstract**. The purpose of the research is to develop a method of manufacturing individual dental abutments for a variety of dental implants. System of industrial X-ray microtomography Phoenix V|tome|X S 240 has been applied for creation of highly accurate model of the dental abutment. Scanning of dental abutment and the optimization of model was produced. The program of milling the individual abutment with a standard conical neck of hexagon was produced for the five-axis milling machine imes - icore 450i from the materials titanium and zirconium oxide.

## Introduction

A beautiful smile is today an integral part of our lives, to realize it is often necessary to resort to the help of dental implants. Implant placement is almost impossible without the abutment.

The abutment is a connecting link between the dental implant and the crown of the tooth. The most convenient and efficient are the individually fabricated abutments, which providing more comfort in their use and with such abutments unable to repeat the tooth shape and to achieve the highest esthetic similarity with the planting of the native teeth.

Functionally, the abutment can be divided into two parts: a conical neck hexagon, which response for fixing the abutment to the implant, and the segment for the individual superstructures, which is fixed on the collet. For install the individual abutment on the standard implant, it is necessary to save a standard conical neck of hexagon, and produce segment for individual superstructures with regard to the shape of the dental crown.

Thus, an individual abutment must have a standard conical neck of hexagon and segment for individual superstructures, which is modeled for each case. Therefore, the aim of this research was to develop a methodology of manufacturing individual dental abutments for various standard implants. To achieve this goal it is necessary to solve the following tasks:

1. Perform an X-ray computed tomography of the abutment and estimate the accuracy of the scanned model.
2. Optimize the model and reduce the file size to simplify the work with the model.
3. Creating a strategy for milling on five axis milling machine.

## Main part

IOP Conf. Series: Materials Science and Engineering **240** (2017) 012068 doi:10.1088/1757-899X/240/1/012068

To create a fixing of individual abutment to standard implants required to save attaching point of the abutment to the implant.

To achieve the goal, we use a standard metal abutment in two perspectives (Fig. 1). In this abutment we are interested in the upper part, which have the shape of a hexagon because it is the standard and is necessary to preserve its size and shape.

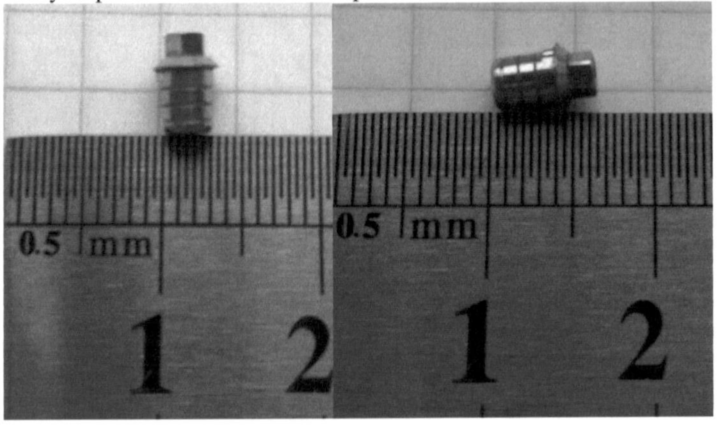


Fig.1. Abutment.

We refused from manual measurement and design 3d models in software such as KOMPAS-3D and SolidWorks due to the small size of the abutment and the presence on the surface of the channels grooves of small quantities which can't be measured using calipers. The method of using an optical scanner such Rexcan 450 also had to be abandoned due to the fact that the abutment is made from metal and the surface is glossy; it becomes necessary to apply a special spray that gives a matte surface. Channels with shallow depth on the surface can become clogged by particles of spray, as a result we get the model does not satisfy the accuracy.

Computer tomography method is most suitable for our purposes, it allows getting the most accurate 3D model of the outer surface, and most importantly providing the exact size and shape of the inside of the through hole. A large load on the locking screw arises in case of violation or changing form of inner hole for screw connection, which can lead to premature destruction.

X-ray computed tomography (CT) - the method of non-destructive research of internal structure of the object is based on measuring the attenuation of X-rays by various parts of the object, differing in density, composition, and thickness. A complex computer processing of data for a variety of its two-dimensional projection is used to generate three-dimensional images of the internal structure of the object.

To create a 3D computer model of the abutment used industrial X-ray microtomography system Phoenix V|tome| X S 240 is equipped with two X-ray tubes: microfocus with a maximum accelerating voltage of 240kV / 320W power and nanofocus maximum accelerating voltage of 180kV / 15W power.

Creating 3D models takes place on the basis of a large number of radiographs, more than a thousand, taken from different angles due the rotating manipulator. Projections set converted into the 3D model by software datos|x reconstruction. Software VG Studio MAX 2.1 and Avizo Fire 7.1 are used for the data visualization and analysis on the elements of three-dimensional image.

Nanofocus tube was used for obtained samples. Scanning was performed at an accelerating voltage of 110kV and current of 100mA. Picture resolution is 6.5 mkm (volume of one voxel). Voxel is the volumetric pixel, in this case, the model consisted of cubic elements of the size of 6.5 * 6.5 * 6,5 mkm. CT method imprecision is 0.1 microns, which is acceptable for parts of our size. The main advantage of the method of X-ray CT, consists in that it was possible to obtain not only the outer surface of the model, but the exact shape and dimensions of the inside (Fig. 2) without cutting model

IOP Conf. Series: Materials Science and Engineering **240** (2017) 012068 doi:10.1088/1757-899X/240/1/012068

and any other destructive effects. High accuracy of scanning allows creating snug fit the abutment to the implant, thereby to protect from infection in cracks and in the junction. Also to protect from infection penetration which can lead to rejection of the implant.

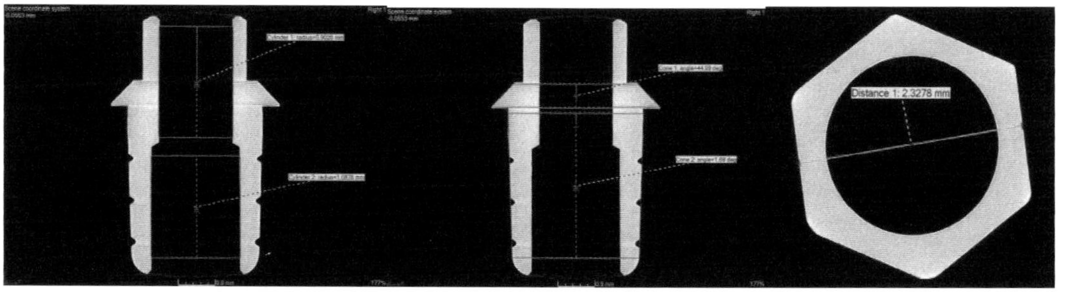


Fig.2 Form and scales of 3D model of abutment in a cut after CT.

The resulting model had STL file format and consisted of 472,275 polygons. Model must be optimized for comfortable work in CAD system by reducing the number of polygons, while maintaining the size, without loss of accuracy and functionality of the model. This operation was carried out in the program 3ds Max. Was used the operation ProOptimizer which reduce the number of polygons by replacement of group polygons to one that is closest to the original group by location peaks. Because of this was able to reduce the number of polygons by 50 times, from 472275 to 9444. The model is angular (Fig. 3), it is possible to fix by smoothing model.

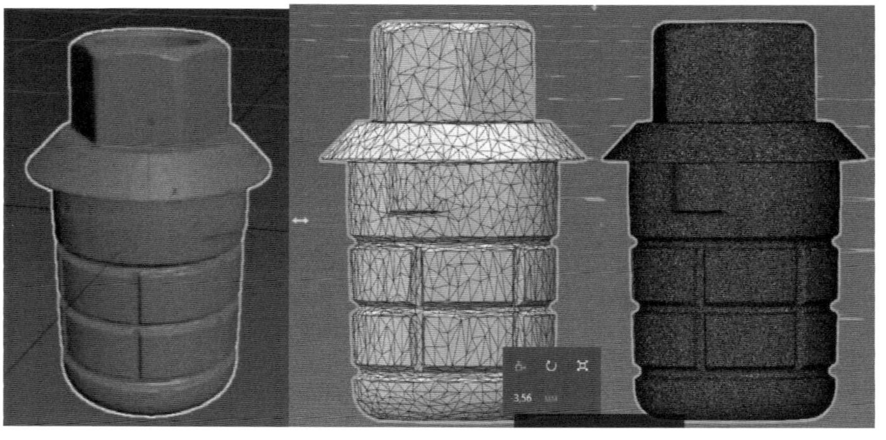

a b c

Fig.3. a) The result of reducing the number of polygons by 50 times, angular (rough) surface. b) The final result of the optimization model. c) Exist model.

Smoothing effect works by combining the process of changing the angle between the polygons on a more obtuse angles and increasing the number of polygons, using for example Bezier curves. Was used the operation Smooth. Model becomes smoother with increasing smoothing coefficient, optimal value 13. Model deformation observed with increasing values of smoothing coefficient.

The file size is reduced from 86 MB to 461 KB (Fig. 2 b, c). This will speed up the subsequent process of the work, because most of the CAD softwares, while working with files of the large volume required spending more time to process the model.

All operations to reduce polygons and volume of file, have not affected on the functionality of the abutment model. Comparing of the distance between the control points have been conducted to verify the accuracy of the size of the abutment after the stage of optimization models. We compared the distance between the two control points on the model before optimization and after optimization. Positioning the control points are indicated in Figure 4 a, b, c, the maximum imprecision was 3 microns.

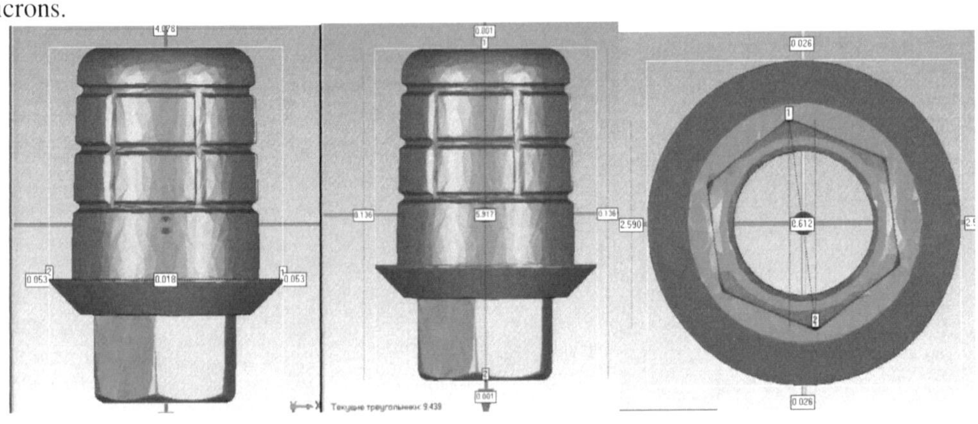


a b c

Fig. 4 a) The distance between the points horizontally. b) The distance between the points along the vertical. c) The dimension of the hexagon.

Fabrication of the abutment from titanium and zirconium oxide occurs by milling on a 5-axis milling machine imes - icore 450i. Preparation milling strategy occurs in the program SUM 3D.

While writing strategies taken into account the features of processing of materials. For titanium necessary to maintain a small contact area by reducing the milling step on the vertical axis because titanium is the low thermal conductivity material. In this metal is only a small part of the heat away with shavings, most of the heat is transmitted to the milling cutter. The important point is cutting an arc, which allows saving the resource of milling cutter. The effect is comparable to a hammer blow to the cutting edge in the direct cutting of milling cutter. For direct plunging of milling cutter was used circular walkways. Due to the small size of the abutment, milling cutter with diameter 0.6 mm was used. Therefore, heat transfer from the workpiece increased by the milling with the milling cutter which does not exceed 70 percent of the groove diameter. With a smaller gap significantly increases the risk of restricting access of coolant to the instrument and retention of the shavings.

When working with zirconium dioxide should take into account the fragility of the material and reduce the milling step in all directions and cutter speed. Of zirconium dioxide feature is to have the effect of shrinkage during baking in the oven, after which attains a high strength. This feature should be considered while writing strategy. During registration of the disc is required to specify the shrinkage factor, which is written on the disc package and is usually from 1.15 to 1.25. Imprecision of five-axis milling machine imes - icore 450i equals 15 microns.

## Conclusion

With reverse-engineering was possible to obtain high precision abutment. It is possible to produce customized abutments to install on a standard implant by this developed technique. Abutment is manufactured by milling of titanium and zirconium oxide on the machine-tool imes - icore 450i. The maximum imprecision in the production of the abutment by reverse engineering was 18.1 microns.

## References

[1] Hamid R. Shafie 2014 Clinical and Laboratory Manual of Dental Implant Abutments Wiley-Blackwell
[2] Kashapov L, Rudyk A, Kashapov R 2014 IOP Conference Series: Materials Science and

Engineering Volume 69 Issue 1 012016
[3] Kashapov R, Korobkina A, Platonov E, Saleeva G 2014 IOP Conf. Series: Materials Science and Engineering 69 (2014) 012026

IOP Conf. Series: Materials Science and Engineering **240** (2017) 012069 doi:10.1088/1757-899X/240/1/012069

# The simulation method of chemical composition of vermicular graphite iron on the basis of genetic algorithm

**L R Yusupov[1], K V Klochkova[2], L A Simonova[2]**

[1] Engineering-Economic College, NChI K(P)FU, Naberezhnye Chelny, Russian Federation
[2] Automation and Control Department, NChI K(P)FU, Naberezhnye Chelny, Russian Federation

KVKlochkova@kpfu.ru

**Abstract**. The paper presents a methodology of modeling the chemical composition of the composite material via genetic algorithm for optimization of the manufacturing process of products. The paper presents algorithms of methods based on intelligent system of vermicular graphite iron design

Progress in industry is inextricably linked with the development and wide deployment of advanced materials: volumetricv nanomodified materials, composite materials, steel and cast iron for special purposes. Cast iron with vermicular graphite (CGI) has a number of specific properties that place it at the number of new promising structural materials for various purposes castings [1, p. 170].

Currently there is no complete theoretical description of the chemical mechanism of the process of graphite forms formation, as there is no satisfactory explanation of the influence of various spheroidizing and graphitizer modifiers on the final performance properties.

In this regard, modeling of optimal chemical composition on the basis of input information about the required physical and mechanical properties will reduce the number of experimental studies and will allow producing castings of the required quality CGI, without a large number of production operations, which will ultimately have a positive impact on the cost of manufactured parts.

Modeling of technological process based on the determination of optimal chemical composition in the final stage of detail processing and find the optimal process parameters to obtain a given chemical composition. Criteria of optimality in solving this problem are quality indicator (at the stage of finishing) and cost parameters (at all other stages).

The structural model of formation of technological process on the basis of the inverse problem [2, p. 72-78] (finding the optimal composition on physical and mechanical specified value) is presented in figure 1.

When modeling the following tools of knowledge-base intelligent system of CGI design are used [3, p. 2] (figure 2):

– rule base;
–precedents database;
– cascade neural network.

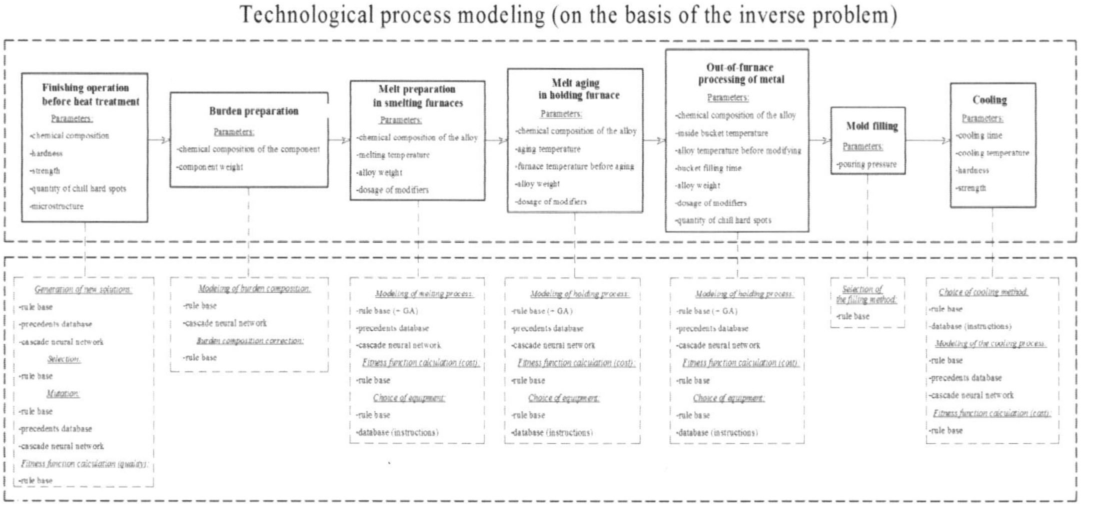


Figure 1 – Structural model of technological process formation on the basis of the inverse problem

*The rule base* serves as a tool to determine various parameters of chemical composition and provides control of the restrictions and requirements imposed on the CGI design process.

Using *precedents database* the response of the system (without model) on the output parameters of technological process is checked. On the basis of the obtained parameters the precedent is composed and the availability of similar precedents in the precedent database is being searched.

*Cascade neural network* is the main tool for predicting the physical-mechanical properties according to the parameters determined by the parameter of chemical composition, which are determined by the rule base.

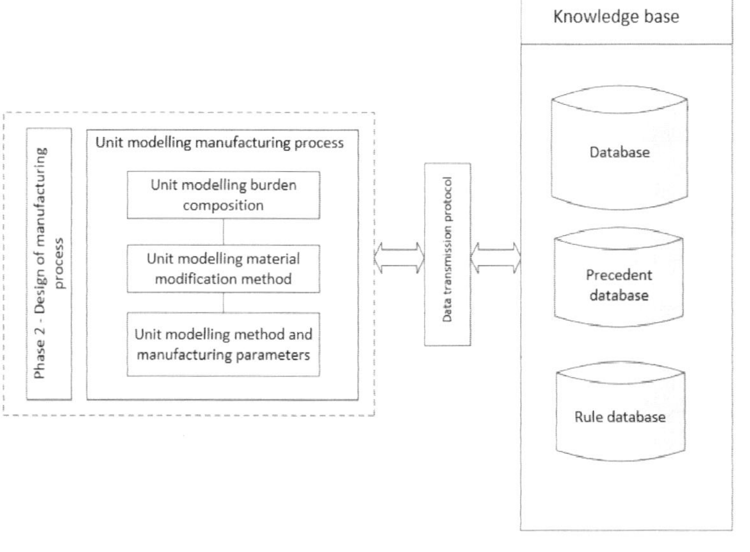


Figure 2 – Block diagram of the knowledge base and function block of technological process modeling

As mentioned above, when modeling chemical composition using genetic algorithm, all elements of the intelligence system are addressed: rule base, database, precedents database, cascade neural network (figure 3).

In the first stage of generating initial population, the system runs a search of precedents depending on the casting type for which you want to predict chemical composition. The found results are presented in the form of a set of chromosomes. In the case when the results of the experiments are not found, the generation of new chromosomes is implemented. Next, the function of chromosomes suitability is evaluated, and the elite selection is made. For initial population 4 chromosomes must be generated.

The chromosomes for the parent pairs are selected from a set of initial population by random selection.

With the selected pair of parents the operation of mutation is made. Mutation can occur both randomly and directionally. When directed mutations, the direction of mutation (change of process parameters) is derived using the rule base and cascading neural networks.

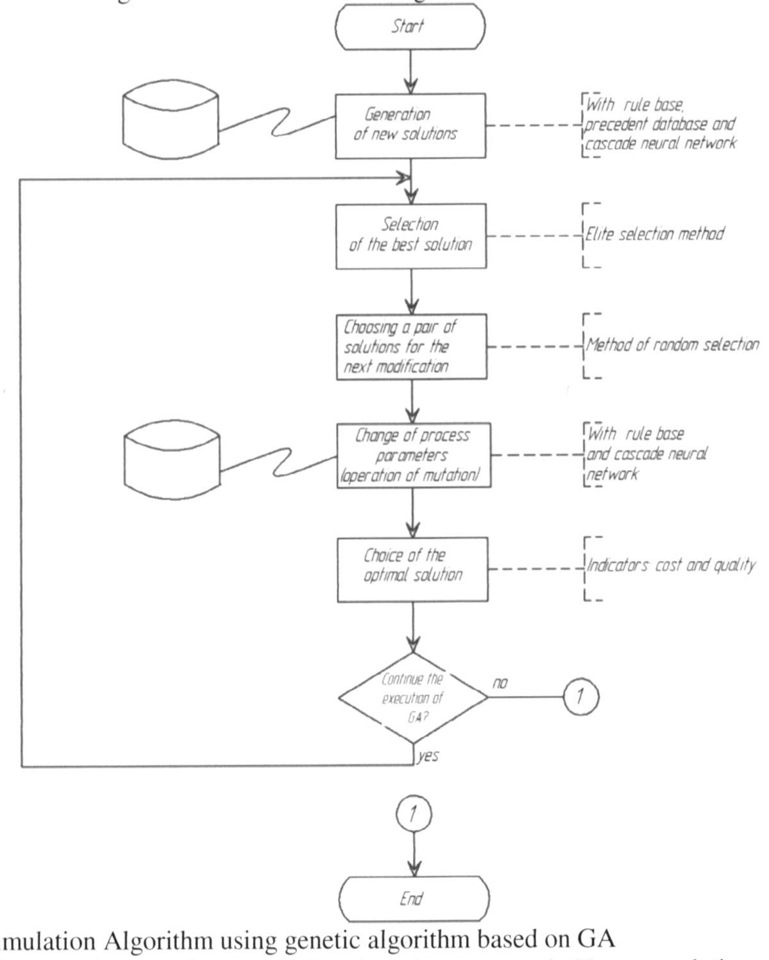


Figure 3 – simulation Algorithm using genetic algorithm based on GA

After mutation mechanism the population is being renewed. New population will contain 2 chromosomes of parents and 2 randomly selected chromosomes from the previous generation (the duplicate chromosomes are excluded). Increasing the number of iterations the chance of elite selection during the update of the population increases. The algorithm of the mutation mechanism is presented in figure 4.

IOP Conf. Series: Materials Science and Engineering **240** (2017) 012069 doi:10.1088/1757-899X/240/1/012069

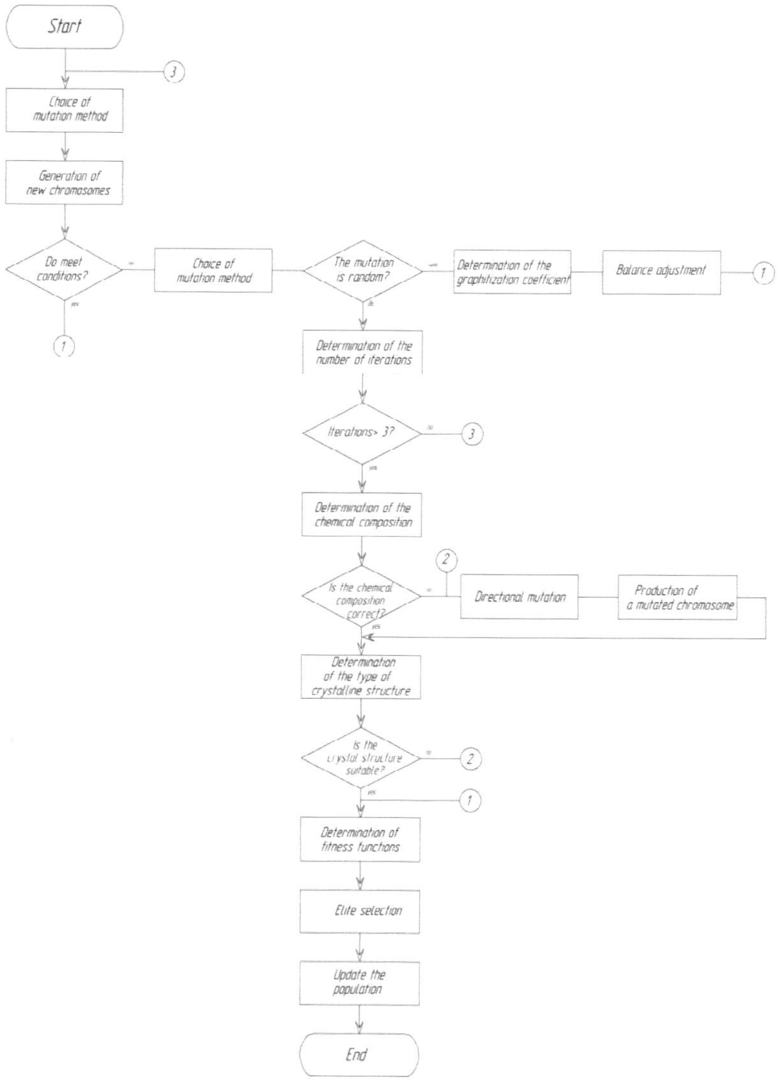


Figure 4 - Algorithm of the mutation mechanism

The proposed method can be used to solve optimization problems, problems of choosing the rational parameters of technological processes at different stages of processing, choosing the optimal solution in the precedents database and for finding the optimal chemical composition on the required physical and mechanical properties.

## References

[1] Girshovich N G 1978 *Handbook of cast iron casting* (L. mechanical engineering) 758p.

[2] Klochkova K V, Petrovich S V, Simonova L A and Yusupov L R 2015 Stages of vermicular cast iron properties modeling in the intelligent design system *IOP Conf. Series: Materials Science and Engineering* Vol 86 Issue 1 **012015**

[3] Klochkova K V, Petrovich S V, Simonova L A and Yusupov L R 2015 Stages of vermicular cast iron properties modeling in the intelligent design system *IOP Conf. Series: Materials Science and Engineering* Vol 86 Issue 1 012015

[4] Klochkova K V, Petrovich S V and Yusupov L R 2014 Model formation of knowledge base of intelligent system design of cast iron with vermicular graphite *Student scientific magazine "Faces of science"* Volume 2 No 2 pp 73-8
[5] Panov A G, Kornienko A E, Degtyareva N G and Zinoviev Yu A 2013 Homogenizing modification of foundry melts of graphitizer cast iron: a monograph *Nizhny Novgorod: Nizhegorodskiy state technical university named after Alekseeva R E* 191 p
[6] Bolshakov V I, Dubrov Yu I, Tkachenko A N 2010 *The Problem of identifying the quality characteristics of materials based on expert systems* D.: PDABA No 1 pp 46-9

IOP Conf. Series: Materials Science and Engineering **240** (2017) 012070 doi:10.1088/1757-899X/240/1/012070

# Automated road marking recognition system

**R R Ziyatdinov, R R Shigabiev and D N Talipov**

Naberezhnye Chelny Institute of Kazan Federal University, Naberezhnye Chelny, the Russian Federation
(The Russian Federation, 423810, Naberezhnye Chelny, Mira Avenue, 68/19)

rust.kfu@gmail.com

**Abstract**. Development of the automated road marking recognition systems in existing and future vehicles control systems is an urgent task. One way to implement such systems is the use of neural networks. To test the possibility of using neural network software has been developed with the use of a single-layer perceptron. The resulting system based on neural network has successfully coped with the task both when driving in the daytime and at night.

## 1. Introduction

Modern vehicles meet high requirements for driving control and safety comfort. For this purpose, vehicles are equipped with various systems that enhance driving safety on the road. One of such systems, which belongs to the make-up systems of active safety is the Lane Keeping Assist System. It is designed to help the driver to stick to the selected lane, preventing traffic accidents. The system shows high efficiency when driving on federal highways where drivers are characterized by inattention, craving to sleep because of the monotony of movement. One of the main functions of the Lane Keeping Assist System is the function of road marking recognition.

Today, the trend in the automotive industry has been the development of prototypes of self-driving vehicles, allowing to refuse from the driver in perspective. Such vehicles require recognition systems of ambient conditions including a road marking recognition systems.

## 2. Recognition of road markings

The task of road marking recognition in real time is reduced to forming a sequence of frames or video using camera and subsequent processing of the images. In images recognition neural networks are frequently used [1, 2]. A neural processing element can be considered as a simple device that performs a set of given simple functions. These devices combining neural processing elements by connections can be assembled into a single network. In this paper for finding markings we used a single-layer perceptron, which, despite its simple architecture, is able to solve quite complex problems [3]. Herewith the resulting model is quite simple and easy to implement using standard software tools [4-7].

For simple neural architectures, methods and objectives, you can use almost any programming language, but for complex projects most useful languages are object-oriented programming. When choosing a programming language such factors as the target platform, language flexibility, performance, prevalence and other factors were taken into consideration. As a result, the common programming language Python was selected, together with a library of algorithms of computer vision, image processing and numerical universal algorithms Open CV [8].

IOP Conf. Series: Materials Science and Engineering **240** (2017) 012070 doi:10.1088/1757-899X/240/1/012070

The algorithm of the road marking recognition system includes the following steps:

1. Pre-processing of images.
2. Detection of road marking lines.
3. Error filtering.
4. Approximation of the obtained values

Operations of images recognition in images of certain objects usually are preceded by image processing to create conditions that increase the efficiency and quality of the extraction and recognition of the unknown objects or the objects under study. Pre-treatment methods depend on the research objectives and are quite diverse. In this paper we used one of the most commonly used algorithms for image pre-processing - Canny algorithm. Canny algorithm is still one of the best detectors. In addition to special individual cases it is difficult to find an algorithm that would work much better than the Canny algorithm.

Before applying the algorithm the original image was converted to the grayscale. This can reduce computational costs. This action is typical for many image processing techniques. After a conversion, we obtain a binary image for further actions.

In the next step the neural network detects the road line. Each synapse in its domain is responsive to the road line.

After a neural network recognizes lines, false positives are possible or unnecessary lanes can be determined, for example, road markings of the neighboring lanes. Mistakenly found data will further affect the quality of the determined line, and therefore a control system in general. That is why it is necessary to remove them by filtration.

In the last step the data are approximated. In this paper, to simplify the task a linear approximation by the method of least squares was used. The results of road marking recognition on the captured video from the DVR in the daytime and night-time are shown in Figures 1 and 2 respectively.

Figure 1. Marking recognitions in the daytime

Figure 2. Marking recognitions at the night-time

## 3. Conclusion

Creating systems that enhance driving safety on the road, undoubtedly, is one of the important tasks. Currently, this topic is particularly relevant in connection with interest all over the world to create unmanned vehicles, where a high reliability and performance regardless of external conditions is required from these systems.

The use of neural networks for image recognition allows to define a road marking in real time and thus to apply it for automated Lane Keeping Assist System in modern and future unmanned vehicles.

## References

[1] Ziyatdinov R R, Mukhametzyanov V R and Nabiullina G I 2015 Application Of Neural Networks For The Diagnosis Of Depth Sucker Rod Pumps *International Journal of Applied Engineering Research* (IJAER) Volume 10 Number 24

[2] Zhuravlev Y I 1978 On the algebraic approach to solving the problems of recognition and classification *Problems of Cybernetics* V 33 pp 5-68

[3] Uosserman F 1992 *Neurocomputing equipment: Theory and Practice* Trans. from English M.: Mir

[4] Jones M T 2011 *Programming Artificial Intelligence in applications* Trans. from English. Osipov AI - M.: DMK Press 312 p

[5] Devyatkov V V 2001 *Artificial Intelligence Systems: Proc. manual for schools* M.: Publishing House of the MSTU Bauman 352 p

[6] Luger D F 2003 *Artificial Intelligence: Strategies and methods for solving complex problems,* 4th edition.Trans.from English. - M.: Williams" Publishing House

[7] Russell S, Norvig P 2006 *Artificial Intelligence: A Modern Approach*, 2nd ed ..: Trans. From English. "Williams" Publishing House 1408 p

[8] Bradski G, Kaehler A 2008 *Learning OpenCV. Computer Vision with the OpenCV Library* Publisher: O'Reilly Media

IOP Conf. Series: Materials Science and Engineering **240** (2017) 012071 doi:10.1088/1757-899X/240/1/012071

# Analysis and development of methods for obtaining metallic powders for selective laser melting

**R N Kashapov[1,2], L N Kashapov[1] and N F Kashapov[1]**
[1] Engineering Institute, Kazan Federal University, Kazan, 420008, Russia
[2] Laboratory of Radiation Physics, Kazan Physical-Technical Institute of the Kazan Scientific Center of the Russian Academy of Sciences, Kazan, 420029, Russia

E-mail: kashramiln@gmail.com

## Annotation

The paper compares the existing methods for obtaining metal powder compositions and reveals their shortcomings. A technique for evaluating the properties of powders for suitability for use in selective laser melting (SLM) processes is proposed. Analysis of methods for obtaining powders has shown the need to investigate the possibility of using a plasma-electrolyte process in the production of consumables for SLM technology.

## Introduction

Additive technologies, in particular the method of selective laser melting (SLM), are increasingly used in industry [1]. This is primarily due to the appearance of new materials for 3D printing of metal and ceramic products. [2]. In installations of selective laser melting, powder materials are used which must satisfy the following requirements: have a certain dispersion composition, the particles must be spherical. At the moment, several technologies are known for the production of metal powders for SLM: gas, water, plasma with a rotating electrode, plasma in crossed flows, and high-frequency discharge of atomization. In the first case, the molten metal is sprayed in a supersonic jet of an inert gas. In the second case, the spray is produced in a liquid stream. In the third one, an arc discharge is ignited between the electrodes, rotation of the melted electrode around its axis takes place, which leads to spattering of the metal. In the fourth case, the metal rod is sprayed in crossed plasma streams. In the fifth case, the pre-ground metal particles are passed through the installation with a high-frequency induction or capacitive discharge, and as a result, they take the form of a sphere. Disadvantages of these technological productions: a high dispersion of particle sizes from 1 μm to 200 μm, which requires additional cleaning and separation, the difficulty of transition of the plant to the production of another type of material, high energy consumption of the process, the difficulty of controlling and extracting particles of the nanometer range, the use of expensive equipment. The concept of additive manufacturing involves individual product geometry, as well as the ability to use a wide range of materials [3]. Now the main limitations are the shortcomings in the production of powders and the difficulty of a rapid transition from one type of metal to another. Thus, the next stage in the development of the SLM process is the creation of technologies for the production of micro and nanopowders, which are devoid of the above disadvantages. One solution is to use a gas discharge with liquid electrodes [4,5,6], this method is simple and does not require expensive equipment. Therefore, the aim of the work was to determine the possibility of its applicability by comparing the plasma-electrolyte method with the existing atomization technologies for powders.

## Main part

Due to the fact that powders used in additive technologies are used in conditions other than standard technologies (powder metallurgy, powder coating) it is necessary to determine which of their properties are the most important and to choose the methods for measuring these parameters. Methods for determining the properties of powders used in standard technologies are known: determination of bulk density, powder flowability according to ASTM B213, measurement of the natural bevel angle. Of interest is the paper [7], which describes the device FT4 Powder Rheometer® (Freeman Technology) allowing to determine a wider class of dynamic and volumetric characteristics of the powder. The authors of the paper showed the effectiveness of studying the powder properties by the "shear cell" methods, measuring the possibility of passing a powder rotating through the flow, measuring the fluidity of the powder through a screw feeder, aerating to create a uniform powder distribution, determining the cohesion of powder particles, measuring the effect of air flow velocity on powder motion, determination of density, compressibility, permeability and air release. Also, an important conclusion is made in the paper about the deterioration of the dynamic properties of the powder after use in additive production machines. Thus, monitoring the properties of the powder is a factor in obtaining the product without internal defects.

Attention should be drawn to the paper [8], in which a comparative analysis of powders 17-4PH obtained by gas and water atomization is carried out. The morphology of particles about 3 μm in size is similar and has a spherical shape. However, for the powder obtained by the water atomization method, with an increase in the particle size to about 20 μm, the particle sphericity is disturbed. This, in turn, does not allow using it in laser melting technologies. Therefore, the main methods of powder production for SLM and LMD are methods of gas atomization, plasma and electric arc with a rotating electrode. These methods make it possible to control the size of the powder obtained while maintaining its sphericity.

It is of interest to obtain a powder in a gas discharge with liquid electrodes. Gas discharges with liquid electrodes have been studied quite well since the middle of the 19th century, and have found wide application in practice: the formation of surface microrelief, heat treatment of products, the formation of functional coatings, the cleaning of metal surfaces, water sterilization, nanoparticle production etc. Burning discharge can occur both on the metal anode and cathode, while the electrode can be immersed in the liquid, and may be above it. The combination of all these conditions with the variation in the shape of the current and the magnitude of the applied voltage provides a variety of options for searching for micron-size powder production regimes. Our starting point for the studies was a survey paper [9], which describes the main methods for obtaining metal nanoparticles and their oxides. Proceeding from this, the set of possible variants of studies was greatly reduced, and the experimental setup was optimized.

Figure 1 shows the functional scheme of the experimental setup, which consists of an electric power supply system 1, an electrolytic bath 2, an electrode system 3, an oscilloscope 4, an additional resistance 5, a voltmeter 6, an ammeter 7 and a thermocouple 8. With the help of the electrode system, the distance between the anode and the electrolyte solution was monitored. With the help of an oscilloscope 4, the shape of the applied voltage and current was monitored, the voltmeter and ammeter were used to measure the voltage and discharge current. The electric power supply system represents a high-voltage DC power source for creating and maintaining electric discharge combustion with smooth regulation of the output voltage in the range from 0 - 3 kV and current 0-10A. Voltage and discharge current were measured by means of two digital universal measuring devices MMH-930 and APPA 109N, the relative error of measurement is 0.8%.

IOP Conf. Series: Materials Science and Engineering **240** (2017) 012071 doi:10.1088/1757-899X/240/1/012071

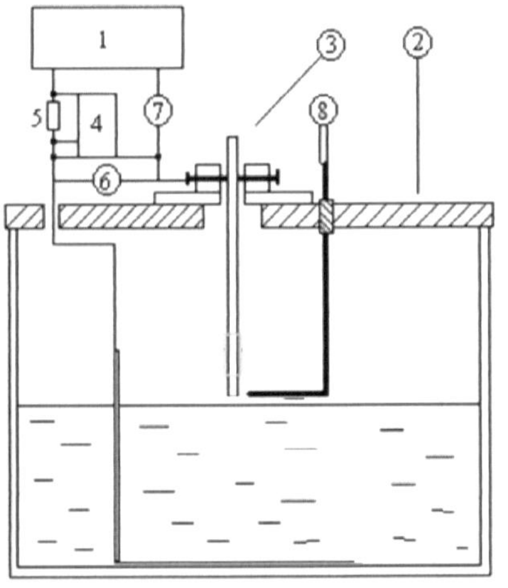

Fig. 1. The scheme of the experimental setup

Burning gas discharge occurs between the metal anode made of steel grade S17400 / 630 and an electrolytic cathode. The anode is a metal cylinder with a diameter of 5 mm, located above the surface of the electrolyte at a height of 1 to 5 mm. As the electrolytic cathode, aqueous solutions of NaCl and $Na_2CO_3$ with a concentration of 0.1-1% by weight were used.

When certain current and voltage values are reached, a process of spraying the metal anode is observed, most of the powder enters the electrolytic cathode and crystallizes. In parallel, there is a process of evaporation of the liquid electrode. The resulting powder was washed with deionized water and dried in a drying oven. The morphology of the powder was studied using a scanning electron microscope, Carl Zeiss EVO 50. The dispersion composition was determined by sieving with a sieve set of sieves of from 10 to 300 microns. The average sieving time was 30 minutes. The microhardness was measured with a PMT-3M device, using the Vickers method. The powder was mixed with cyanoacrylate and pressed into tablets, which were then ground and polished. The resulting microsections were examined at a load of 50 gauss, a holding time of 25 s and 5 prints. Analysis of the chemical composition was determined with the aid of a scanning electron microscope attachment of an energy-dispersive electron-probe spectrometer Oxford Instruments Inca X-act.

The main parameters of plasma-electrolyte production of steel powder are voltage, current, discharge power, used for heat generation, physical and chemical properties of the metal anode. The power put into the discharge is determined by the current-voltage characteristic, on the basis of which it is possible to estimate the energy contribution. Figure 2 shows the current-voltage characteristic of the plasma-electrolyte process for three interelectrode distances.

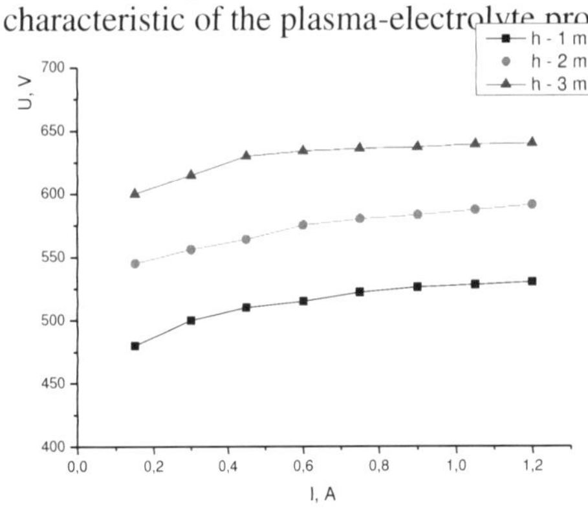

Fig. 2. Volt-ampere characteristics of the plasma-electrolyte process

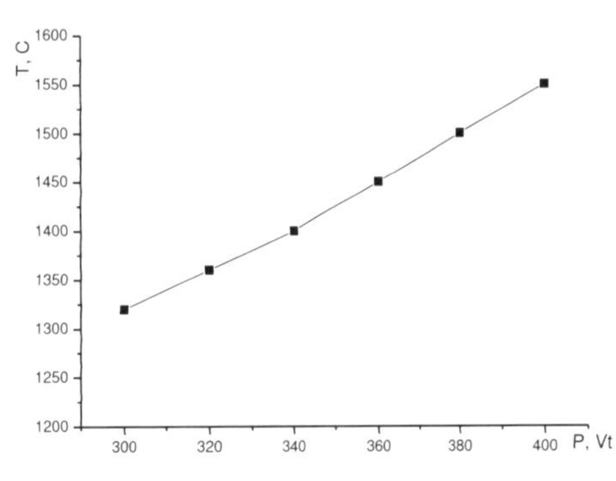

Fig. 3. Dependence of the anode temperature on the power

IOP Conf. Series: Materials Science and Engineering **240** (2017) 012071 doi:10.1088/1757-899X/240/1/012071

The significant effect on the discharge combustion process is exerted by the electrolyte concentration, it is established that the electric field strength decreases with increasing concentration. Therefore, for stability of the results, solutions with a concentration of less than 1% were used.

Figure 3 shows the linear dependence of the change in anode temperature with increasing discharge power.

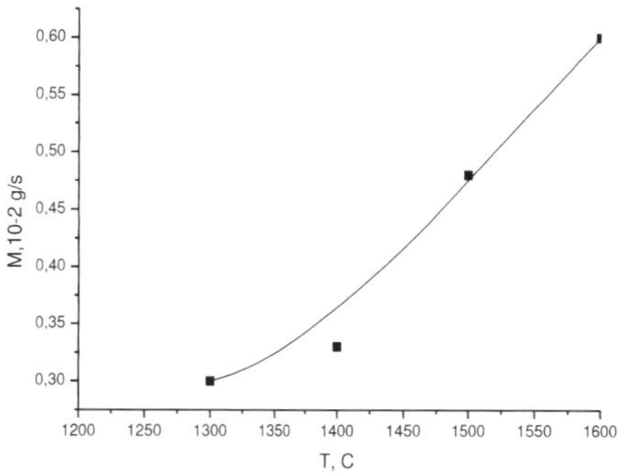


Fig. 4. Dependence of powder performance on anode temperature

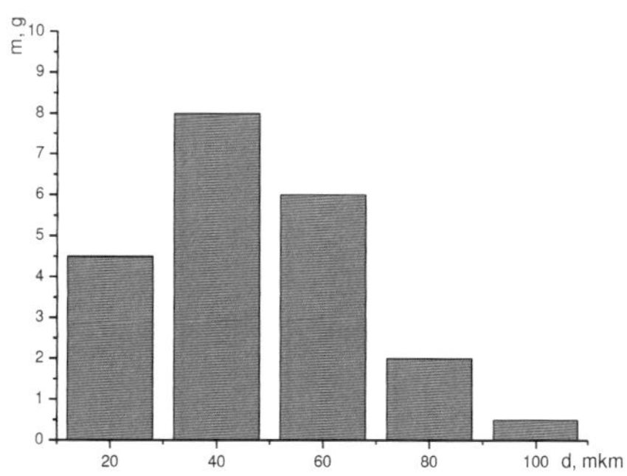


Fig. 5. Powder distribution histogram

The formation of the powder occurs when the surface temperature of the anode reaches the appropriate solidus temperature. When the liquidus temperature is reached, the electrode is melted and large drops of metal are formed. Proceeding from this, the powder can be produced in the temperature range of the liquidus and solidus anode. An increase in temperature will lead to an increase in the productivity of the powder, but with an increase in the particle size of the powder. Figure 4 shows the dependence of the powder performance on the anode temperature. It has a nonlinear dependence and increases with increasing temperature. A linear increase in the average particle size with increasing anode temperature is established. Figure 5 shows the histogram of the granulometric composition of the obtained powder. The largest amount of powder is obtained by a size smaller than 40 µm. Figure 6 shows SEM images of the resulting powders. The powder has a spherical shape, the smallest particle size is 0.5 µm. The resulting powder is suitable for use in selective laser alloying plants. However, the need for further research related to the increase of the process productivity by increasing the anode area definition and dynamic characteristics of the powder.

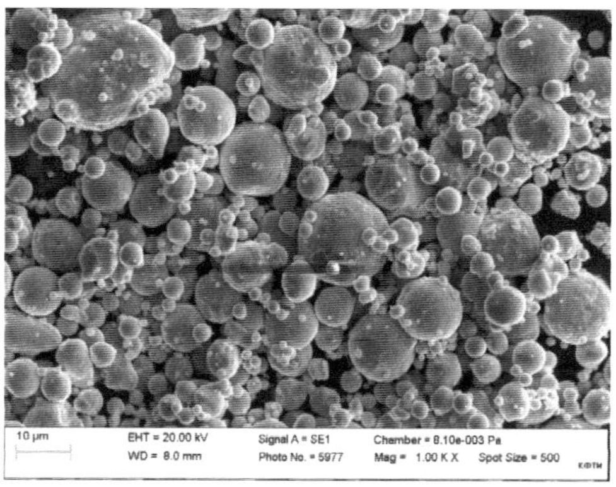


a

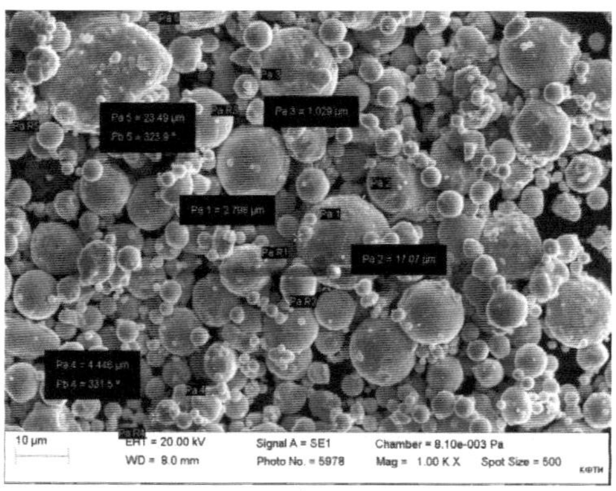


b

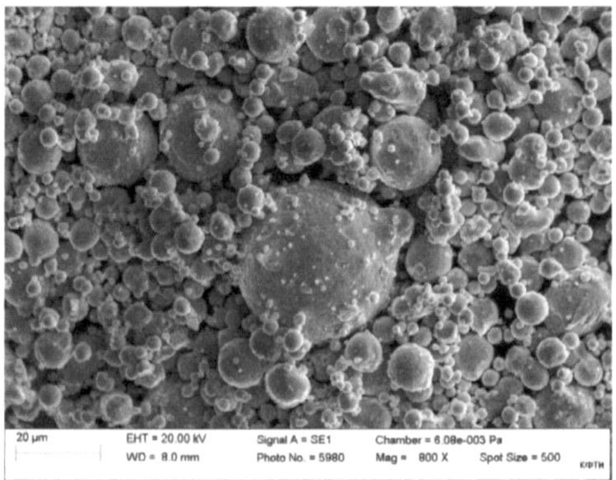


c

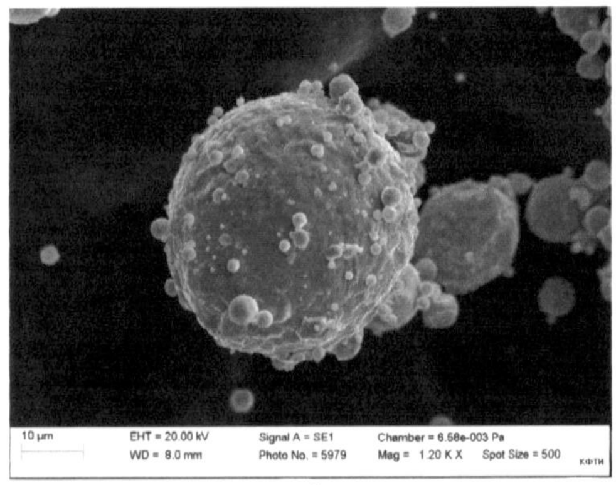


d

Fig. 6. SEM images of steel powder obtained in the plasma-electrolyte process

## Conclusion

Studies of the combustion of a gas discharge between a metal electrode S17400/630 and liquid electrodes have shown the possibility of obtaining a spherical powder with a particle dispersion from 0.5 to 40 µm. Analysis of the obtained SEM photographs shows similarity with the powder obtained by the gas atomization method. The influence of the gas discharge parameters on the process productivity and the size of the particles obtained is determined. In the voltage range from 500 to 800 V, particles smaller than 40 µm in size with a productivity of 1 • 10-2 g / s occur. Above this range, the electrode is melted and the process for obtaining the powder ceases. The location of the metal electrode affects the physicochemical processes occurring in the discharge and the mechanism of particle formation.

This work was financed by grant MK-1861.2017.8 allocated by the Council on grants to the President of the Russian Federation for state support of young Russian scientists and for state support of the leading scientific schools of the Russian Federation.

This work was funded by the subsidy allocated to Kazan Federal University for the state assignment in the sphere of scientific activities (11.1300.2017/4.6).

This work was financed within the framework of state support of leading universities in order to increase their competitiveness among the world's leading scientific and educational centers.

## References

1. Journal of Alloys and Compounds Volume 691, 15 January 2017, Pages 316-322
2. International Journal of Heat and Mass Transfer Volume 104, 1 January 2017, Pages 665-674.
3. Scripta Materialia Volume 126, 1 January 2017, Pages 41-44.
4. L. Kashapov, N. Kashapov and R. Kashapov, Journal of Physics : Conference Series. Volume 479, Issue 1, 2013, Article number 012011
5. D. Denisov, N. Kashapov and R Kashapov, IOP Conference Series: Materials Science and Engineering, Volume 86, Issue 1, 26 June 2015, Article number 012005.
6. L. Kashapov, N. Kashapov, R Kashapov and D. Denisov, Journal of Physics: Conference Series. Volume 669, Issue 1, 14 January 2016, Article number 012029
7. J. Clayton, Metal Powder Report. Volume 69, Issue 5, September–October 2014, Pages 14–17
8. B. Hausnerova, B. Mukund, D. Sanetrnik, Powder Technology, 312 (2017), p. 152 – 158.
9. T. Abdul Kareem, A.Anu Kaliami, Ionics (2012) 18: 315 - 327.

IOP Conf. Series: Materials Science and Engineering **240** (2017) 012072 doi:10.1088/1757-899X/240/1/012072

# Method of control position of laser focus during surfacing teeth of cutters

**V V Zvezdin[1], R M Hisamutdinov, R R Rakhimov[1], I H Israfilov[1] and R F Akhtiamov[2]**

[1] Kazan (Volga) Federal University (KFU), Naberezhnye Chelny, Russia
[2] Research and production association ROSTAR, Naberezhnye Chelny, Russia

irmaris@yandex.ru, rafisih88@mail.ru

**Abstract**. Providing the quality laser of surfacing the edges of teeth requires control not only the energy of the radiation parameters, but also the position of the focal spot. The control channel of position of laser focus during surfacing, which determines the parameters of quality of the deposited layer, was calculated in the work. The parameters of the active opto-electronic system for the subsystem adjust the focus position relative to the deposited layer with a laser illumination of the cutting edges the teeth cutters were calculated, the model of a control channel based on thermal phenomena occurring in the zone of surfacing was proposed.

## 1. Introduction
The main role in the development of the industry belongs to tool production, which determines the level of the productive forces of the entire industry. To meet the needs of mechanical engineering, metalworking and other industries in the metal cutting tools necessary to significantly improve the quality, improve operating techniques. The quality of the tool depends on the design, material, manufacturing technology, the type of surface thermal treatment. Modern technologies of tool manufacture are: the use of powder metallurgy; automation of technological processes, the use of automatic line laser technological complexes; application of methods of thermal and thermo-chemical treatment, wear-resistant coating; extension of the scope of thermal processing methods. These areas, ultimately, contribute to the solution of the main tasks of increasing efficiency of production and quality.

## 2. Automatic control system of the laser technological complex
Stabilization of energy parameters in the zone of interaction of laser radiation (LR) to the metal during surfacing and improving the accuracy of positioning its focus with respect to the cutting edges of teeth cutters, is the best way of building automation systems, laser technological complex (ACS LTC). This leads to the challenge in meeting the requirements for the quality of the technological process (TP) surfacing [1, 2].

Tracing the trajectory LR accurately focus movement provides non-contact methods, in particular, based on measuring as the light flux of the thermal radiation of metal, as well as reflected from the

IOP Conf. Series: Materials Science and Engineering **240** (2017) 012072 doi:10.1088/1757-899X/240/1/012072

plane cutters teeth [3]. The solution of the control focus position LR is it possible with the help of photodetectors line (PL) [8].

On the basis of the energy balance:

$$W_{\mathrm{LR}} + W_{Boost} + W_{\mathrm{Lightpressure}} = W_{\mathrm{Melting}} + W_{\mathrm{Mechanical.}} + W_{\mathrm{Reflections.}} + W_{\mathrm{Kinetic}},$$

where $W_{\mathrm{LR}}$ - energy is supplied to the treatment zone LR, J.; $W_{Boost}$ - energy shielding gas pressure, J; $W_{Light\ \mathrm{pressure}}$ - mechanical energy of the impact on the metal surface of LR, J.; $W_{Melting}$ - the energy expended on melting metal; $W_{Mechanical}$ - the energy of the mechanical oscillations in the metal, J; $W_{Reflections}$ - reflected energy from the interaction zone LR metal, J; $W_{Kinetik}$ - the energy expended on the flight of the molten metal particles, J.

It can be stated that $W_{Reflections}$ is about 10% of the $W_{\mathrm{LR}}$ energy as the metal temperature exceeds the melting point and the co-absorption coefficient LR melt is 0.9.

Providing high accuracy of the trajectory of the laser head, it requires a clear boundary illumination sites photodetectors, to this end, we make the calculation laser illumination channel. Area illumination diameter of 10 mm is formed in the channel laser illumination at a distance of 150 mm from the lens by a semiconductor laser $\lambda = 0{,}84$ m with an angle of divergence of the laser beam is then 40 ° [4].

In laser illumination channel illumination area of this size is relatively large and so it does not require such a high quality in image forming optical systems, the lens is made of a single lens [5-6].

The distance from the front surface to the first focus lens is 22.3 mm. The calculated wavelength is 0.84 microns. The focal distance is 25.1 mm.

Select channel optical parameters of the photodetector for the reflected total radiation requires accounting for such requirements and constraints as:

-the linear dimension of the subject area (plane cutters teeth) quality image which is to be formed in the photodetector plane - 10 mm;

- aberration circle diameter (maximum) - 10 microns;
- image size PL element in the plane of the teeth cutters - 20 microns;
- the number of sensitive elements in the line PL (minimum) - 500;
- the distance from the lens to the plane of the cutter teeth - 150 mm;
- adequacy of PL exposure level of sensitive elements of the illuminated laser tooth plane (power rating, see below);
- use AF type DALSA line with the size of the sensor 14x14 microns or 13x13 microns.

The required focal distance f 's lens and the distance from the lens to the PL to obtain a first approximation of geometric optics formulas for thin lenses.

$$s' = f' + x' = 150,$$

$$\Gamma = x'/f' = f/x = s'/s = 1{,}5.$$

here s = 100 мм, f' = 60 мм.

These requirements for the optical system PL channel implemented in three lens. The design parameters of the lens are shown in Fig. 1.

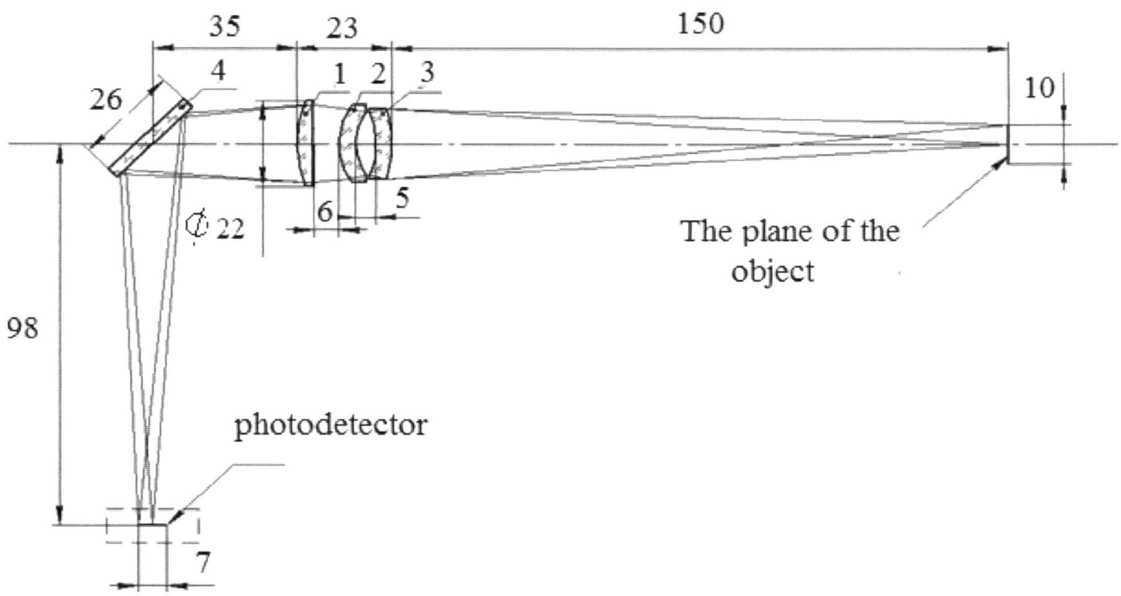


**Figure 1.** Progress beams and structural parameters of the channel OP:
1, 2, 3 - lens; 4 - mirror.

The calculated wavelength is 0.84 microns. The focal length of 71.3 mm. The distance from the focus F to the front surface of the first lens, 85.5 mm. Distance from the PL to the first surface of the lens 133 mm. The distance from the last surface of the lens to the back focus F '= 42,8 mm.

Flat-breaking mirror installed between PL and lens will reduce the size. With similar mirrors can be optimized size of the PL optical channel. Flat mirror does not affect the image quality [7-8].

Calculation of indicators of the quality of the lens, made in the pro-gram «Zemax». Diagram of scattering spots was created in the plane of the tooth for the field of views corresponding to a deviation from the optical axis 0, 3, 8 and 5 mm. Scattering circle diameter does not exceed 5 microns.

Energy assessment will show sufficient irradiance PL sensor elements at the chosen parameters of the laser optics channel and photodetector channel [4].

Initial data for calculation:

- power laser, $P_l = 0.2$ W;
- laser lens channel transmission $\tau_l$, not less than 0.8;
- threshold exposure of photodetector at $\lambda = 0,84$ m, $N_{thr} = 13 \cdot 10^{-12}$ W ·s/ cm$^2$;
- exposure of photodetector saturation at $\lambda = 0,84$ m, $N_p = 23 \cdot 10^{-9}$ W ·s/ cm$^2$;
- transmission of lens of photodetector channel with mirrors, $\tau_{ph} = 0.8$;
- coefficient of the metal reflectance in the tooth plane, $\rho = 0,7$;
- laser spot area at the tooth plane $S_l = 0.8$ cm$^2$;
- half aperture angle of lens of photodetector channel, $\alpha = 2,86$ °;
- bodily aperture angle lens of photodetector channel
$\Omega = 2 \cdot \pi \cdot (1 - \cos \alpha) = 0,0078$ cp;
- linear increase lens in the direction from the photodetector to the tooth plane, $\Gamma = 1,5^x$.

The calculation is as follows. Power illumination (irradiance) of the laser in the plane of the tooth:

$E_s = P_l \cdot \tau_л / S_л = 0,2 \cdot 0,8 / 0,8 = 0,2$ W / cm$^2$.

Energy brightness laser spot:

$B = E_s \rho / \pi = 0,2 \cdot 0,7 / 3,14 = 4,45 \cdot 10^{-2}$ W / (cm$^2$·sr).

Irradiation in the plane PL:

$E_{ph} = B\, \Omega\, \tau_{ph}\, \Gamma^2 = 4,45 \cdot 10^{-2} \cdot 0,0078 \cdot 0,8 \cdot 1,52 = 0,6 \cdot 10^{-3}$ W / cm$^2$.

In the case of a standard charge accumulation photodetector $t_n = 25$ microseconds threshold irradiance of photodetector

$E_{thr} = N_{thr} / t_n = 13 \cdot 10^{-12} / 25 \cdot 10^{-6} = 0,5 \cdot 10^{-6}$ W / cm$^2$.

This implies that the signal / noise ratio will be

$E_{fp} / E_{thr} = 0,6 \cdot 10\text{-}3 / 0,5 \cdot 10^{-6} = 1200$.

Limit charge accumulation time must not exceed

$t_{max} = Np / E_{fp} = 23 \cdot 10\text{-}9 / 0,6 \cdot 10\text{-}3 = 38 \cdot 10^{-6}$ c.

## 3. Conclusion

It can be concluded that the calculation of the energy assessment of laser illumination of the weld shows sufficient irradiance sensor elements of the photodetector line (PL) at the chosen parameters of the laser optics channel and photodetector channel.

The presented method of controlling the position of the laser focus relative to the plane of the tooth with considering calculated geometry can significantly improve the quality of welding.

## References

[1] Saubanov R R, Rakhimov R R, Zvezdin V V and Israfilov I Kh 2011 A method for nanostructured surface modification layer parts concentrated streams of energy *Proceedings of the 13th Int. Scient.* (Conf Section 3 St-Petersburg Univ) ed Polytechnic University pp 389-394

[2] Khisamutdinov R M, Zvezdin V V, Saubanov Ruz R, Israfilov I H, Rakhimov R R and Spirin A A 2016 Study of processes of steels surfaces modification with highly concentrated energy flows *IOP Conf. Ser* (Materials science and engineering) Volume 669 Issue 1 **012024**

[3] Zvezdin V V, Khamadeev A V, Zagiriv R G and Shangaraev I R 2008 Positioning of laser radiation with respect to the weld as an indicator of the quality of the technological process *Bulletin of KNRTU. Tupole* number 3 Kazan National ResearchTechnical University. Tupolev– KAI (*Kazan*) pp 17-19.

[4] Rakhimov R R, Zvezdin V V, Israfilov I H, Nabiullina G I Saubanov R R 2014 Modeling of the laser heat treatment *News of the Tula State University* (Technical science) number 11-2 pp 476-484

[5] Zvezdin V V, Israfilov D I, Portnov S M, Saubanov R R, Rakhimov R R and Zvezdina N M 2015 Automatic control system of high-precision welding of preparations by the laser radiation at influence of the plasma torch *Proceedings of the higher educational institutions (Physics)* number 9-3 Volume 58 ed Siberian Physico-Technical Institute (*Tomsk*) pp 51-54

[6] Zvezdin V V, Khamadeev A V, Fardiev R K, Israfilov D I and Bashmakov D A 2007 Influence of the mode composition of laser radiation on the heat affected zone in metals *Bulletin of KNRTU. Tupolev* number 2 ed Kazan National ResearchTechnical University. Tupolev - KAI (*Kazan*) pp 84-85

[7] Aleev R M, Ivanov V P and Ovsyannikov V A 2004 *Non-scanning thermal imaging devices. Basic theory and calculation* ed Publishing House of Kazan (University Press)Volume 57 number 3 p 228

[8] Filachev A M Taubkin I I and Trishenkov M A 2005 PHOTOELECTRONICS (Physical bases) (Moscow: Physics and mathematics book) p 384

IOP Conf. Series: Materials Science and Engineering **240** (2017) 012073 doi:10.1088/1757-899X/240/1/012073

# Management of laser welding based on analysis informative signals

**V V Zvezdin[1], R R Rakhimov[1], Ruz R Saubanov[1], I H Israfilov[1] and R F Akhtiamov[2]**

[1] Kazan (Volga) Federal University (KFU), Naberezhnye Chelny, Russia
[2] Research and production association ROSTAR, Naberezhnye Chelny, Russia

irmaris@yandex.ru, rafisih88@mail.ru

**Abstract**. Features of formation precision weld of metal were presented. It has been shown that the quality of the welding process depends not only on the energy characteristics of the laser processing facility, the temperature of the surface layer, but also on the accuracy of positioning laser focus relative to seam and the workpiece surface. So the laser focus positioning accuracy is an estimate of the quality of the welding process. This approach allows to build a system automated control of the laser technological complex with the stabilization of the setpoint accuracy of of positioning of the laser beam relative to the workpiece surface.

## 1. Introduction

The most promising method of welding is laser welding. Laser welding is carried out by local melting and subsequent crystallisation of metals at their butt surfaces, resulting in strong adhesion occurs at interatomic interactions of welded parts [1].

Automation of the process of laser welding parts in mechanical improve efficiency of technological process (TP). The most expedient is to provide an automatic control system (ACS) Laser technological complexes (LTC) with the stabilization and optimization of energy and time characteristics to obtain the required output parameters TP [2].

## 2. Development of automatic control system

The structure of information support of ACS include models of individual dynamic modules, as well as models of acting on the signal and noise. This requires adequate mathematical models of TP on the basis of experimental data. However, due to the complexity of the physical processes occurring in laser welding of metals, they do not give a complete picture of events taking place in the area of laser radiation (LR) to the metal [3, 7].

According to the results of experimental research on welding of different metals are widely spread metrics TP quality (the roughness of the weld, the lack of fusions, burning and shells, etc.).

The photographs (see. Fig. 1, 2, 3) are seen strongly pronounced weld defects for steel with given parameters of pulse LR. Fig. 1 seen along the length of the uneven roughness. Fig. 2 shows the burning metal. These results are obtained welding by continuous LR without stabilization LTC settings. When welding with pulse LR (Fig. 3) appeared sealing rings, leading to unevenness in depth microhardness.

**Figure 1.** Photo of weld by continuously LR without stabilization LTC settings.

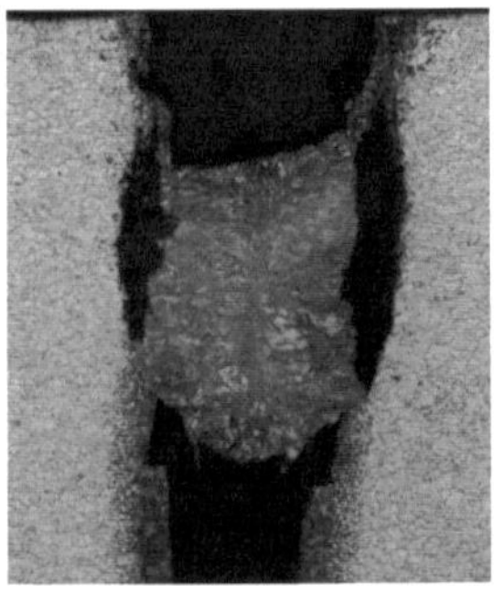

**Figure 2.** Fig. 2. Photo of incision a weld by continuous LR without stabilization LTC settings with burning metal.

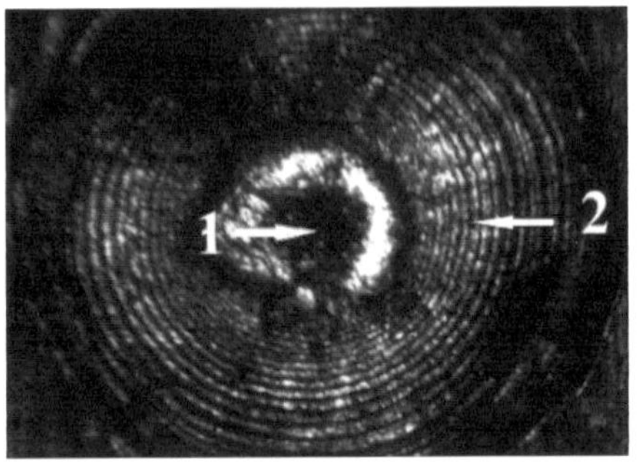


**Figure 3.** Photo steel surface after exposure to pulsed LR. 1 - impact zone (Φ 0.5 mm, energy LR - 12 J, pulse duration of 3 ms.). 2 - sealing ring.

This is explained by the instability of LTC settings, and the variation of the surface layer of welded metal.

Positioning accuracy LR with respect to joint is the main parameter that characterizes the quality of the welding. For fast-flowing laser welding process is necessary to provide automated management LTC with feedback in parameters measured in real time the progress of TP [4].

The experimental setup was developed, a block diagram of which is shown in Fig. 4.

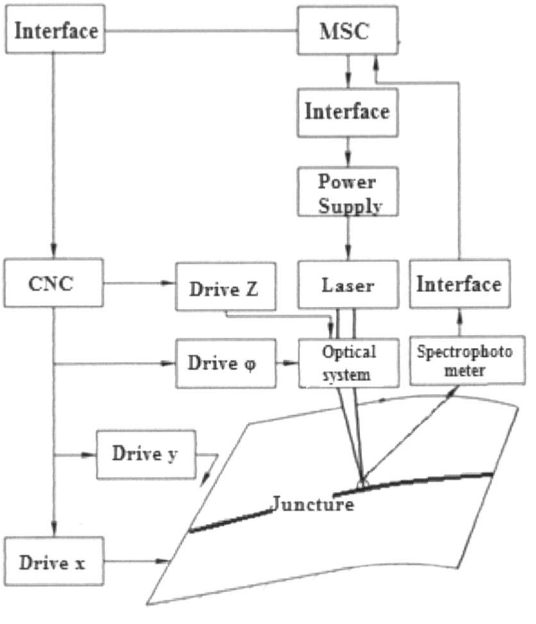


**Figure 4**. Block diagram LTC

Management parameters of LTC and their stabilization by due to negative feedback on the measured parameters of the zone interaction of LR with the metal. This is achieved by analyzing the position of LR focus with respect to weld and LR angle to a normal plane surface of the workpiece [5]. Structure ACS LTC is a system which is characterized by a large number of feedbacks and is nonlinear.

Certain assumptions allow linearize their transfer functions have been taken in justifying the choice of the mathematical model of the links. Calculations and studies of the properties of ACS made for the linearized system. The analysis in the field of development of ACS research shows the effectiveness of multiple-relay systems with feedback by informative parameters measured in real-time to control the LTC. The analysis was conducted by ACS developed models in MathCAD. The transfer function of the open-loop system has the form [6]:

$$W_r = \frac{K}{(T_1 \cdot p+1)\cdot(T_2^2 \cdot p^2 + 2\cdot T_2 \cdot \varsigma \cdot p+1)\cdot(T_3 \cdot p+1)\cdot(T_4 \cdot p+1)} \qquad (1)$$

where $K = \prod K_i = 0{,}0005$ .

The main element of the spectrophotometer is much area line of photodetectors (Fig. 5).

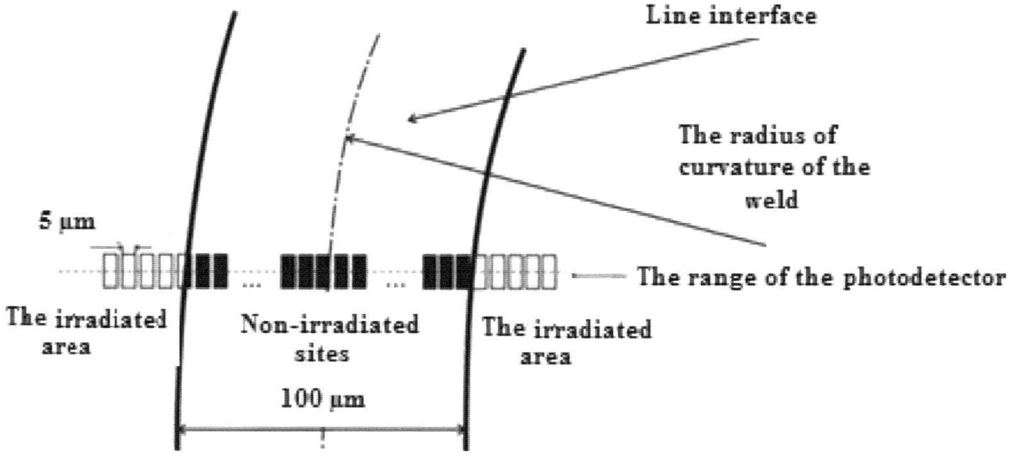


**Figure** 5. Position much area line of photodetectors with respect to the weld on a curved surface of the part to be welded elements

Technological process of metals welding is provided by a careful fitting edges across the seam length (less than 100 microns) and high positioning accuracy LR. Protection joint surface from oxidation is carried out with a mixture of helium and argon fed through a special nozzle.

## 3. Conclusion

Experimental research effects of LR to the metals shown possibility of optimizing the energy parameters LTC for welding with obtain the desired quality metrics, which leads to lower energy consumption. Developed ACS LTC satisfies the requirements on positioning accuracy of LR with respect to seam and provide stable welding quality indicators. Locality of the welding zone ensures the rational design of welded components and minimizes the residual deformation.

## References

[1] Grigor'yants A G, Shiganov I N and Misyurov AI 2006 Technological processes of laser machining *Proceedings Handbook for Universitiee* ed. AG Grigoryants, Publishing house of the BMSTU (*Moscow*) p 664

[2] Khisamutdinov R M, Zvezdin V V, Saubanov Ruz R, Israfilov I H, Rakhimov R R and Spirin A A 2016 Study of processes of steels surfaces modification with highly concentrated energy flows *IOP Conf. Ser* (Materials science and engineering) Volume 669 Issue 1 **012024**

[3] Zvezdin V V, Khamadeev A V, Zagiriv R G and Shangaraev I R 2008 Positioning of laser radiation with respect to the weld as an indicator of the quality of the technological process *Bulletin of KNRTU Tupolev* number 3 ed Kazan National ResearchTechnical University.

IOP Conf. Series: Materials Science and Engineering **240** (2017) 012073 doi:10.1088/1757-899X/240/1/012073

Tupolev – KAI (*Kazan*) pp 17-19
[4] Zvezdin V V, Aleev R M, Saubanov R R, Galiev R M and Rakhimov R R 2011 Method of measuring informative parameter based on optical - physical methods of research *Intelligent systems in production* (*Scientific and practical journal*) number 1 (17) pp 231-237
[5] Zvezdin V V, Grigor'yants A G and Israfilov I H 2012 Control method of laser technology metal welding *Scientific journal* (*Moscow*) number 1 (17) pp 231-237
[6] Saubanov R R, Zvezdin V V, Zvezdina N M, Rakhimov R R and Zagiriv R G 2015 Optoelectronic method research area laser hardening metals *Socio-economic and technical systems: research, design, optimization. (Electronic Materials* vol 1*)* ed Naberezhnochelninsky Institute KFU number 4 (67 ) pp 59-69
[7] Denisov D G, Kashapov N F and Kashapov R N 2015 The appearance of shock waves in the plasma electrolytic processing *Iop conference series* (Materials science and engineering) **012005**

# Formation of cracks in the selective laser melting of objects from powdered stainless steel 17-4 PH

**R N Kashapov[1,2], L N Kashapov[1] and N F Kashapov[1]**
[1] Engineering Institute, Kazan Federal University, Kazan, 420008, Russia
[2] Laboratory of Radiation Physics, Kazan Physical-Technical Institute of the Kazan Scientific Center of the Russian Academy of Sciences, Kazan, 420029, Russia

E-mail: kashramiln@gmail.com

## Annotation:

In the work, the process of selective laser melting of thick-walled objects of molds for wax models of 17-4PH steel powder on a ProX 300 was studied. The microstructure of the surface has been studied, the formation of cracks has been revealed and the possible reasons for their formation and propagation have been proposed. Analysis of selective laser melting of thin-walled objects revealed no cracks. To prevent the occurrence of cracks, it is necessary to warm the working platform to 200 ° C.

## Introduction

Selective laser melting (SLM) is one of the main directions of additive technologies, it is possible to create metal objects of complex geometry, prototypes, moving objects, mesh structures and functional tools that can't be produced by standard technologies of "cutting" [1]. Another advantage of this technology is the ease of operation, namely the use of a 3D model, which modeled in computer-aideddesign (CAD) or obtained by scanning in the STL format. This file is downloaded to the software of the SLM-installation, where the support simulation is performed, where laser scanning parameters, layer thickness and other parameters are specified. The resulting structure, containing the object itself and supports, is subjected to the "slicing" procedure, namely it is divided into a plurality of layers, each layer will be located at a certain height and contain information about the object profile. All these procedures are simple, and an engineer working with CAD can easily master them, thus eliminating the need for an installation operator, as it is necessary when working with a machine. Another important factor is the speed of getting a finished object with a complex profile geometry that exceeds the standard technologies. All these factors significantly shorten the period of mastering of a new type of object in the factory. However, despite all these advantages, there are limitations that hinder the widespread use of SLM in the industry, one of them is the worst mechanical characteristics, the presence of internal stresses, pores leading to cracking, and high surface roughness [2].

These shortcomings are eliminated by post-processing. There are different methods to improve the properties of the object, each of which is aimed at eliminating any one defect. Thus, the gas-statistic method [3] is designed to reduce porosity, it is realized by placing the object in a gas atmosphere with a temperature of 1000° C and a pressure of 3000 atmospheres. Another interesting method is laserre-melting [4], which can be used both in the process of selective melting and at the final stage of surface treatment of the product. This also helps to eliminate the presence of pores and increase the hardness of the samples. From the viewpoint of decreasing the surface roughness, the electrochemical polishing method [5] or more advanced methods of plasma electrolyte treatment [6,7,8] are used. Also, a mandatory procedure is heat treatment to relieve internal stresses.

IOP Conf. Series: Materials Science and Engineering **240** (2017) 012074 doi:10.1088/1757-899X/240/1/012074

All the above said so far does not allow to produce products with high quality requirements, for example, aviation products, but it allows to be integrated into the existing technological process. Namely, to manufacture molds for the production of wax models intended for subsequent casting. Molds are massive composite objects with a wall thickness of up to 30 mm, when they are melted, strong stresses may arise, which will lead to the formation of cracks. Therefore, the aim of the work was to study the process of selective laser melting in the sintering of molds from 17-4 PH powder and to determine the causes of the formation of cracks.

## Main part

In the experiments was used a selective laser melting system ProX300 3D Systems (Fig. 1, a). Figure 1 b shows the diagram of the inner chamber, the length of which is 1500 mm, the height of 800 mm. The chamber is hermetic and allows creating an atmosphere of inert gas, in the chamber to the left is a metal roller for applying powder, it has the possibility of horizontal movement throughout the chamber. In the middle there is a box with a powder inside which a movable piston is located, on the right is a construction zone consisting of a movable piston with a platform and a laser mounted above it. The cycle for the construction of one layer consists of the following steps: lifting the central piston by 60 μm, lowering the working platform of the construction by 40 μm, moving the roller through the central box, moving the excess powder to the working platform of construction and scanning the surface of the powder with the laser. During the scan, combustion products can be generated, which are carried away by the flow of gas into the dust container.

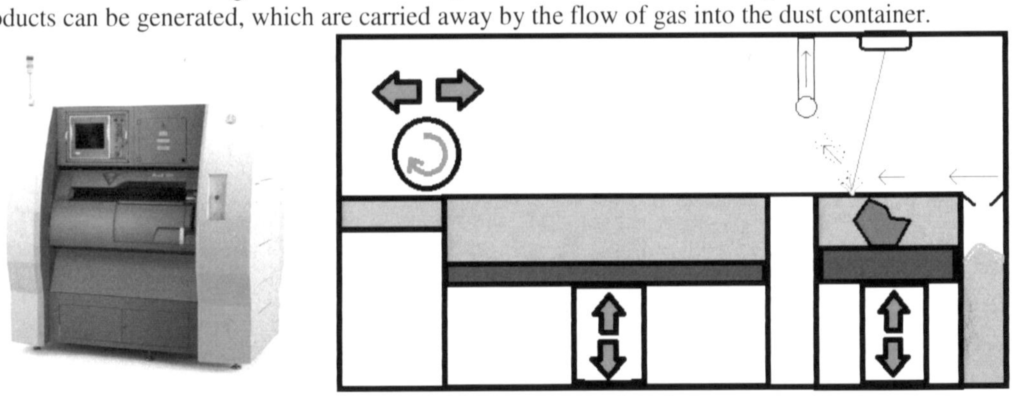

a b

Fig. 1. a - the appearance of the ProX 300, b - the scheme of the working chamber ProX 300.

The ProX 300 is equipped with a fiber Nd: YAG laser power of 500 W, with a wavelength of 1080 nm, the laser spot area was 20 μm. The size of the construction zone is x - 250 mm, y - 250 mm, z - 300 mm, the minimum product size x = 100 μm, y = 100 μm, z = 20 μm. During the melting of the product support, the laser power is 100 W, when the product itself is melted -150 W, the scanning speed is 380 mm/s. Stainless steel powder 17-4PH, dispersed from 0.5 to 40 μm, made by 3D Systems, was used in the experiments.

For melting, 5 virtual models of mold parts were chosen, they all have vertical side walls, and the forming part is in the upper end face. This choice was associated with the construction of solid supports in the lower part of the object to eliminate the effect of geometry on the formation of a crack. Figure 2 shows a series of photographs of prototypes obtained by selective laser melting. All samples have similar defects, namely the formation of cracks near the base. The growth of existing cracks during the cooling of prototypes was observed. In Figure 2b, there is a crack at the support and product boundary that extends into the object. In photos 2a and 2c, cracks are also observed, but in the support itself, where the angle of $90^0$ is formed. The presence of cracks in the support is not a negative result, if it does not lead to distortion of the geometry of the product. It was decided to implement controlled education and the spread of cracks in support. Figure 2d shows the separation

IOP Conf. Series: Materials Science and Engineering **240** (2017) 012074 doi:10.1088/1757-899X/240/1/012074

of support from the work platform. Realize this situation was obtained by uneven melting of the powder.

a

b

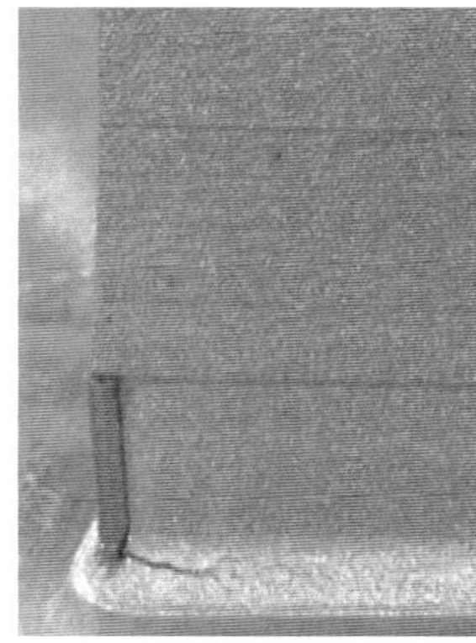
c

d

e

f

g

h

i

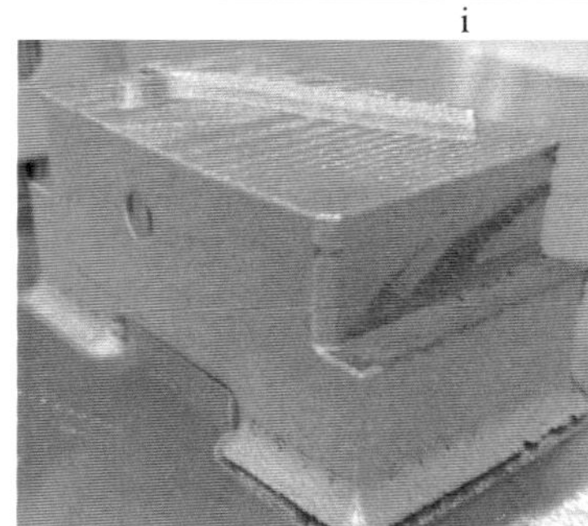

j k

Fig. 2 Prototypes of mold elements.

Namely on the right side of the object, the support starts to melt with the first layer, and in the left part only from the third layer. Thus, we have obtained in the left part of the support a layer with a large porosity and poor adhesion to the platform. The second method of controlled crack propagation was implemented on the same object, only by forming in a virtual model geometric voids in the form of triangles. As the experiment showed, this approach was successful (Figure 2e, 2f, 2g-right side). The third method of controlled formation and propagation of the crack was to stop the melting process for 3 hours, during which time the formed layers cooled to 350 C, and after that the SLM process was resumed. The results of this experiment are presented in Figures 2g and 2i, a smooth divergence of the two layers is observed. Figures 2j and 2k show a sample of the experiment of melting an object of small overall dimensions (600 * 400 * 300 mm), combining the first technique and the second one. As can be seen, the formation of cracks is absent, there is a separation of support without distortion of the geometry. Figure 3 shows photomicrographs of the structure of the obtained samples.

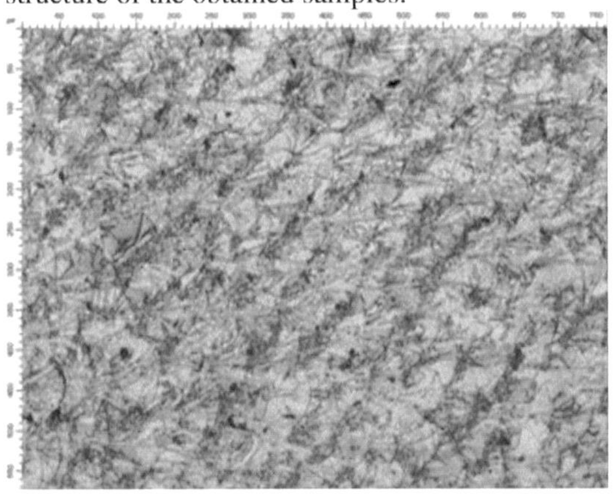

Fig.3a. Microsection along the sample (x100)

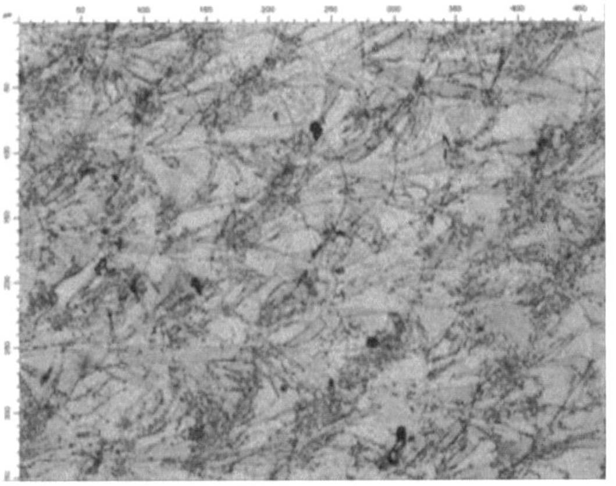

Fig.3b. Microsection along the sample (x200)

In Figures 3, at various magnifications, the presence of pores and inclusions is observed, which are the sources of the development of cracks.

The solution to the problem of the development of cracks can be found in the preheating of the platform to a temperature of 300 ° C, but in the ProX 300 this possibility is absent. Preheating allowed to reduce the amount of internal stresses. To reduce pores in this case, only one method is suitable - laser re-melting, which allows to remove pores during melting.

## Conclusion

As a result of studies of selective laser melting of thick-walled objects of mold elements from metal powder 17-4 PH, the development of cracks in the installation ProX 300 has been revealed. Four methods for solving this problem are proposed:

1. Controlled detachment of the support base of the object from the platform.
2. Controlled development of a crack inside support by creating internal triangular cavities.
3. Controllable rupture of two layers as a result of a temporary shutdown of the installation.
4. Elimination of crack formation due to preheating of the working platform of construction.

All proposed methods can be successfully implemented, except for the fourth, which will require the upgrade of the ProX 300.

It has been established that only one method is suitable for fighting pores in this case - laser re-melting.

This work was financed by grant MK-1861.2017.8 allocated by the Council on grants to the President of the Russian Federation for state support of young Russian scientists and for state support of the leading scientific schools of the Russian Federation.

This work was funded by the subsidy allocated to Kazan Federal University for the state assignment in the sphere of scientific activities (11.1300.2017/4.6).

This work was financed within the framework of state support of leading universities in order to increase their competitiveness among the world's leading scientific and educational centers.

**References**

1) S. Youssef, E. Maire, R. Gaertner, Acta Mater. 2005, 53, 719.
2) J. Ramos-Grez, D. Bourell, International journal of materials and product technology, 2004, 21, 297 -316.
3) K. Kempmen, Master thesis, Catholic University of Leuven, Heverlee, Belgium, 2010.
4) E. Yasa, J-P. Kruth, Procedia Engineering 19 (2011), 389 -395.
5) B.G. Pyka, A.Burakowski, G. Kerckhofs, M. Moesen, S. Van Bael, J. Schrooten and M. Wevers, Advanced Engineering materials 2012, 14, №6, 363-370.
6) L. Kashapov, N. Kashapov and R. Kashapov, Journal of Physics : Conference Series. Volume 479, Issue 1, 2013, Article number 012011
7) D. Denisov, N. Kashapov and R Kashapov, IOP Conference Series: Materials Science and Engineering, Volume 86, Issue 1, 26 June 2015, Article number 012005.
8) L. Kashapov, N. Kashapov, R Kashapov and D. Denisov, Journal of Physics: Conference Series. Volume 669, Issue 1, 14 January 2016, Article number 012029

**Institute of Physics**
Dirac House, Temple Back
Bristol BS1 6BE UK

ISSN: 1757-8981
ISBN 978-1-5108-4892-4

# Mechanical Science and Technology Update (MSTU 2017)

Journal of Physics: Conference Series Volume 858

Omsk, Russia
27 – 28 February 2017

IOP
Institute of Physics